W9-ASF-269

The Story of Physics

The Story of Physics

Lloyd Motz
and
Jefferson Hane Weaver

Plenum Press • New York and London

Library of Congress Cataloging in Publication Data

Motz, Lloyd, 1910–
The story of physics / Lloyd Motz and Jefferson Hane Weaver.
p. cm.
Includes bibliographical references and index.
ISBN 0-306-43076-2
1. Physics. I. Weaver, Jefferson Hane. II. Title.
QC21.2.M65 1989 88-33655
530–dc19 CIP

First Printing – April 1989
Second Printing – October 1989

Grateful acknowledgment is made to the AIP Niels Bohr Library for permission to reproduce most of the portraits that appear throughout the book.

The following individuals, organizations, and collections were the original sources for certain portraits obtained through the Bohr Library: the E. Scott Barr Collection (Lagrange), the Margrethe Bohr Collection (Bohr), the Burndy Library (Becquerel, Huygens, Newton), the Fermi Film Collection (Fermi), Hale Observatories (Hubble), the W. F. Meggers Collection (Curie, Planck), the National Portrait Gallery, London (Faraday), the *Physics Today* Collection (Rayleigh), Francis Simon (Schrödinger), Stanford University (Hofstadter), the University of Texas at Austin (Wheeler), WGBH/Boston (Feynman), and the Zeleny Collection (Thomson).

Grateful acknowledgment is made to the AIP Meggers Gallery of Nobel Laureates for permission to reproduce the following portraits: Dirac, Heisenberg, and Lorentz.

Plenum Press is a Division of Plenum Publishing Corporation
233 Spring Street, New York, N.Y. 10013

Printed in the United States of America

To Minne and Shelley

Preface

The author of the story of a science must be guided by the constraints and demands imposed on him by the definition of "science" itself. A science is more than a body of knowledge expounded in original papers and collected in books; it is the active pursuit of this knowledge by a dedicated group of people (scientists) who are devoted to this "great adventure" by an inner drive that they cannot deny. Since physics as an intellectual activity is the search for the fundamental laws of nature, it is the basic science from which all others are derived; no phenomena in the universe are foreign to the physicist. But the physicist goes beyond the mere knowledge of facts because his ultimate concern is deducing from these facts basic laws that enable him to correlate what appear to be disparate phenomena and to predict future events. An excellent current example of this concern is the astrophysicist's description of the evolution of stars (for example, the sun) from their present states to their ultimate demise as white dwarfs, neutron stars, or black holes. The astrophysicist performs this task by applying the known physical laws to stellar interiors to discover their dynamic processes.

Since knowledge of natural phenomena or even of natural laws alone is not science, we have presented the story of physics here not only as the growth of a body of facts but also as the emergence and evolution of nature's laws from facts, which could have stemmed only from a remarkable intellectual synthesis of fact and fancy (speculation). In connection with this theme, we emphasize again the important distinction between knowledge and science. Every living thing in the universe, even a single cell, has the knowledge necessary for life, which is far beyond anything that we know consciously. Our eyes (or the cells in our eyes) know far more about the laws of optics than we do, and if we had to tell the organs in our bodies how to operate, we would quickly die. But despite all their cleverness, the cells in our bodies are not scientists, nor, as another example, are bees scientists, even though they know that they can keep their hives cool by rapidly vibrating their wings; they are arguably technicians, but not scientists.

Proceeding with this idea from cells, insects, and lower animals to ourselves, we note that each of us, even the most untutored in science, learns a great deal about the laws of nature without even being conscious of it. In walking, running, balancing ourselves, and avoiding all kinds of natural dangers, we constantly apply our subconscious knowledge of the laws of motion, the law of gravity, the laws of thermodynamics, vectorial concepts, and symmetry and conservation principles. Keeping in mind this distinction between knowledge *per se* and science, we begin our story of physics with the ancient Greeks, because their written records show that they were involved in the deliberate pursuit of knowledge (the beginning of science) as the pathway to an understanding of the universe. In that sense, and in accordance with our prescription, they were certainly scientists, although not very successful ones.

Since our book is not a history of physics, we do not explore all the facets of "Greek physics," but present only those salient features that influenced, whether in the right or the wrong way, the thinking of the scientists who followed. The discoveries of Pythagoras, Euclid, Archimedes, Aristarchus, Hipparchus, and Ptolemy are most notable in this respect, but to have done more than describe the works of these remarkable philosophers briefly, yet in sufficient detail to be understandable, would have expanded this book beyond its intended domain.

The reader who is not interested in the Greek contributions to physics can begin this book at Chapter 3, which deals primarily with the astronomical works of Nicolaus Copernicus, Tycho Brahe, and Johannes Kepler, whose derivation of the laws of planetary motion from Brahe's observational data is one of the great intellectual syntheses of the post-Copernican era. Comparing this achievement with that of the greatest of the ancient Greeks shows clearly the vast difference between the speculations of the Greeks (which were experimentally or observationally unsupported) and the sound observational basis of Kepler's deductions.

We single out Galileo Galilei's concept of inertia and Sir Isaac Newton's laws of motion for special emphasis because they are such a great departure from the thinking of the Greeks and the philosophies of the pre-Renaissance scholastics. This is most evident in the rapid post-Newtonian development of physics, which, in a relatively few years, laid the basis for all of classical physics and modern physics, even though modern physics, stemming from the quantum theory and the theory of relativity, departs drastically from Newtonian physics in certain fundamental features. However, the conservation principles, the symmetry principles, and the least action principles of classical (Newtonian) physics developed in the 18th and 19th centuries by the classical mathematical physicists were carried over to modern physics with certain crucial changes.

In our description of the development of classical physics, we emphasize such principles and show that they form the thread that connects one group of concepts with another (for example, the dynamics of particles with thermodynamics) and also define the continuity in the evolution of physics. Since this continuity is not broken by the rise of modern physics—that is, the quantum theory and the theory of relativity—we carefully show why the quantum theory was necessary and how it emerged from classical physics.

The transition from the quantum theory to the quantum mechanics (matrix mechanics and wave mechanics) as developed by Louis de Broglie, Erwin Schrödinger, Werner Heisenberg, Paul Dirac, and Max Born produced a much greater revolution in our thinking than did Max Planck's introduction of the quantum concept itself, for what that transition brought with it (that is, correctly predicted) was phenomena that defy physical understanding. We have therefore emphasized the physical features of the quantum mechanics as much as possible while pointing out the features that have to be accepted at this point without question.

Given the rapid development of high-energy particle physics during the last quarter of a century, no story of physics would be complete without a discussion, however brief, of the important theoretical and experimental features of current particle physics. We have therefore included a discussion of this topic in Chapter 19.

Lloyd Motz
Jefferson Hane Weaver

Contents

The Story of Physics

CHAPTER 1

Greek Physics

Everything of importance has been said before by somebody who did not discover it.

—ALFRED NORTH WHITEHEAD

In writing a book that is not a history of physics but rather a story of the continuity of the ideas, observations, speculations, and syntheses that constitute the body of knowledge that we now call physics, we had to leave out aspects of this subject that rightly belong in a history. With this point in mind, we approached the Greek contributions to this story with the knowledge that whatever we included or did not include, the result would not be entirely satisfactory. Though Greek physics cannot be properly integrated into Newtonian physics, Greek philosophy and Euclidean geometry still influence our thinking; for this reason, we have included in this story of physics the features of Greek thinking that we believe to be pertinent to physics.

Physics, as we understand and practice it today, was unknown to the ancient Greeks, and we may speculate where society would be today if Newton's laws of motion and his law of gravity had been discovered by Aristotle or Archimedes. This is not to say that the Greek philosophers and mathematicians were not scientists in a very general and amorphous way; they were indeed, as evidenced by their keen observations of the heavens, their endless speculations, and their mathematical creations. Wherein, then, does their science differ from Newtonian science? Precisely in the absence of physical principles or laws that enable one to predict future events from current observations or, more generally, to correlate many apparently disparate phenomena in the universe.

A few examples will illustrate and illuminate this very important difference. No matter how much the Greek astromoners learned about the motions of the planets observationally, that information alone could not enable them to predict or understand the periodicity of the tides, the behavior of freely falling bodies, or the revolution of two neighboring stars (a binary) about a common point. Newton's laws, on the other hand, permitted him and the physical scientists who followed him to correlate and explain planetary motions, tides, and other phenomena as manifestations of the same physical force, namely, gravity. In the same way, without a deep understanding of pressure, the Greeks could not apply Archimedes's buoyancy principle to the explanation of general atmospheric phenomena as the Newtonians did. Greek science was thus entirely empirical and without basic laws.

But we still owe much to the ancient Greeks for their mathematics, their observational astronomy, and their range of speculations. Although mathematics is not physics, an important branch of mathematics, geometry, in which the Greeks were great experts, is so intimately related to physics that the study of Greek geometry is essential to the proper study of physics. Geometry is important to physics because the laws of motion of bodies can be only expressed in a geometrical context. This is also true of such phenomena as the spatial interrelationships of bodies and the empirical description of the motion of a body. If we had no geometry, we could not formulate physical laws that are useful precisely because they enable us to correlate disparate *spatial* events.

Today, we know that three kinds of geometries exist, Euclidean (flat space), hyperbolic (negatively curved space), and elliptical (positively curved space), but the Greeks knew only of Euclidean geometry, to which contributions were made not only by Euclid but also by Pythagoras and Eudoxus. Pythagoras (560–480 B.C.) founded a school of philosophers that lasted some 200 years and greatly influenced Greek thinking. Little is known about the details of Pythagoras's life, but he is believed to have spent much of his earlier years in Egypt and Babylonia learning mathematics. Forced to leave his lifelong home at Samos, he settled in Croton, Italy, in 530 B.C. and founded his school of philosophy. Although Pythagoras's teachings were influential throughout southern Italy, his antidemocratic views generated strong opposition that ultimately forced him to flee in 500 B.C. to Metapontum, where he spent his remaining years.

To the Pythagoreans, number was everything; they believed that all phenomena in nature could be explained in terms of numerical relationships, but they gave no recipe for discovering these relationships, and so their numerical philosophy was sterile. However, like all basic principles, the Pythagorean numerology had great heuristic value in that it prompted the Pythagoreans to seek symmetries and harmonies in all natural phenomena. This search led them to the discovery that the harmony of musical sounds depends on the regularity of the intervals between the pitches of harmonious sounds.

They generalized these ideas to propose a universal harmony to account for the apparent motions of the planets, which they associated with musical notes of different pitch. This theory, called the "harmony of the spheres," influenced even Johannes Kepler, who tried, in his early speculations, to represent the motions of the various planets by different octaves in the musical scale.

Today, Pythagoras is best known for his famous geometric law or theorem that expresses the length of the hypotenuse of a right triangle in terms of the lengths of the other two sides. This simple relationship, which Pythagoras established for a right triangle on a plane, has been generalized to any number of dimensions and to non-Euclidean geometry. So generalized, it is the basis of the geometrical interpretation of the laws of nature. Indeed, Pythagoras's theorem, in its most general form, is the starting point of Albert Einstein's general theory of relativity and all modern attempts to unify the laws of nature as manifestations of space–time geometry.

Euclid, of course, is famous for his *Elements,* contained in 13 volumes of definitions, postulates (axioms), and theorems, which summarize all the mathematical knowledge of ancient Greece. Its influence was tremendous, and Euclidean three-dimensional geometry was accepted, for hundreds of years, as the correct geometrical framework on which to formulate the laws of nature. Newtonian mechanics and James Clerk Maxwell's electromagnetism incorporated Euclidean geometry into their theoretical structures. The break with Euclidean geometry occurred when the great 19th-century geometers like Carl Friedrich Gauss, Nikolai Lobachevski, and Georg Riemann began to challenge Euclid's fifth postulate, which states that given a line and a point outside it, only one line can be drawn through the point parallel to the given line. The denial of this axiom led to modern non-Euclidean geometry, from which so much modern theory has evolved.

The vast difference between modern physics and Greek physics is perhaps best indicated by our present atomic theory and the Greek atomism that stemmed from Democritus and his school of philosophers. Democritus proposed the very attractive hypothesis that all matter consists of indivisible particles (atoms) differing in many ways (for example, in size, mass, color) that combine with each other to form all the matter we see in the universe. Since the Greek atomists gave no prescription or mathematical formulas for calculating any properties of matter or predicting any phenomena, their atomic theory remained useless and sterile.

On the other hand, modern atomic theory based on the electromagnetic interactions of the electrically charged constituents of atoms is a precisely formulated discipline that has evolved out of a synthesis of mathematics and basic physical principles. As such, it enables physicists to calculate atomic and molecular phenomena with incredible accuracy. The Greeks knew about electricity and magnetism, but they never connected electrical and magnetic phenomena with the atoms of Democritus.

Of all the Greek philosophers who concerned themselves with physical phenomena, Archimedes was the most notable and was the closest to what we now consider a scientist to be. Archimedes (287–212 B.C.), the son of the astronomer Phidias, was born at Syracuse and was good friends with King Hieron, the local ruler. He spent part of his youth in Egypt learning mathematics from the immediate successors of Euclid. He then returned to Syracuse, where he remained for the rest of his life.

Archimedes combined theory and experiment in a manner similar to scientific procedure today, but no body of basic scientific principles resulted from his work. He attempted to do for science what Euclid had done for geometry: to show that scientific knowledge can be deduced as theorems from a set of self-evident propositions. But little is known about Archimedes's axioms or the theorems he deduced from them.

That Archimedes was a great experimentalist, an inventor, and a keen student of nature is indicated by his discoveries and his mathematical treatises. Not having a well-equipped laboratory to do actual experiments, he must have carried out the kind of thought experiments that characterize all great scientists. He is most famous for his discovery of the principle of buoyancy (the Archimedes principle), and he probably also knew the law of the spatial reflection of light from mirrors. His inventions ranged from the water screw to a planetarium and the astronomical cross-staff with which he made accurate celestial observations. He demonstrated his mathematical skill by showing how to deduce geometrically the number pi (the ratio of the circumference of a circle to its diameter) to any desired accuracy. He did this by approximating the circumference of a circle with the perimeter of a circumscribed or inscribed many-sided regular polygon. By allowing the number of sides to grow without limit and equating the perimeter of such a polygon to the circumference of the circle, one obtains an infinite series for pi.

Archimedes also wrote his *Sand-Reckoner* to demonstrate that very large finite numbers and the infinite itself are indeed different, as shown by the opening sentences: "There are some, King Gelon, who think that the number of the sand is infinite in multitude: and I mean by the sand not only that which exists about Syracuse and the rest of Sicily but also that which is found in every region whether inhabited or uninhabited. And again, there are some who, without regarding it as infinite, yet think that no number has been named which is great enough to exceed its multitude." Archimedes calculated how many grains of sand would fit in a poppy seed, then how many poppy seeds would be needed to equal the diameter of a finger, and so on out to a distance of some 10,000 stadia (one stadium is 607 feet) to arrive at the number of grains of sand he believed would be needed to fill the entire universe. More important than his ease in dealing with these large numbers was his classification of them by orders and periods.

Archimedes died at the age of 75 when Syracuse finally fell to Rome after a brutal siege prolonged by Archimedes's ingenious defensive devices. According to Herbert Westren Turnbull's book *The Great Mathematicians,*[1] the Roman commander, Marcellus, had ordered that Archimedes be taken alive because he "uses our ships like cups to ladle water from the sea, drives off our sambuca ignominiously with cudgel-blows, and by the multitude of missiles that he hurls at us all at once, outdoes the hundred-handed giants of mythology!" Although Archimedes's efforts in defense of his city were extraordinary, he saw them as no more than applications of mechanics, a subject that paled in importance in comparison with his beloved geometry. So devoted was Archimedes to his subject that when the city fell and the Roman legions were pouring through the breached gates, Archimedes continued to puzzle over a mathematical diagram drawn in the sand and was killed by a Roman soldier. Alfred North Whitehead viewed the death of Archimedes as a monumental event because "[t]he Romans were a great race, but they were cursed by the sterility which waits upon practicality." In Whitehead's opinion, "the Romans were not dreamers enough to arrive at new points of view, which could give more fundamental control over the forces of nature." In short, "no Roman lost his life because he was absorbed in the contemplation of a mathematical diagram."

We finally come to Aristotle (384–322 B.C.), Plato's most famous student, who predated Archimedes by some hundred years. Born in Stagira in Chalcidice, Aristotle's philosophy governed human thinking for nearly two millennia in fields ranging from physics and meteorology to biology and psychology. His father was the court physician at Macedon and probably contributed to Aristotle's early interest in biology and the

Aristotle (384–322 B.C.)

classification of sciences. Orphaned at an early age, Aristotle joined Plato's Academy in 367 B.C. and spent the next 20 years studying under the master, who "recognized the greatness of this pupil from the supposedly barbarian north, and spoke of him once as the *Nous* of the Academy—as if to say, Intelligence Personified."[2] After the death of his teacher in 347 B.C., Aristotle spent several years wandering among several of the nearby Greek kingdoms before returning to Macedon to tutor the young prince who would one day be known as Alexander the Great. After his return to Athens, Aristotle founded his school, the Lyceum, which attracted many students and—unlike Plato's Academy, which was devoted to mathematics and political philosophy—emphasized biology and the natural sciences.[3] Aristotle's belief that observation was essential to the study of science prompted him to collect "a natural history museum and a library of maps and manuscripts (including his own essays and lecture notes), and organiz[e] a program of research which *inter alia* laid the foundation for all histories of Greek natural philosophy, mathematics and astronomy, and medicine.''[4] ''If we may believe Pliny, Alexander instructed his hunters, gamekeepers, gardeners and fishermen to furnish Aristotle with all the zoological and botanical material he might desire; other ancient writers tell us that at one time he had at his disposal a thousand men scattered throughout Greece and Asia, collecting for him specimens of the fauna and flora of every land.''[5]

Aristotle viewed mathematics as the key to providing a model for organizing science. This impression was probably formed while he was at Plato's Academy, where mathematics and dialectic discussions geared toward examining the assumptions made in reasoning were most heavily studied. Aristotle viewed the structure of science as ''an axiomatic system in which theorems are validly derived from basic principles, some proprietary to the science ('hypotheses' and 'definitions,' the second corresponding to Euclid's 'definitions'), others having an application in more than one system ('axioms,' corresponding to Euclid's 'common notions').'' His attempt to use mathematics as a tool for generalization, however, necessitated that the dialectic so favored by Plato be assigned a supporting role, to be called forth when mathematics could not free science of its regress and circularity.[6]

Although Aristotle is deservedly praised for his classification system, which exercised such a strong influence on the development of biology, his contributions to physics were undistinguished. His *Physics* was something of a metaphysical mishmash that purported to grapple with so-called ''ultimate topics'' ranging from infinity and time to motion and space. It did provide a valuable historical record because Aristotle recounted the views of earlier pre-Socratic philosophers. However, his purpose was not so much to call attention to the contributions of his predecessors as to enable him to refute and disparage their opinions. While *Physics* offered little in the way of astronomical knowledge and explicitly rejected the Pythagorean belief that the sun is at the center of the universe, Aristotle's meteorological speculations about the continual process of change in the world were inspired: ''[T]he sun forever evaporates the sea, dries up rivers and springs, and transforms at last the boundless ocean into the barest rock; while conversely the uplifted moisture, gathered into clouds, falls and renews the rivers and the seas.''[7] However, Aristotle was unable to synthesize his observations, discern underlying patterns in nature, and thereby formulate a useful theory about the physical world.

He did try to develop a theory of motion that would explain the kinematical behavior of all observable objects from the stars down to terrestrial bodies. He was misled in his analysis of the motions of bodies by his belief that a body can be kept in motion only if the body is in direct contact with a "continually operating mover." If the mover did not maintain contact with the body, the body stopped moving instantaneously; Aristotle had no notion of the concept of inertia, so he failed to discover the laws of motion.

To explain why phenomena happen, Aristotle introduced his doctrine of causes, which reduced all causes to four basic ones that he labeled "material," "formal," "efficient," and "final." We mention these labels here merely to show how far removed Aristotle's thinking was from the modern concept of causality. That Aristotle was a keen observer is evidenced by his geological discoveries and his biological classification schemes. These contributions are still noteworthy and valid.

Taken as a whole, then, Greek physics is not very significant; its greatest value lies in demonstrating how fruitless a putative exact science is if it does not have a sound theoretical foundation supported by a powerful mathematics. The Greeks discovered many interesting facts about nature, but their science did not progress because they had no principles to guide them in constructing a science with its own seeds of growth. We believe, however, that we can learn something important from the Greeks, for modern physics is in danger of developing into a body of theories without facts. Though the Greeks had no mathematical formalism to develop a strong theoretical base for their physics, they were ingenious and clever in their speculations. Today, something similar prevails in the most advanced stages of physics; elementary particle physics is drowning in a sea of formalism. Paper after paper, each with a welter of recondite mathematical equations, but with no numerical deductions, appears in the most prestigious journals of physics today. The absence of numbers at the ends of these papers is a clear symptom of the ill health of theoretical physics today, for it shows that theoreticians are discussing a fanciful universe, rather than the real one.

CHAPTER 2

Greek Astronomy

Except the blind forces of Nature, nothing moves in this world which is not Greek in its origin.

—SIR HENRY JAMES SUMNER MAINE

We make no apology for including Greek astronomy in our story of physics because astronomy today, more than ever in the past, is accepted as a branch of physics. We need only consider the interrelationships between high-energy physics and cosmology, stellar evolution, and nuclear physics or those between the structure of galaxies and hydrodynamics to see how closely these two branches of knowledge are related. In a sense, the story of physics properly begins with Greek astronomy, because the Greeks were the first to try to understand and explain the movements of the stars and planets in the evening sky. The idea that astronomical objects obey unseen forces (which we now associate with physical laws) probably originated, albeit in a murky form, with the Greek astronomers. Although their beliefs now appear to us as antiquated, the Greeks tried to use their mathematics to understand what they imagined to be the "geometrical" structure of the universe. Despite their preference for conceptual mathematics (as opposed to its practical applications), several Greek astronomers showed the usefulness of mathematical techniques by calculating such things as the diameter of the earth and the number of grains of sand in the known universe. These feats suggested that mathematics could be useful to astronomers by providing a means for quantifying disparate physical phenomena. However, the idea of using mathematics to provide a common language for the sciences was not seriously considered by the Greeks. Consequently, they were unable to move beyond the relatively sterile considerations of observational astronomy and formulate useful physical laws to explain their observations.

Though the Greeks developed no body of astronomical laws or principles to guide them in their astronomical studies, they accumulated a vast amount of observational data and proposed various models of the universe that enormously stimulated post-Grecian astronomy. The Greeks contributed far more to astronomy than to physics, primarily because contemplating the night sky was much easier and far more pleasant than trying to analyze the content and structure of matter. The heavens, by their very beauty and mystery, invited study and wonderment; their appeal to the very philosophically minded Greeks could not be resisted. As the study of astronomy flourished, the study of physics languished.

Although Pythagoras and his followers placed the earth at the center of the universe, they pictured it as a sphere and as moving around the circumference of a small circle once every 24 hours. They thus accounted for the apparent daily rotation of the starry heavens and the daily motion of the sun and the moon. It did not occur to them that the earth rotates on its own axis. The Pythagoreans believed the center of the earth's circle to be occupied by a central fire that illuminated the sun and the moon. Since this primitive model could not account for all the observations, the Pythagoreans had to embellish it with many other attributes that had no physical basis.

Despite the commonsense view that the earth was stationary and the heavens circled around it, there were a few individuals who were not convinced that the geocentric (earth-centered) theory of the universe was correct. Aristarchus of Samos (310 B.C.-230 B.C.) is believed to have been the first person to offer a heliocentric (sun-centered) theory of the universe. He was born on the island of Samos and attended the Lyceum originally founded by Aristotle, where he studied under Strato, the so-called "physical philosopher." Aristarchus preferred mathematics and indeed became known among his contemporaries as "the mathematician," although that designation may have been given to distinguish him from several others having the same name.[1]

Aristarchus took the first big step in the development of our modern picture of the heliocentric solar system. Only his book *On the Size and Distances of the Sun and Moon* has been preserved, and it is clear from this book that he made the first serious attempt to measure the relative distances of the sun and moon from the earth. It was certainly such considerations that finally led him to a heliocentric model of the solar system that greatly influenced Nicolaus Copernicus, whom we may speak of as the modern Aristarchus. Aristarchus reasoned that when the moon is in its first quarter (in quadrature), at which time exactly half its visible disk is illuminated, the line from the center of the earth to the center of the moon is at right angles to the line from the moon's center to the sun. If, then, the distance from the moon to the sun were about equal to the moon's distance from the earth, the line from the earth to the sun would form an angle of about 45° with the line from the earth to the moon. But, as Aristarchus discovered, this is not so. This angle is very nearly 90°, which means that the sun's distance from the moon, and hence from the earth, is very many times greater than the moon's distance from the earth. But since the moon and sun appear to be equal in size, this means that the sun is actually many times larger than the moon and the earth. From this discovery, Aristarchus argued that it is much more reasonable to have a small body like the earth revolve around a large one like the sun than vice versa.

The Roman architect Vitruvius wrote that Aristarchus was a man who possessed not only extraordinary mathematical talents but also the ability to apply his talents to the solution of practical problems; Vitruvius credited Aristarchus with having invented the *skaphe,* a sundial "consisting of a hemispheric bowl with a needle erected vertically in the middle to cast shadows."[1] Although it is unclear what caused a distinguished mathematican like Aristarchus to concern himself with the relative positions of the bodies in the solar system, it is probable that preparing his book "gave him an appreciation of the relative sizes of the sun and earth and led him to propound a heliocentric system."[1]

Although the boldness of Aristarchus's thinking was applauded by some of the ancient commentators, the heliocentric model was not widely accepted. Archimedes, for one, argued that the heliocentric model proposed by Aristarchus was mathematically flawed because it seemed to suggest that "the ratio that the earth bears to the universe, as it is commonly conceived, is equal to the ratio that the sphere in which the earth revolves, in Aristarchus's theme, bears to the sphere of the fixed stars."[2] Archimedes's criticism was not accurate, however, because he assumed that Aristarchus was making a mathematical statement when in fact he was simply contrasting the smallness of the planet earth with the immensity of the heavens. More fatal to the durability of Aristarchus's theory was its dependence on mystical speculations about the universe that proved to be of decreasing importance as the more talented Greek mathematicians turned their attention to astronomy. The exacting demonstrations and calculations of Apollonius of Pera, Hipparchus, and Ptolemy were based on a geocentric orientation.[2] As these newer cosmological theories, with their epicycles and eccentrics, were more mathematically intricate than the older heliocentric theory, they seemed to possess greater intellectual validity and, consequently, gradually attracted the attention of the Greek astronomers and mathematicians.

Herakleides, who lived in the 4th century B.C. and therefore antedated Aristarchus by almost one century, proposed that the earth moves "not progressively but in a turning manner, like a wheel fitted with an axis from west to east round its own center." This idea of a revolving earth was a bold departure from the teachings of Plato and Aristotle (that the heavens rotated around a fixed earth), with whom Herakleides studied; he was probably also a Pythagorean, for he is said to have attended the Pythagorean schools. In the literature of other Greek philosophers of that period, Herakleides is reported to have suggested that Venus revolves around the sun, rather than around the earth, because the distance of Venus from the earth, as indicated by its changing apparent brightness, varies considerably.

Following Aristarchus, a few hundred years elapsed before any serious attempts were made to construct a physical model of the solar system that correctly reproduced the apparent motions of the sun, moon, and the known planets. Observational astronomy still flourished, particularly in Alexandria, where a school of observers, supported and encouraged by the Ptolemaic dynasty, accurately determined the positions of the planets and stars, using graduated instruments. Concerned with regulating the calendar, they also followed the apparent motions of the moon and the sun. Their observations later helped Hipparchus and Ptolemy to develop their theory of epicycles to account for the apparent motions of the planets in their geocentric model of the solar system.

Theoretical astronomy during that time was not altogether neglected; Apollonius of Perge (262–200 B.C.), one of the great early Greek mathematicians, who spent most of his life in Alexandria, developed the geometry of the apparent retrograde motions of the planets, which Ptolemy incorporated *in toto* into his geocentric theory. Apollonius's theory of conic sections also influenced Johannes Kepler, some 1300 years later, to consider conic sections (the ellipses) as planetary orbits around the sun. Apollonius became known as "the great geometer" for his extensive writings in mathematics, which were particularly distinguished by their clarity and absence of technical terminology. By employing a form of pure geometry, he derived the properties of conics and showed how to find the shortest and longest distance between a given point to a conic. According to Herbert Westren Turnbull's *The Great Mathematicians,* Apollonius also learned "to work with what is virtually an equation of the sixth degree in *x* and *y*, or its geometrical equivalent—in its day a wonderful feat.'' Apollonius is also credited with having invented a quick method for approximating the number pi in his *Unordered Irrationals*, a work that hints at what Turnbull believes was a primitive theory of uniform convergence.

The ever-increasing accuracy in navigation and knowledge of the geometry of the earth, impelled by the growth of commerce and the expansion of terrestrial exploration during this period, stimulated considerable geographical activity. That the earth is spherical was already known to, or surmised by, the Pythagoreans, and Aristotle had estimated the circumference of the earth to be about 40,000 miles. That the earth is spherical was immediately obvious to the ancient navigators, who noted that the positions of the stars relative to the horizon changed as they sailed north or south; constellations that never rose when navigators went far north did rise and set when they sailed toward the south.

It remained for the great geometer Eratosthenes of Alexandria (276–194 B.C.), however, to make the first accurate determination of the size of the earth's circumference; he was then the librarian of the great museum in Alexandria. Eratosthenes's procedure was similar to the modern geodetic methods of determining the ''lengths of a degree'' on the earth's surface. The ''length of a degree'' on the earth's surface is defined as the distance one must walk along a great circle (a circumference of the earth) for the direction of a plumb line (the direction of the vertical) to change by 1°. This distance is about 69 miles, so that the earth's circumference (complete length of a great circle) is 69 × 360° or about 24,840 miles.

Eratosthenes did not measure the length of a degree, but noted that at noon on the longest day of the year, the sun at Syene was directly overhead (it cast no shadow at the bottom of a deep well). But at the same time, the gnomon (a vertical rod) at Alexandria cast a shadow of a length that indicated that the sun there was $7\frac{1}{4}°$ south of (below) the zenith. This means that the direction of a plumb line (the vertical) at Syene and at Alexandria differ by $7\frac{1}{4}°$ (that is, their latitudes differ by $7\frac{1}{2}°$). But the distance, along a great circle, between Alexandria and Syene is about 500 miles, and $7\frac{1}{4}°$ is approximately 1/50th of 360°. Hence, Eratosthenes found the earth's circumference to be about 25,000 miles (more accurately, 24,500 miles), which is remarkably close to the modern value.

Hipparchus (*ca.* 190–120 B.C.) was undoubtedly the greatest of the earliest Greek astronomers; his observational and theoretical work set the stage for Ptolemy, whose book *The Almagest* was the European astronomer's textbook for the next 1500 years. Only one of Hipparchus's numerous books survives. Written in 140 B.C., it predates his discovery of the precession of the equinoxes and his star catalogue. According to Ptolemy, who, in his *Almagest,* described Hipparchus's work completely, Hipparchus began his celestial observations in 161 B.C. He had studied the astronomical work of the Babylonians as well as that of the early Alexandrians, which led him to his most famous discovery: the westward precession of the equinoxes.

To explain this discovery, we first define the equinoxes. Astronomers use two imaginary great celestial circles in their description of the rising and setting of the stars and the apparent motions of the sun, moon, and planets: the celestial equator (the great circle in which the plane of the earth's equator cuts the sky) and the ecliptic (the imaginary circle along which the sun appears to move eastwardly among the constellations from day to day). Since the planes of these two great circles are tilted about $23\frac{1}{2}°$ with respect to each other, the celestial equator and the ecliptic intersect at two diametrically opposite points on the sky called the ''vernal equinox'' and the ''autumnal equinox.'' As the sun in its apparent motion moves eastwardly along the ecliptic, it coincides with each of these equinoxes once a year—on about March 21 with the vernal equinox when spring begins (hence ''vernal'') and on about September 21 with the autumnal equinox. These points are called the equinoxes because on those two days, the sun rises at 6 A.M. exactly in the east point and sets exactly in the west point at 6 P.M. and the day and night are exactly equal in length.

Hipparchus, using the early observations of the times of the rising and setting of the stars and the beginning of spring, noted that spring seemed to begin somewhat earlier each year when compared to the rising of the fixed stars. To put it somewhat differently, Hipparchus discovered that the fixed stars seemed to shift their positions eastwardly with respect to the vernal equinox by a small amount each year, which he estimated roughly to be about 1° per 75 years. The precise figure, as measured today, is 50.2619″ per year, so that the seasonal year (the interval between two successive coincidences of the sun with the vernal equinox) is 20 minutes shorter than the earth's period of revolution around the sun (the sidereal year). This phenomenon is called the ''westward precession of the equinoxes.''

The cause of this precession was unknown then, but we now know that it arises because the earth is not a perfect sphere but an oblate spheroid with flattening at the poles and an equatorial bulge. Owing to this fact, the earth's axis of rotation does not point in the same direction at all times because the sun and moon pull on the equatorial bulge (that is, exert torque on the bulge) in such a manner as to make the axis rotate westwardly completely around once every 26,000 years. Put differently, the north celestial pole is not a fixed point in the sky; it moves westwardly in a circle around the pole of the ecliptic.

Hipparchus was undoubtedly the outstanding astronomer of his day and advanced science in general, and astronomy in particular, more than any other astronomer before him had done. His greatest contribution was to introduce precise measurements, mathematics, and careful reasoning into the analysis of astronomical data. The concept of stellar magnitudes that astronomers use today to designate stellar brightnesses goes back to Hipparchus, who arranged stars in classes according to their apparent brightnesses. From an analysis of the duration of a lunar eclipse, he deduced a fairly accurate value for the moon's size and its distance from the earth. He is also credited with having invented the science of the measurement of triangles, which is now known as trigonometry, although the details remained to be worked out by Ptolemy. While his own astronomical investigations convinced him that Aristotle's belief that the celestial bodies move around in circles was erroneous, he did not try other curves such as the ellipses that are now known to represent the planetary orbits. Instead, he and his successors experimented with combinations of circles known as epicycles and deferents, hoping to arrive at a model of the solar system that would explain the movements of the planets across the evening sky. So great was his influence that his efforts to develop mathematically the geocentric theory effectively scuttled meaningful discussion about Aristarchus's heliocentric doctrine for almost 16 centuries.

While Hipparchus was certainly the greatest astronomical observer of the ancient Grecian period, much of his work would have been lost if Ptolemy (100–170 A.D.) had not written his *Almagest*. Although Ptolemy lived in Egypt, his Latin name *Claudius Ptolemaeus* indicated that he possessed Roman citizenship, which may have been granted to an ancestor by the emperor Claudius or Nero.[3] Although Ptolemy was an accomplished mathematician, he is best remembered for his astronomical work, which synthesized much of earlier Greek astronomy with a thoroughness comparable to Euclid's treatise on geometry: "Ptolemy assumes in the reader [of the *Almagest*] nothing beyond a knowledge of Euclidean geometry and an understanding of common astronomical terms; starting from first principles, he guides [the reader] through the prerequisite cosmological and mathematical apparatus to an exposition of the theory of the motion of those heavenly bodies which the ancients knew (sun, moon, Mercury, Venus, Mars, Jupiter, Saturn, and the fixed stars, the latter being considered to lie on a single sphere concentric with the earth) and of various phenomena associated with them, such as eclipses."[3]

Ptolemy's book is a complete story of ancient astronomy, with great emphasis on its Alexandrian period, which was dominated by Hipparchus. In his *Almagest*, Ptolemy developed the complete theory of epicycles to explain the apparent motions of the sun, moon, and planets in a geocentric solar system. Ptolemy not only expounded the work of Hipparchus but also in many instances completed and extended it. Thus, he greatly improved the magnitude classification of stars, which was used until 1850, when it was replaced by a precise magnitude scale introduced by Pogson. The importance of the *Almagest* lay not so much in its astronomy as in the methodology it introduced into science: careful observations combined with mathematics. That no one could present a better way than epicycles to account for the motions of the planets attests to Ptolemy's great mathematical skill and power of reasoning.

CHAPTER 3

Pre-Galilean Science

The civilization of one epoch becomes the manure of the next.

—CYRIL CONNOLLY

Three great figures dominated scientific thinking in the century between 1500 and 1600: Nicolaus Copernicus, Tycho Brahe, and Johannes Kepler. We might well have devoted this chapter entirely to them, but that would have left the reader wondering why a gap of about 15 centuries separates Greek astronomy from the beginning of modern astronomy. For that reason, we first give a brief discussion of science during the Middle Ages to show the vast difference between the dogmatic scholasticism (non-science) of that period and the remarkable discoveries of Copernicus, Brahe, and Kepler. The medieval scholastics such as Hugo of Saint Victor and Thomas Aquinas sought to understand the world by uniting faith and reason into a single intellectual framework, while the early Renaissance scientists, particularly Kepler, tried to make sense of the universe by searching for mathematical patterns (that could be quantified) amid their astronomical observations. Although the scholastics developed impressive proofs for the existence of God, among other things, they could offer no external evidence to support their conclusions. By contrast, Kepler, after 30 years of labor, derived three simple mathematical relationships that could describe the motions of the planets in the sky. His repeated testing of theory against observation provided an early model for what we now call modern science and offered an empirical approach toward the world that continues to be relevant to this day.

Very little science was pursued in the long interval between Ptolemy and Nicolaus Copernicus, a time variously referred to as the Middle or Dark Ages, although technology did develop, which, of course, aided the growth of the science that came later. Thus, navigational instruments, mechanical clocks, gunpowder, firearms, spinning and weaving, metallurgy, and paper making advanced rapidly. The jewel in the crown of this technology was the invention of the printing press by Gutenberg at Mainz in 1436. Although the invention of a new technology implies the application of scientific principles, the medieval inventors did not start from basic scientific laws; in fact, they produced their inventions without even knowing the laws or principles involved, in contradistinction to Galileo, who constructed his telescope by a careful application of the law of optical refraction.

But medieval Europe was not entirely devoid of pure science, as evidenced by the researches in the 13th century by Roger Bacon, a Franciscan of Oxford, who stated that the "true student should know natural science by experiment" and should reject the untested opinions of fallible authorities. This was a very bold position to take, since it challenged the scholasticism of Aristotle, St. Augustine, and the revered theologians Albert Magnus and Thomas Aquinas. Bacon's threat to accepted authority was considered dangerous and heretical, and so he was reprimanded and forced to give up his study of optics, mechanics, and fluid dynamics with the stricture from his Franciscan superior that "the tree of science cheats many of the tree of life, or exposes them to the severest pains of purgatory."

Although Aristotelian concepts dominated the thinking of the Middle Ages, the Aristotelian concepts of motions were meeting ever-increasing resistance and criticism. Thus, William of Ockham in the 14th century argued that a body in motion does not require the physical contact of a "mover" to keep it moving and rejected the idea that heavenly bodies (for example, planets) are kept moving in their orbits by a choir of angels, suggesting instead that God might have endowed these bodies with motion initially (a kind of divine concept of inertia). "It is vain," he argued, "to do with more what can be done with fewer." Thus was born the principle of "Ockham's razor," which has been applied so often in choosing one among various theories.

This concept of an initial divine impetus to account for the observed motions of bodies became quite popular, and Nicholas of Cusa, bishop of Brixen in the 15th century, accepted the concept of the rotation of the earth as an impetus conferred on it at its creation. He expressed his acceptance in the statement that wherever a person may be—on a planet, a star, or the earth—he will always believe himself to be motionless at the center of the universe and all other bodies to be moving.

The supreme genius of the late Middle Ages and the early Renaissance was, unquestionably, the 15th-century Italian artist, inventor, and scientist Leonardo da Vinci, who conducted research in hydraulics, mechanics, and geology. He was an architect and prolific inventor who produced hundreds of designs for all kinds of machines and instruments—designs that indicated a profound knowledge of the basic physical and engineering principles that applied to his architectural work and inventions. But Leonardo left behind no evidence that he had discovered or formulated any basic physical laws, so that, in a sense, despite his great genius, he contributed very little to physics.

The development of physics during the period from Leonardo to Galileo was due primarily to the efforts of Copernicus, Brahe, and Kepler. Although they dealt primarily with astronomy, their discoveries greatly influenced Galileo and Newton, who may be considered the first physicist of the modern era. Of the three, Nicolaus Copernicus (1473–1543) most symbolized the new spirit of inquiry that appeared in 16th-century Europe as the flourishing of arts and culture caused a few brave individuals to begin reconsidering the philosophical validity of the officially sanctioned geocentric (earth-centered) cosmos.

Copernicus was born in Poland, the son of a prosperous merchant. His father died when Nicolaus was ten, so he was taken in by his uncle, Lucas Watzelrode, who became bishop of Ermland in 1489.[1] Little is known of his early education, but in 1491, Copernicus did enter the University of Cracow, where he studied mathematics. His election as a canon of the cathedral chapter of Frauenburg was aided by the efforts of his uncle, and the office provided Copernicus with an ample guaranteed income that made it possible for him to continue his education outside the country. Copernicus enrolled at the University of Bologna in Italy in 1496 to study law, but soon developed an interest in astronomy. In November 1500, he made his first observation of a lunar eclipse in Rome and gave a well-received lecture about it to a large gathering of scholars.[2]

Although Copernicus was given permission by his chapter to attend medical school, he preferred to pursue graduate studies in law. In 1503, Copernicus received a doctoral degree in canon law from the University of Ferrara and returned to Varmia, where he remained for the rest of his life. Copernicus continued his astronomical studies privately while devoting most of his energies to the service of his chapter. In 1513 Copernicus purchased some 800 building stones and a barrel of lime to construct a little roofless tower for making observations of the evening sky.[3] Extensive studies of the sun, the moon, and the stars convinced Copernicus that the geocentric theory was wrong, but he dared not publicly voice his opinions. Instead, he wrote his *Commentariolus* to document his own

Nicolaus Copernicus (1473–1543)

view that the sun, not the earth, is stationary, and circulated the manuscript among his closest friends. Although the influence of the heliocentric (sun-centered) theory of Aristarchus was apparent in Copernicus's thinking, Copernicus deliberately chose not to credit the Greek with having originated the idea. Regardless of the source of his inspiration, however, the *Commentariolus* was notable not only for challenging the intellectual monopoly long enjoyed by the stale Ptolemaic cosmology but also for explaining the apparent daily rotation of the heavens, the yearly journey of the sun through the ecliptic, and the apparent alternation of retrograde and direct motion by the planets.[4]

This is not to say that Copernicus was ready to throw out the last vestiges of the Ptolemaic system. Like Ptolemy, he believed that the planets move in circular orbits around the sun: "Copernicus considered that Ptolemy's system was, 'not sufficiently absolute, not sufficiently pleasing to the mind,' because Ptolemy had departed from the strict letter of the Pythagorean preconceptions.''[5] He believed that most astronomers were overly preoccupied with trying to explain the movements of the planets within an increasingly unworkable geocentric framework and were not interested in explanations that would challenge the prevailing orthodoxy. Copernicus knew the ancient and Ptolemaic cosmologies thoroughly and rejected them because they were unnecessarily complicated and, from his point of view, incomplete. In his words, ''they [his predecessors] are found either to have omitted something essential, or to have admitted something extraneous and wholly irrelevant.'' He was sure that the complexity of epicycles that Ptolemy had introduced to account for the apparent motions of the sun, moon, and planets arose because the earth's motions (for example, rotation, revolution) had been assigned to these bodies; if the earth were given back its ''true motions,'' the lunar, solar, and planetary motions would be greatly simplified. He saw that if the sun were placed at the center of the solar system and the earth were rotating around its own axis and moving around the sun with the other planets, most of the Ptolemaic epicycles could be discarded and the solar system would acquire a beautiful symmetry with respect to the sun that the Ptolemaic system lacked.

Copernicus's insistence on circular orbits for the earth and the planets meant that he had to introduce epicycles of his own to account for certain irregularities and asymmetries in the apparent motion of the sun and the planets, which were not fully explained until Kepler's great astronomical discoveries. Copernicus was not an observational astronomer, but tried to deduce his cosmology from as few simple axioms as he could; these axioms, however, included many of the ancient Greek misconceptions about the nature of motion, such as that the motions of celestial bodies must be circular and uniform. The argument for circular orbits was purely theological and was accepted by Copernicus in accordance with his belief that the ''perfection'' of heavenly bodies required that they move about in ''perfect'' orbits—namely, circles. The ''perfection'' that Copernicus referred to *is* what physicists now call symmetry, which plays a very important role in physics today. A circle in a plane has maximum symmetry in the sense that it has the same appearance no matter from what direction in the plane we look at it. The symmetry of an ellipse, however, is smaller than that of the circle.

Copernicus's heliocentric solar system caught on very slowly, primarily owing to the theological objections to it but also because a spinning, moving earth could not be reconciled with the feeling of complete rest that people have on the earth. But as mounting quantities of celestial data were accumulated, the ever-increasing number of epicycles required to account for the data made the geocentric model of the solar system seem more and more unwieldy and repugnant.

Another consequence of setting the earth in motion around the sun was that it forced Copernicus to consider the immense distances that seemed to separate the sun and planets from the stars: "[As] compared with the earth, the heavens are immense and present the aspect of an infinite magnitude, while on the testimony of the senses, the earth is related to the heavens as a point to a body, and a finite to an infinite magnitude."[6] By drawing on the atomic theory of Democritus and Lucretius, Copernicus envisioned a universe of infinite magnitude in which the earth is part of a system of planets:

> [Atoms] can be multiplied to such an extent that in the end there are enough of them to combine in a perceptible magnitude. The same may be said about the position of the earth. Although it is not in the center of the universe, nevertheless its distance therefrom is still insignificant, especially in relationship to the sphere of the fixed stars.[7]

Copernicus never openly declared that the universe is infinite, because he was reluctant to cast the sun loose in an infinite void. His conservatism in this regard is curious, because removing the earth from its central position had not bothered him greatly, even though he was unable to condemn the sun to the same fate. As the naked eye cannot detect the very small parallax displacement of the stars produced by the annual motion of the earth, the evidence seemed to favor an infinite universe. Copernicus resolved his philosophical objections by brushing the matter aside and refusing to characterize the universe as either infinite or finite.

His book *De Revolutionibus* was published in 1543, on the day of his death, but a century elapsed before his heliocentric theory was considered seriously by more than a few bold thinkers, and this was due to the work of the three great intervening scientists, Brahe, Kepler, and Galileo, who died in 1643. Copernicus's astronomical observations were scanty—no more than 27—but Brahe more than made up for that in precision and quantity.

Born in Denmark, Tycho Brahe (1546–1601) was one of ten children of Otto Brahe, a privy councillor who later became governor of Helsingborg Castle. Young Tycho went to live with his uncle and was tutored at home from the age of seven.[8] In 1559, he began three years of

Tycho Brahe (1546–1601)

study at the Lutheran University of Copenhagen, where he concentrated on Latin and classical literature. He was greatly influenced by mathematical lectures on Pythagorean harmonics and emerged as a staunch Aristotelian.[9] Brahe also studied the sciences, especially the classical Greek treatises on physics and mathematics, and devoted much of his time to learning about astrology. A chance observation of a solar eclipse in 1560 fueled his interest in astronomy and prompted him to begin making his own naked-eye observations of the heavens.

In 1562, Brahe went to the University of Leipzig at the urging of his family to study law. He was accompanied by a tutor, Anders Vedel, who had been hired to help Brahe with his legal studies and to prevent him from pursuing his interest in astronomy, which Brahe's uncle thought to be a waste of time. Brahe was not to be denied, however, and while his tutor slept, he often sneaked outside to study the stars.[10] He spent what money he could save on astronomy books and instruments. After finishing at Leipzig in 1565, Brahe returned to Copenhagen and became involved in a dispute with another Danish nobleman that resulted in a piece of Brahe's nose being sliced off during a duel. For the rest of his life, Brahe wore a silver and copper covering over the wound, and when his tomb was opened in 1901, "a bright green stain was found on the skull at the upper end of the nasal opening."[11]

Finding little reason to remain in Denmark, Brahe spent a year at the University of Basal and then moved on to Augsburg, where he began to make detailed astronomical studies using devices of his own invention, including a massive wooden quadrant some 19 feet in diameter, a portable sextant, and a 5-foot globe, to map the stars of the constellations.[12]

On November 11, 1572, he observed a brilliant new beacon of light in the constellation Cassiopeia, and through a series of measurements of its position relative to nearby stars (which showed no change over the next two years), he concluded that he was indeed looking at a new star and not a companion moon as he had originally suspected.[13] He kept careful notes on his observations of the "nova," notes that became an invaluable source of information to future scientists and historians.

Observational astronomy at that time was in a deplorable state, with no one concerned with observational accuracy. Brahe saw correctly that little progress could be made in astronomy without a systematic, continuous course of observations of the stars and planets. Fortunately, he was invited by Frederick II of Denmark to set up an observatory on the Danish island of Hveen in the Copenhagen Sound; he furnished it with the best naked-eye instruments available and, from 1576 to 1597, collected a vast body of very precise data. A superb naked-eye observer, he greatly increased observational accuracy; his work was a decisive factor in Kepler's discovery of the three laws of planetary motion. The accuracy of Brahe's observations, in fact, was down to the very limits of the accuracy of the human eye. Brahe also amassed an extensive personal library of astronomical manuscripts and compiled a vast body of precise observations. Believing that science can progress only within a hospitable theoretical framework, however, he clung to a modified geocentric theory that was of little value to his observational work.

In 1597, Frederick II died, and Brahe was forced to look elsewhere for financial support. Having fallen out of favor with the Danish court, Brahe considered resettling in Holland before friends convinced him to travel to Prague for an audience with the emperor. The emperor was sympathetic and arranged for Brahe to receive financial support and the use of the castle of Benatky northeast of Prague, where Brahe built a laboratory and an observatory.[14] However, Brahe was never able to fund his efforts properly. His poor health coupled with cramped working conditions prevented him from making many more important observations.

During the last year of his life, while at Prague in 1599, Brahe invited the young German astronomer Johannes Kepler to be his assistant. Kepler had already been converted to Copernicanism and enthusiastically accepted the invitation, for he saw Brahe's observational data as just what he needed to prove mathematically the truth of the Copernican doctrine. Although Kepler's mathematical talents complemented Brahe's powers of observation, the two often quarreled bitterly. However, each recognized his dependence on the other, and so a tenuous partnership was maintained until Brahe's death. Kepler, inheriting Brahe's data and position in Prague in 1601, began his momentous work on the orbits of the planets, which he completed nearly 30 years later.

Throughout his life, Johannes Kepler (1571–1630) was preoccupied with divining harmonic relations in the universe. His three laws of planetary motion, the product of many years of tedious calculations, were a by-product of his unceasing search for mathematical relationships that could explain the movements of celestial bodies, particularly the planets. Kepler's enthusiasm for the Copernican theory was tempered by his awareness of its shortcomings; its preference for circular planetary orbits continually resulted in discrepancies between the predicted and observed positions of the planets.

Although Kepler possessed a formidable intellect and incredible powers of concentration, he was a sensitive man who did not easily accept criticisms. His somewhat erratic personality was probably shaped by his unhappy childhood. His father, Heinrich, was an ill-tempered mercenary who fought in several military campaigns and finally abandoned the family in 1588.[15] His mother, Katharina, was a meddlesome and embittered woman who was unable to deal with the family's poverty; Kepler later became embroiled in an exhausting and financially disastrous three-year legal dispute defending his mother against charges of witchcraft and in so doing left his own family destitute.[16]

Young Kepler received his early education at Leonberg and excelled

Johannes Kepler (1571–1630)

in Latin. He enrolled at the Adelberg monestary in 1584 and, two years later, began his preparatory education at Maulbronn. In 1589, he began to study astronomy at the University of Tübingen. Kepler's interest in the Copernican theory was nurtured by his professor, Michael Maestlin.[17] After obtaining his master's degree in 1591, Kepler began to study theology, but the death of Georgius Stadius, a teacher of mathematics at the Lutheran school in Graz, opened a vacancy that Kepler was called on to fill.[18] Although Kepler had not initially shown much interest in mathematics, he accepted the position and gave up any further intentions of becoming a clergyman.

Aside from his teaching and tutoring duties, Kepler devoted much of his time to constructing an astrological calendar that made several predictions about peasant uprisings and harsh weather that later occurred, thus enhancing his local reputation.[19] He produced several more editions of his calendar over the next few years, but found himself increasingly uneasy with the unscientific basis of his predictions. Although he was publicly inclined to dismiss the scientific usefulness of astrology, he was inevitably drawn to it because of his belief that there is some sort of mysterious relationship between the activities of human beings and the motions of the stars and planets. However, his calendars did provide a steady source of income, and his astrological skills were later instrumental in his obtaining an appointment as an imperial mathematician.[20]

Kepler's astrological interests inspired his first cosmological work, *The Mystery of the Universe,* which appeared in 1597: "He sought for mathematical harmonies between the orbits of the planets in the Copernican system, finding that the five regular solids could be fitted between the spheres of the planetary orbits."[21] Kepler's conclusion that the five regular solids roughly corresponded to the known orbits of the planets convinced him that the universe has a geometrical basis and that there could be only five planets because there are only five regular polyhedrons. This conclusion clearly had no basis in fact, but his stubborn insistence on its truth shows the degree to which geometrical considerations motivated his thinking.

Although Kepler considered this geometrical model to be among his greatest accomplishments, the work that he is best known for, his three laws of planetary motion, began when he received Tycho Brahe's deathbed request that Kepler use Brahe's tables of observations to prove Brahe's own geocentric theory. Kepler himself was concerned primarily with the orbit of Mars, which appeared to be anything but a circle and which therefore strongly challenged the Copernican hypothesis of uniform motion in circular orbits. Convinced that Copernicus was right in every detail, Kepler spent many fruitless hours trying to squeeze Brahe's Martian orbital data into a circle and almost succeeded. At one point, however, the circle he had constructed to fit the data departed from Brahe's observations by 8 minutes of arc. Kepler therefore reluctantly rejected the idea of a circular orbit for Mars, for he knew that Brahe had been too accurate an observer to have made so large an error as 8 minutes of arc.

Kepler then turned to other possible orbits and finally chose the ellipse, to which he had previously given little thought. The ellipse lacks the perfect symmetry of the circle, but it possesses a marked simple symmetry with respect to its largest diameter—that is, its major axis—a symmetry that is immediately evident from the definition of the ellipse as well as from the following prescription for constructing one: Draw a line of a given length between two points A and A' on a piece of paper and mark off two other points F and F' on this line, with F' at the same distance from A' as F is from A. Take a string of length AA' (the length of the line) and pin one of its ends to F and the other to F'. Now place a pencil in the loop of the string and, keeping the string as taut as possible, trace an ellipse on the paper with the pencil point. The points F and F' are called the "foci" of the ellipse, and the length AA' is its "largest diameter" or its "major axis." Various elliptical shapes can be produced by bringing the points FF' closer together (the shape becomes more circular) or separating them further toward A and A' (the shape becomes flatter).

The ellipse is the basis of Kepler's first law of planetary motion, which he stated as follows: Each planet moves in an ellipse around the sun, which is at one of the foci of the ellipse. Since the sun must be at one focus of the ellipse of each planet, all the planetary ellipses must have one focus in common. This statement is the first example in the history of science of a law of physics that applies to the motion of bodies. It is also the first example of the remarkable correspondence between the dynamics of moving bodies and mathematics.

It is instructive to consider the ellipse from another point of view, namely, as a conic section, which is described as follows: It is the outline of the top of a truncated right circular cone produced by a random cut through the cone. The shape of this outline depends on the tilt of the cut with respect to the base of the cone. Since the shape is a circle if and only if the cut is exactly parallel to the base of the cone, the probability of getting a circle by a random cut is zero. This means that the chance of having a circular orbit for a planet is also zero. We shall consider this point more carefully when we discuss Newton's derivation of Kepler's laws. Other conic sections are the parabola and the hyperbola, which are also possible orbits of celestial bodies.

Since the distance of the pencil point from focus F plus its distance from focus F' is always the length of the string, that is, the length of the major axis, the ellipse is often defined as the locus (set or collection) of all points the sum of distances of which from two fixed points (the foci) is constant. This constant, or half of it (the "semimajor axis," as it is called), is an important numerical parameter of the ellipse, for it determines its size.

Another parameter, the eccentricity of the ellipse, is important in studying the geometry of the ellipse, for it determines the shape of the ellipse. The eccentricity, which can range from 1 to 0, is a measure of the amount by which the ellipse departs from circularity. As the two foci F and F' are brought closer to the center of the ellipse (the point halfway between A and A'), the eccentricity approaches 0 and the ellipse becomes more circular—becoming a circle when F and F' coincide at the midpoint. As F and F' are separated, the eccentricity approaches 1 and the ellipse becomes flatter—becoming a straight line when F and F' coincide with A and A', respectively.

Kepler discovered this first law of planetary motion (elliptical orbits) empirically, that is, by trial and error, without knowing the physical significance of the two geometrical parameters that we have just discussed—the size and shape of the ellipse. Their full dynamic significance will be revealed when we show how Kepler's three laws of planetary motion flow quite naturally from Newton's laws of motion and his law of gravity.

That Kepler, without any basic scientific principles to guide him, discovered the geometrical nature of the planetary orbits is amazing, for he had to give up every preconceived notion about "perfect orbits" to do so. The discovery of this first law did not satisfy Kepler's craving for a thorough knowledge of planetary motion, but it did show the way he had to go to get such an understanding. He saw that giving up circular orbits meant rejecting uniform motion as well, and so he began his study of the motions of the planets. His object was to find something in the motion of a planet around the sun that does not change even though its speed does. This search was very fruitful, for it led him to his other two laws of planetary motion—the law of areas and the harmonic law.

Kepler's discovery of his second law—the law of areas—attests to his great genius and to his superb mathematical skill. Since the law of areas does not say anything about the speed of a planet directly, but rather about how a certain geometrical feature of the ellipse associated with the planet's motion changes, Kepler had to have deep insight and intuition to discover the law, which he stated as follows: The line from the sun to a planet (the planet's "radius vector") sweeps out equal areas in equal times.

In discovering this law, Kepler had to go beyond Brahe's observations, for he had to calculate the area of various elliptical sectors or wedges defined by lines drawn from the sun to a planet at different points in its orbit. This entailed extremely tedious arithmetic, algebraic, and trigonometric work, but it finally led Kepler to the famous law of areas, which states that the area of the sector of the ellipse swept out by the line from the sun to the planet in a given time is always the same no matter where the planet may be in its orbit. This law is generally stated as follows: The radius vector from the sun to a given planet sweeps out equal areas in equal times. This remarkable law is the first example of a statement of a conservation principle in science, although Kepler was not aware of the equivalence of his second law to the principle of the conservation of angular momentum, as we explain in our discussion of Newton's law of gravity (Chapter 6).

Kepler announced his third law of planetary motion—the harmonic law—in his book *The Harmony of the World,* which was published in 1618. As in his previous investigations, he had proceeded by his system of trial and error, testing all kinds of numerical combinations of the periods of the planets and their mean distances from the sun. The work was extremely laborious and detailed, but he persisted until he found the correct relationship between the period of revolution of a planet around the sun and its mean distance from the sun, which he stated as follows: The square of a planet's period (time to complete one revolution) is proportional to the cube of its mean distance.

As Kepler stated it, this law means that the square of a planet's period divided by the cube of its mean solar distance is constant and hence is the same for all planets. He considered this discovery his greatest achievement, for it represented the fruit of 16 years of his most productive period, "for which," as he said, "I joined Tycho Brahe and for which I settled in Prague."

When we discuss Newton's law of gravity, we shall see that Kepler's third law, as he stated it, is not quite correct; when deduced from Newton's law, the third law differs from Kepler's version in that the square of a planet's period divided by the cube of its mean solar distance is not constant, but varies slightly from planet to planet. But this variation is too small for it to have been deduced from Brahe's observations, so Kepler's statement of the third law agrees with Brahe's data, but does not agree exactly with the law of gravity.

Although Kepler knew that the planetary motions are controlled by a solar force, he did not identify this force with gravity, and he misunderstood its nature. He thought of it as a magnetic force that decreases with increasing distance, rather than with the square of the distance, and as a force that acts laterally (at right angles to the line from sun to planet), rather than toward the sun. Moreover, he believed that if this force vanished, the planets would stop dead in their orbital paths instead of continuing to move.

Kepler studied optics and designed a telescope that he probably built but never used. He discovered the inverse-square law of the decrease in the brightness of a source of light, for he saw instinctively that light from a faint source spreads out spherically and that the brightness of the source therefore varies inversely as the square of the observer's distance from it. Kepler also investigated the refraction of light and showed that Ptolemy's approximate law of refraction—that the angle of refraction (the angle of bending of the light's path) is proportional to the angle of incidence—holds only for small angles of incidence. However, he did not discover the correct law of refraction, which was discovered in 1621 by Willebrod Snell, a contemporary of Kepler's and a professor of mathematics at Leiden.

In Kepler's *Epitome of the Copernican Astronomy,* which he completed in 1621, he offered his view of how the Copernican theory had been modified since it first appeared and showed how his own substantial contributions had changed cosmology from a speculative and inexact subject to one that was both theoretically unified and mathematically elegant. Kepler's systematic approach toward his subject was anchored by his belief that astronomy consists of five parts: "First, the observation of the heavens; secondly, hypotheses to explain the apparent movements observed; thirdly, the physics or metaphysics of cosmology; fourthly, the computing of past or future positions of the heavenly bodies; and fifthly, a mechanical part dealing with the making and the use of instruments."[22] This book became the most widely read astronomical textbook in Europe at that time, since it contained Kepler's three laws of planetary motion and described the heliocentric theory in unprecedented detail.

Kepler also spent nearly 30 years preparing the *Rudolphine Tables,* which integrated Brahe's observational data and his own research about the motions of the planets to provide the most accurate compendium of planetary orbits available. That the *Tables* did not appear until 1627 was due to its complicated printing requirements as well as its lavish illustrations:

> The printed volume of the *Tabulae Rudolphinae* contains 120 folio pages of text in the form of precepts and 119 pages of tables. Besides the planetary, solar, and lunar tables and the associated tables of logarithms it includes Brahe's catalog of 1,000 fixed stars, a chronological synopsis, and a list of geographical positions. In some of the copies there is also a foldout map of the world, measuring 40 × 65 centimeters; the map was engraved in 1630 but apparently was not distributed until many years later. This work stands alone among Kepler's books in having an engraved frontispiece—filled with intricate baroque symbolism, it represents the temple of Urania, with the Tychonic system inscribed on the ceiling. . . .[23]

Kepler convinced Emperor Ferdinand II to agree to impose a tax on three cities so that Kepler could be paid some 6300 guldens due in back pay for services to the court in exchange for Kepler's agreement to publish the *Tables* in Austria.[24] Although Kepler was able to collect only about one third of the monies due him, he financed the printing of the book in Linz in 1624. Soon after Kepler settled into his new home in Linz, however, the Counter-Reformation exploded in Europe and brought war to Kepler's doorstep. Linz was blockaded and the printing house was burned to the ground. Fearing that he might never see the *Tables* printed, Kepler packed his belongings and moved to Ulm, where he found a new printer and personally supervised the production of the book. Although the continual religious and political upheavels caused Kepler to consider moving to Italy or Holland, he was unwilling to forfeit his salaries as provincial and imperial mathematician.[25] Kepler spent his remaining years traveling among the principalities of central Europe and conducting astrological readings for his benefactors. His last work, *The Dream,* was completed shortly before his death; it was a work of fantasy that is notable because "its perceptive description of celestial motions as seen from the moon produced an ingenious polemic on behalf of the Copernican system."[26] It was not published until after Kepler's death, when his son sold it to a publisher to help pay some of the family debts.

Kepler, in his approach to and practice of science, was closer to the modern scientist than any of his predecessors. He was highly imaginative and speculative, but he was never content with an amorphous, unstructured theory or hypothesis. Theories and hypotheses had to be formulated precisely and in such a way that they could be tested against accurate observations. To Kepler, the observed data were the final criteria by which all theories had to be judged, and he never flinched from applying such judgments to his own theories.

CHAPTER 4

The Physics of Galileo

Here and elsewhere we shall not obtain the best insights into things until we actually see them growing from the beginning.

—ARISTOTLE

Although Galileo Galilei (1564–1642) was a contemporary of Johannes Kepler, the two scientists seldom communicated with each other and had little in common, even though they were most responsible for laying the scientific foundation that made possible Isaac Newton's contributions to the study of mechanics. Galileo was perhaps not as talented a mathematician as Kepler, but his professional interests were more diverse, and he made unparalled use of experiments to illustrate physical phenomena, such as the acceleration of freely falling bodies. Moreover, Galileo was an innovative craftsman who could construct devices such as greatly improved telescopes, which made possible a number of important astronomical discoveries and greatly extended the boundaries of the observable universe.

The son of a Florentine merchant, Galileo was born three days before the death of Michelangelo at a time when science was in its infancy and scholarly inquiry was weighed down by the heavy hand of the papal dogmatists. Galileo's father saw no reason not to encourage the free exchange of ideas and doubtless passed his enthusiasm for unfettered debate in any discipline down to his son. Young Galileo received his early education at the monastery of Vollombrosa near Florence and then studied mathematics at the University of Pisa. After working as a lecturer at the Florentine Academy, Galileo began teaching mathematics at the University of Pisa in 1592. The 18 years Galileo spent on the faculty at Pisa were the most productive period of his life. He conducted a number of experiments that demonstrated the shortcomings of Aristotelian physics, such as the belief that the continuous motion of a body was possible only if it

Galileo Galilei (1564–1642)

remained in contact with the propelling force. Galileo's studies of mechanics showed that a body does not come to a halt when the force propelling it is removed, but instead decelerates at a rate dependent on the amount of friction it encounters; this conclusion led him very close to the concept of inertia. Galileo also argued that all objects would fall at the same rate if there were no atmospheric friction. Although legend has it that Galileo dropped objects of different weights from the top of the Leaning Tower of Pisa to disprove Aristotle's contention that the heavier object would reach the ground first, there is no record that he actually conducted such experiments.

Among Galileo's scientific discoveries are the isochronism of the pendulum, the hydrostatic balance, the principles of dynamics, the proportional compass, and the thermometer.[1] He also made many improvements in the telescope and established himself as a peerless celestial investigator with his observations of the surface of the moon, the moons of Jupiter, and the phases of Venus. His careful study of the rugged landscape of the moon, however, first caused Galileo to run afoul of religious orthodoxy. Through his telescope, he saw what appeared to be "dark seas and bright lands, oceans and continents, mountain peaks swelling in the morning light, valleys lapsed in shadow."[2] The obvious similarity to the geography of the earth was not lost on Galileo, and he did not hesitate to speculate that the earth and moon consisted of the same matter. This conclusion was dangerous because the earth was thought to occupy a central place in the universe and therefore to be made of substances not found elsewhere. His preference for the heliocentric cosmology of Copernicus and his suspicion that the earth is a minor planet in the solar system convinced Galileo that the orthodox hierarchial view of the universe was wrong.

That Galileo's doubts about Aristotelian philosophy represented a challenge to the intellectual integrity of the Church was not lost on those churchmen familiar with Galileo's opinions. The Church itself was not inclined to be particularly tolerant of dissenting views, since it faced the threat of the Protestant Reformation; indeed, too much freedom of thought was thought by many to have contributed to the schism. Consequently, the Church dared not permit any further challenges to its doctrines. Galileo was aware of this attitude, and his enthusiasm for the Copernican system was tempered by his belief that the open advocacy of heliocentricity would invite punishment from the religious establishment, as shown by his 1597 letter to Johannes Kepler:

> . . . for years I have been an adherent to the Copernican view which explains to me the causes of many natural phenomena that remain utterly inexplicable within the limits of the commonly accepted hypothesis. To disprove this latter I have compiled numerous arguments but I dare not bring them to the light of public attention for fear lest my fate become that of our master Copernicus who, although in the esteem of some his fame has come to be immortal, stands in the opinion of infinitely many (for such are the numbers of fools) as an object of ridicule and derision.[3]

Galileo's reticence was further encouraged by the burning at the stake of Giordano Bruno in 1600 for heresy. Bruno's death in the Campo Dei Fiori in Rome had been preceded by six years of imprisonment by Inquisition authorities; his death came with his refusal to recant his philosophical opinions, which centered around his belief that the universe is infinite and has an infinite number of worlds. The absence of a cosmic center as well as the multiplicity of worlds in Bruno's universe posed direct challenges to the geocentric theory; Bruno's unbending advocacy of these blasphemous opinions sealed his fate. One important difference between Bruno and Galileo, however, was that the former had constructed a cosmology that conflicted directly with the picture of the world offered by the Bible, while the latter was not prepared to interpret the conclusions of Copernicus's theory in an anti-Christian manner:

> For all of [Galileo's] life he was convinced that the concept of a stationary sun and moving earth was completely consistent with the Bible, if only the Bible were read properly. There is no evidence that Galileo ever had any struggles with his religious conscience; it appears that he not only started the modern age of science but also was the first representative of that group of scientists who had no difficulty reconciling science with possible supernatural intervention.[4]

Galileo's religious beliefs did not dissuade him from arguing for the Copernican theory, but his efforts to sidestep Church objections failed, and he was censured by the Holy Office in 1616. He was ordered not to hold, teach, or defend the heliocentric doctrine. As Galileo was a more practical man than Giordano Bruno had been and saw little personal advantage to be gained by martyrdom at the stake, he agreed to abide by the terms of the censure action. Galileo published nothing until 1623, when his friend Cardinal Berberini became Pope Urban VIII. Urban was thought by Galileo to be a supporter of the arts and sciences, and while he could not reverse the antiheliocentric decision, Urban "did not forbid discussion of the theory as a speculative hypothesis."[5] When Galileo published his *Dialogues Concerning the Two Chief World Systems* in 1632 in support of the condemned theory, he dedicated it to Urban and prefaced the work with a long statement about his devotion to the Church.

Although Galileo had hitherto been careful to present the Copernican theory only as a possible alternative to accepted doctrine, his preference was made clear in his *Dialogues*. Galileo had many enemies in the papacy. Fearful of the continued erosion of Aristotelian philosophy, they convinced the Holy Office that Galileo had not abided by the terms of his 1616 censure. As a result, Galileo was ordered by the Holy Office in 1632 to go to Rome to face the Inquisition. He arrived there in February 1633 and was examined four months later on charges that the publication of his *Dialogues* had violated the 1616 decree. Galileo was found guilty of the charges and sentenced to jail; he was forced to kneel and recant his belief that the earth moves around the sun. It is alleged that Galileo muttered, "*E pur si muove*" ("Yet it does move") under his breath even as he was promising not to challenge the geocentric doctrine. Galileo was allowed to return to his villa in Florence, but he remained under house arrest for the last eight years of his life. Although he had no choice but to refrain from further public controversy, he did continue his scientific experiments during that time and managed to smuggle out a copy of his *Dialogues* so that it could be translated into Latin.

As noted above, Galileo played a pivotal role in helping to anchor the infant science in a bedrock of experiments and observations while purging it of idle philosophical speculations, especially in the field of mechanics. The science of mechanics grew out of the study of motion, which was developed in two steps—kinematics and dynamics. Kinematics deals with the motions of bodies without inquiring into the causes of such motions; dynamics deals with force as the cause of changes in the state of motion of a body. Mechanics was thus not fully developed until Newton propounded his three laws of motion, without which dynamics could not have evolved, but Galileo had already established kinematics as a science. The study of mechanics stemmed from the need to understand such practical phenomena as the flight of projectiles (for example, arrows, bullets, cannonballs), the propulsion of vehicles, the movements of animals, and the flight of birds. Leonardo da Vinci had studied mechanics intensely, for he considered it the worthiest of all sciences and one that was amenable to mathematical formulation and testing. It is not surprising, then, that Galileo, da Vinci's intellectual descendant, concerned himself more with mechanics than with any other phase of physics. He felt intuitively that until the nature of motion was thoroughly understood, mechanics could not be mastered.

Very little progress had been made in the study of motion since the time of Aristotle, who had introduced what appears to be a very reasonable idea: that bodies move only because they are being pushed or pulled by some kind of force. This was generally accepted as a correct statement because people knew from direct experience that objects on the ground moved only if they were pushed or pulled; physical effort was needed to keep an object moving. This Aristotelian concept led to severe difficulties in the study of the motions of the planets, for they appeared to move without being pulled or pushed. The scholastics and theologians "solved" this problem by simply assigning to each planet an angel who kept that planet moving in its appointed celestial orbit. But opposition to this concept was taking hold, as first expressed by the "impetuists," who, following William of Ockham, declared that God gave each celestial orb an impetus that has kept it going since. This was thus an easy way out of the dilemma without offending the Church, but it added nothing to the understanding of motion. Mechanics hobbled along without progressing beyond the discussion stage until Galileo began to subject it to experimental tests and to formulate a mathematical theory of motion.

After Johannes Kepler, Galileo was the second great scientist who saw the importance of mathematics in the development of the principles and laws of nature; he set himself the task of applying mathematics to the investigation of physical phenomena, arguing that any such phenomena involving measurable qualities could be formulated mathematically. He went beyond this idea, however, and pointed out that the mathematical formulation of a problem permitted one to deduce results, by means of mathematical manipulations, that are not directly observable in the phenomena themselves. This, of course, is the basis of most scientific discoveries today (as it has been from the time of Newton).

Galileo first applied his mathematical skills to the problem of scaling, which is used extensively in all branches of engineering and science today. He noted that the dimensions of the supports (for example, the legs of an elephant) of two similar structures (structures that have the same chemical composition and density) do not scale dimensionally in the same way as the two structures themselves. Thus, if the size of the larger structure is 5 times that of the smaller one, the thicknesses of the supports of the larger one must be more than 5 times those of the smaller one. The reason is that the weight of a structure increases as its volume and hence as the cube of its height (thus by a multiple of 125 in the 5-fold height increase). In contrast, the support strength depends only on the cross-sectional area of the supports and hence increases as the square of their height (thus by a multiple of 25). Thus, the strength of the supports is deficient by a factor of 5 if all the dimensions of the supports are increased by a factor of 5. If the heights of the supports are increased by a factor of 5, the diameters (thicknesses) must be increased by a factor of at least the square root of 125.

To study the motions of freely falling bodies (motions that are not hampered by some kind of resistance), Galileo saw that he would have to take into account the resistance of the air, which hinders free fall. Having no vacuum in which to perform his experiments, he decided to work with small metal spheres of different masses; since the air resistance to the motion of a body depends on the area of the surface of the body facing the flow of the air, a sphere experiences less resistance than a body of any other shape having the same mass. With that settled, Galileo had to devise some way of following the motion of a freely falling body in detail as it falls, which is impossible if the body falls vertically, since things then happen too rapidly. He overcame this difficulty by allowing his metal spheres to roll down very smooth inclined planes.

Galileo wanted to discover how the pull of gravity (which is always vertical) changes the motion of a body and whether the motions of all bodies are affected in the same way. He had dropped stones of different weights from the same height, and it had seemed to him that they had all hit the ground at the same time, but this sort of casual observation was not conclusive enough for him, so he devised the inclined-plane experiment to make precise measurements. A sphere rolling down an inclined plane is not affected by the entire pull of gravity, but only by that part of it that is along (parallel to) the inclined plane. As the slope of the incline is increased, the pull of gravity along the incline increases from a zero value (when the plane is horizontal) to a maximum (the full value) when the plane is vertical. Thus, by decreasing the slope of the incline, Galileo could make the metal spheres roll down as slowly as he desired—slowly enough to permit him to make any measurements he desired and to time the motion carefully. At a slope of 30°, the pull of gravity along the incline is half its full value; at 45°, it is 1/2 the square root of 2 times its full value; and at 60°, it is 1/2 the square root of 3 times its full value.

He made a series of important observations that became the basis of Newtonian mechanics, as formulated in Newton's three laws of motion, the first of which—the law of inertia—was a direct consequence of Galileo's study of motion. Galileo observed that as long as a sphere is rolling down the inclined plane, its speed increases by the same amount in equal time intervals. But once it leaves the incline and moves along the smooth horizontal surface, its speed remains constant. He thus disproved experimentally Aristotle's contention that a force must act on a body just to keep it moving at a constant speed. The inclined-plane experiments demonstrated that the constant force of gravity along the inclined plane increases the speed of the rolling sphere constantly so that the action of a force changes the velocity of a body (the motion along the inclined plane), whereas the absence of a force (motion along the horizontal) means constant velocity. Thus, Galileo associated acceleration (changing velocity) with the action of a force.

Galileo made a number of mathematical deductions from his observations of the rolling spheres. He first showed that the speed of a rolling sphere increases steadily with time and that this rate of increase of speed (the acceleration) is the same for all spheres, regardless of their weights or sizes—that is, all spheres, starting from the top of the same inclined plane, have equal speeds at the bottom. He also proved mathematically that the distance a sphere rolls along the plane is proportional to the square of the time it is rolling and that the square of its speed at any point on the incline is proportional to the distance along the plane of that point from the top of the plane. From these simple mathematical relationships, he concluded that if a body (regardless of weight) were falling freely in a vacuum, its speed would increase by about 32 feet per second every second (the acceleration of gravity). He noted further that the reason the sphere finally stops on the horizontal surface is that this surface is not perfectly smooth; as the smoothness of the surface is increased, the sphere rolls an ever-increasing distance. He therefore concluded that if the surface were perfectly smooth, the sphere would roll on forever (the concept of inertia). He made one more important observation: The speed of a sphere at the bottom of all inclined planes is the same, regardless of the lengths of the planes, provided that the tops of the planes are all at the same height from the ground. Only the height of the plane from the ground determines the speed of the rolling sphere at the bottom of the plane. Simple as these observations and deductions were, they initiated the science of mechanics.

Galileo went beyond his experiments with spheres on inclined planes and applied his kinematical mathematical formulas to the flight of projectiles such as cannonballs—a very important application in those days. He showed that the path of a projectile launched at any angle to the ground must be a parabola because its motion in flight consists of two components, a horizontal motion and a vertical motion. The horizontal speed of the projectile is constant, so that its horizontal distance from the launching point increases with time, but its upward vertical speed diminishes constantly with time, so that its height above the ground varies with the square of the time. By combining the vertical and horizontal components of the projectile's motion, Galileo showed that the projectile's flight path is a parabola and that its range is a maximum if it is launched at an angle of 45° to the ground.

Galileo did not limit his scientific research to mechanics, but applied his very active and inquisitive mind to many different technical and scientific problems, showing in each phase of his work the predictive power of mathematics combined with correct scientific principles. In his study of heat, he developed the first thermometer to measure temperature and used the oscillations of a pendulum to measure time, pointing out that the frequency of the pendulum (the number of oscillations in a given time) depends only on the length of the pendulum, but not, to a good first approximation, on the size of its swing from one side to the other. He used a pendulum to measure pulse rates and pointed out the importance of such measurements in diagnostic medicine.

Today, Galileo is most famous for his astronomical telescope, which altered astronomy from a hit-or-miss science to a precise observational discipline with an accuracy that exceeded by a great factor the best efforts of naked-eye astronomy. Astronomy before the telescope depended on such instruments as the quadrant, the armillary sphere, and the parallactic ruler, with which Tycho Brahe had made his remarkable observations. With the advent of the astronomical telescope, naked-eye observational astronomy and its instruments soon disappeared. Galileo made four important telescopic observations that convinced him beyond any doubt that the Copernican cosmology was correct and scholasticism and Ptolemaic astronomy were wrong: (1) The moon's surface is cratered and highly irregular, thus negating the idea that celestial bodies are "perfect"; (2) the phases of Venus and those of the moon are similar, proving that Venus revolves around the sun and not around the earth; (3) four moons (satellites) revolve around Jupiter, illustrating in miniature the Copernican model of the solar system; and (4) the Milky Way consists of numerous points of light, which Galileo correctly interpreted as very distant stars.

CHAPTER 5

Newton and His Physics

The Nature of Theory

We think of Euclid as of fine ice; we admire Newton as we admire the Peak of Teneriffe. Even the intensest labors, the most remote triumphs of the abstract intellect, seem to carry us into a region different from our own—to be in a terra incognita of pure reasoning, to cast a chill on human reasoning.

—WALTER BAGEHOT

Sir Isaac Newton (1642–1727) was born on Christmas Day, three months after the death of his father, a yeoman farmer after whom Isaac was named, to the former Hannah Ayscough at Woolsthorpe, Lincolnshire. The baby was frail and sickly, but he somehow managed to survive and grow stronger, even though he never enjoyed excellent health. Isaac did not have a happy childhood because before he reached the age of two years, his mother married a wealthy minister named Barnabas Smith and, leaving Isaac to be raised by his grandmother, moved to the nearby village where Smith lived to help him raise his three children. Isaac was separated from his mother for nearly nine years until the death of his stepfather in 1653, and it is almost certain that her absence severely affected the development of his personality. It undoubtedly shaped his attitudes toward women; he had little to do with them throughout his life. He never married and, apart from a youthful romance, seems to have focused his attentions solely on his work and, to a lesser extent, on his critics: "The acute sense of insecurity that rendered him obsessively anxious when his work was published and irrationally violent when he defended it accompanied Newton throughout his life and can plausibly be traced to his early years."[1]

Newton's early childhood gave little indication as to his mental

Sir Isaac Newton (1642–1727)

powers. He was a curious boy and an average student in the grammar school at Grantham. He spent more time daydreaming in class than he did studying his lessons. Newton preferred his own company more than that of others and seldom played games and sports with other children. He was temperamental and high-strung, but he was also introspective and shy. Newton did display some mechanical ingenuity and ''constructed mechanical devices of his own design such as kites, sundials, waterclocks and so on.''[2]

After the death of Newton's stepfather, his mother invited him to come manage the considerable property she now owned. It was a short career, as Newton proved to be completely incompetent at running the estate; he was unable to get along with the farmhands and showed little interest in agricultural matters. Fortunately, Newton's maternal uncle convinced his mother that Newton should be sent back to the grammar school at Grantham to study Latin and arithmetic to prepare for the rigors of a university education. Newton did well enough in his subjects to be admitted to Trinity College, Cambridge, where he matriculated in 1661 at the age of 18.

Although Newton attended Trinity College after Nicolaus Copernicus, Galileo Galilei, and Johannes Kepler had made their great contributions to modern science, the education offered by Cambridge, like that offered by most other universities at the time, was steeped in Aristotelian doctrine. Little discussion was devoted to the heliocentric system of Copernicus or the mechanics of Galileo. Instead, Newton and his fellow students had to learn about the works of Aristotle and Plato and the familiar but increasingly unrealistic geocentric vision of the cosmos. However, Newton was drawn to the works of physical philosophers such as René Descartes, who "had begun to formulate a new conception of nature as an intricate, impersonal, and inert machine."[3] The influence Descartes had on Newton was enormous because, unlike Aristotle, Descartes "viewed physical reality as composed entirely of particles of matter in motion" and "held that all the phenomena of nature result from their mechanical interaction."[3] Newton also came under the influence of Isaac Barrow, a mathematician who was the first to recognize Newton's brilliance. Barrow encouraged Newton's interest in mathematics and directed his attention to the study of optics. During his last two years at Cambridge, Newton mastered his mathematics while continuing to study the works of the Renaissance scientists and philosophers; he also began to formulate the concepts that formed the basis for his own unparalleled contributions to science. Because he devoted so much of his efforts to his private study, however, his official academic career was undistinguished. "When Newton received the bachelor's degree in April 1665, the most remarkable undergraduate career in the history of university education had passed unrecognized" because Newton "had sought out the new philosophy and the new mathematics and made them his own, but he had confined the progress of his studies to his notebooks."[3]

The outbreak of the plague in London in 1665 caused Newton to leave Cambridge and return to his home at Woolsthorpe, where he spent the next two years contemplating the ideas about space and time and motion that he had first considered while at the university: "By the time of his return to Cambridge [in 1667] it is tolerably certain that he had already firmly laid the foundations of his work in the three great fields with which his name is forever associated—the calculus, the nature of white light, and universal gravitation and its consequences."[4] He also discovered the binomial theorem, and it "was during this time that he examined the elements of circular motion and, applying his analysis to the moon and the planets, derived the inverse square relation that the radially directed force acting on a planet decreases with the square of its distance from the sun—which was later crucial to the law of universal gravitation."[5]

In those two extraordinary years at Woolsthorpe, Newton carried the work of Galileo and Kepler to its logical conclusion and formulated the physical laws needed to explain the dynamics of a mechanistic universe; Newton's scientific achievements dominated science and philosophy for the next two centuries. Although it is difficult to understand how such intellectual feats could be accomplished by so young a man in so short a time, the clues to Newton's brilliance lay in his unparalleled powers of concentration.

> His particular gift was the power of holding continuously in his mind a purely mental problem until he had seen straight through it. I fancy his pre-eminence is due to his muscles of intuition being the strongest and most enduring with which a man has ever been gifted. Anyone who has ever attempted pure scientific or philosophical thought knows how one can hold a problem momentarily in one's mind and apply all one's powers of concentration to piercing through it, and how it will dissolve and escape and you will find that what you are surveying is a blank. I believe that Newton could hold a problem in his mind for hours and days and weeks until it surrendered to him its secret. Then being a supreme mathematical technician he could dress it up, how you will, for purposes of exposition, but it was his intuition which was pre-eminently extraordinary—"so happy in his conjectures," said de Morgan, "as to seem to know more than he could possibly have any means of proving."[6]

Newton's theory of gravitation was based on his theory that "the rate of fall was proportional to the strength of the gravitational force and that this force fell off according to the square of the distance from the center of the earth."[7] His observation of an apple falling from a tree to the ground while at Woolsthorpe led Newton to conclude that the earth was pulling on the apple and the apple on the earth. Although the notion that the earth attracts objects near its surface was not new, Newton was the first to speculate that the force that caused the apple to fall to the ground is the same force that keeps the moon in its orbit around the earth and the earth in its orbit around the sun. His inverse-square law shows mathematically how the attractive forces between two bodies depend on their masses and their distance from each other. Of equal significance was Newton's conclusion that there is nothing special about the attractive force of the earth; this force can be found to emanate from all bodies in the universe. Newton not only unified and completed the mechanics of Galileo and Kepler but also showed that the dynamic motions of the universe can be described by basic mathematical relationships that are valid anywhere in the universe. The demonstrated utility of mathematics gave natural philosophy (which physics was then called) a self-contained theoretical foundation that it had never previously possessed.

The second of Newton's great achievements was his experiments with light and the formulation of his corpuscular theory of light. He had experimented with prisms while at Woolsthorpe and observed that a ray of light passing through a prism is "refracted, but different parts of it are refracted to different extents, and the beam that falls on the screen is not merely a broadened spot of light, but band of consecutive colors in the familiar order of the rainbow: red, orange, yellow, green, blue, and violet."[8] When the light was passed through a second prism, the different colors recombined into white light. These experiments led Newton to conclude that white light consists of all the colors of the rainbow; the apparent immutability of the component colors of white light led him to formulate his corpuscular theory of light: "He held that individual rays (that is, particles of given size) excite sensations of individual colours when they strike the retina of the eye."[9] Although his colleagues later generally accepted his theory that light consists of tiny particles, there were a number of dissenters such as Christian Huygens who argued that light consists of waves. Newton responded that if light were undulatory, it would bend into the shadows in the same way that sound "bends" and can be heard around a corner. Newton was correct, but it was only years later that more precise experiments showed that light does bend and thus have wavelike properties. However, the corpuscular theory of light was to be revived in some sense in the 20th century when Albert Einstein suggested that light consists of discrete particles called "photons." In any event, the debate may never be conclusively resolved, because light appears to have both corpuscular and undulatory attributes.

After Cambridge reopened in 1667, Newton was elected to a fellowship at Trinity College. Two years later, his mentor, Issac Barrow, resigned his position and recommended Newton as his successor. Although Newton had not yet published anything about his discoveries, he began to give lectures in optics. He also continued his experiments with light and built the first reflecting telescope; it aroused great interest in the

Royal Society and led to Newton's election to that body in 1672. This honor prompted Newton to offer a paper on optics that was severely attacked by Robert Hooke. Even though Hooke was a leader of the Royal Society and considered himself to be an expert in optics, the condescending nature of his review enraged Newton, who was unable to accept any criticism of his work and abhorred controversy: "Less than a year after submitting the paper, he [Newton] was so unsettled by the give and take of honest discussion that he began to cut his ties, and he withdrew into virtual isolation."[9]

Newton's disputes with Hooke and Huygens about the nature of light were eventually overshadowed by the debate over the discovery of calculus, which began in 1684 with the publication by Gottfried Wilhelm von Leibnitz of his paper on the subject. Newton's delay in publishing his own papers on the calculus until 1704 confused the matter of authorship even further. While both men were outwardly friendly toward each other in public, each privately encouraged his supporters to disparage the work of the other. Although Leibnitz probably arrived at the calculus independently, it is now clear that Newton actually developed the calculus before Leibnitz had begun to study mathematics. In any event, the dispute began to be cast in nationalistic terms, with people who knew nothing about the work of either man arguing passionately over whether an Englishman or a German had discovered it. The matter was never conclusively resolved during Newton's life, and while many of his colleagues were willing to give him the benefit of the doubt (owing largely to his scientific reputation), the more convenient notation offered by Leibnitz was used by most scientists and mathematicians on the Continent and has survived to the present day.

By the mid-1680s, Newton had made his major discoveries in optics and mechanics, but aside from a few papers on optics, he had published very little about his work, especially his law of gravitation. His strained relations with many of his contemporaries in the Royal Society periodically caused Newton to become disgusted with science and to devote his attention to other subjects that had always intrigued him, such as religion and mysticism. His most important work, the *Principia,* might never have been written had it not been for the argument by his old antagonist Robert Hooke that the motions of the planets could be explained by the inverse-square law of attraction. Hooke was unable to prove his theory, however, and Edmond Halley, a friend of Newton, took the problem to him and asked him how the planets would move if the force of attraction between them and the sun diminished as the square of their distance from the sun. Newton answered that the planets would travel in elliptical paths and, when asked by Halley why he believed that to be so, responded that he had calculated the orbits. Halley's request that Newton demonstrate his work caused Newton to begin working on a book explaining his theory of gravitation, as well as his three laws of motion. He finished the manuscript in 18 months, and it was published in 1687 at Halley's expense as the *Philosophiae Naturalis Principia Mathematica*. Written in the form of a series of densely worded geometrical axioms and proofs, it remains the greatest and most influential scientific work ever written. It offered a vision of a universe wound up by the hand of God and left to run down on its own with all dynamic motions governed by the law of gravitation. The *Principia* brought Newton worldwide fame and ensured his unequaled reputation in the scientific community: "And the Newtonian scheme was based on a set of assumptions, so few and so simple, developed through so clear and so enticing a line of mathematics that conservatives could hardly find the heart and courage to fight it."[10]

Even as Newton became the living symbol of the Age of Reason, he strayed from science and began an arduous, though unsuccessful, effort to show how base metals could be transmuted into gold and wrote lengthy though completely useless treatises about chemistry. A confirmed Unitarian who kept his religious views to himself to keep his job at Cambridge, Newton also wrote more than one million words speculating about the meanings of mystical biblical passages and decided that the earth was about 5000 years old based on the number of generations in the Bible.

In 1692, Newton suffered a nervous breakdown that may have been brought on by sheer exhaustion. Although Newton recovered fully from the illness, it did force him to retire for nearly two years. It did mark the end of his scientific research, however, though he spent some time during the next decade assembling materials about his theory of light published two decades before that appeared as the *Opticks* in 1704. The delay in publication was due to Newton's refusal to publish his treatise on light until after the death of Hooke in 1703. Newton was elected president of the Royal Society in 1703, and he continued to occupy that position until his death. He was also elected a member of Parliament in 1689, but he never made a speech during his several years in office and spoke up only once to ask that an open window be closed.[11]

In 1696, Newton was made a warden of the Mint, and three years later, he assumed the chief post of master of the Mint. Although Newton maintained a professional affiliation with Cambridge until 1701, his appointment to the Mint effectively ended his academic career, as he moved to London to take up his official duties. Although some commentators have argued that this career change robbed the scientific world of its most towering figure for the last quarter century of his life, it appears that Newton was ready for life outside the university and wanted to enjoy his celebrity in the social whirl of London. The appointment was not simply a symbolic honor, because Newton spent much energy implementing the recoinage scheme of Lord Halifax. He was also knighted by Queen Anne in 1705, an honor never before conferred on a scientist. Newton received a steady stream of visiting dignitaries and oversaw the publication of the *Opticks* and two subsequent editions of the *Principia*. He spent the last years of his life basking in the unprecedented public adulation that continued until the day he died at the age of 84: "When he died in 1727 the greatest honours were accorded to him: his body lay in state in the Jerusalem Chamber, his pall was supported by the Lord High Chancellor, two dukes and three earls—which meant something in those days—and the place allotted for his monument had been previously refused to the greatest of our nobility. It was a great occasion—the first and the last time that national honours of this kind have been accorded to a man of science or, I believe, to any figure in the world of thought, learning, or art in England."[12]

Newton's legacy to modern science is rivaled only by the work of Albert Einstein, although it can be argued that Newton's ideas were more revolutionary for their time than those of Einstein in his lifetime. It is interesting that both men made their chief scientific contributions while in their twenties and in a comparatively short period of time. To delineate how Newton's work changed modern science and its concepts about space, time, and motion, we begin when Newton, then a student at Cambridge, returned to his home at Woolsthorpe, where he made a series of discoveries in mathematics and physics that became the foundation of modern physics. He ended his recital of his various discoveries, which included the calculus, the law of gravity, and his corpuscular theory of light, by saying, "All this was in the two years 1665 to 1666 for in those years I was in the prime of my age for intuition and minded mathematics and philosophy more than anytime since."[13] He did not discuss the laws of motion in this statement of his early achievements, but it is clear from his discussion of the gravitational action of the earth on the moon that he must have understood these laws thoroughly and have formulated them even earlier.

Although Galileo had laid the foundation of mechanics in his study of motion and had emphasized the importance of acceleration, it remained for Newton to establish mechanics as a precise science by stating his three laws of motion, which have contributed as much to his fame as his law of gravity. The importance of these laws cannot be overemphasized, for they elevated mechanics from a semiempirical, semimathematical science to a complete mathematical discipline, as rigorous as geometry. Before analyzing these laws of motion and showing their great predictive power, we must define the expression ''law of nature.''

If science were merely the collecting of observational data with no attempt to organize such data into a meaningful intellectual structure in which disparate parts are correlated to each other in some precise way, it would have little appeal to the probing, curious mind. But science is far more than the random gathering of data, as it also encompasses the drive to discover the causal relationships among the individual bits of data that we are constantly aware of as we observe the universe all around us. The stream of data to which we are constantly subjected flows past perhaps the majority of people without stirring their curiosity or desire to know the significance of the data or how the data may be understood in terms of basic interrelationships that govern all phenomena. Scientists are among those whose curiosity is most stimulated by this data stream and, among scientists, physicists are those who seek its explanation—or the basic natural laws, as we call them—that direct its flow.

To be more specific about the nature of a law, we first define an event, which we may accept as the most elementary phenomenon that can occur. To this end, we first introduce the idealized concept of a ''point particle'' (a bit of matter that has substance—to be defined later—but no dimensions). Clearly, a point particle can have no real existence, but it is useful in our intellectual pursuit of the understanding of the nature of an event, which we now define as the coincidence of a point particle with a given point of space at a given time. Here again, we have introduced concepts—a ''point of space'' and ''at a given time''—that are not well defined, but we will not probe their meanings any further at this stage in our narration.

As we shall see, not all the concepts that enter into the laws of physics are defined, but in formulating the laws of nature, the physicist introduces as few undefined concepts as possible and builds his law on these undefinables as a base. In any case, though these basic concepts are undefinable, the physicist uses and works with them only if he can introduce an operational way of measuring them, so that measurement replaces definition.

With this understood, we return now to the relationship of ''the event'' to ''the law'' and consider a series of events associated with a given particle. Even if the particle were fixed with respect to us, as observers and formulators of laws, it would still define a series of events over a span of time, for we would have to assign to it a series of different times for its coincidence with a fixed point of space. But since this special case of a ''series of events'' is not fruitful for the discovery of a law, we allow the particle to move from point to point and connect all these points (or events) by a curve of some sort. We call this curve the ''orbit'' or ''path'' of the particle. A law, then, is a universal statement that enables us to determine the orbit of such a particle under all circumstances. A simple example is the path of a particle thrown into the air. As we shall see, the law—discovered by Newton—that enables us to calculate the path of such a projected particle incorporates the basic features of all such laws. We note that the determination or discovery of the orbits of particles was and is the principal goal of physicists, for the knowledge of particle orbits gives us a deep insight into the nature of the structures in the universe, from the nuclei of atoms to galaxies, and permits us to check the truth of the laws we seek.

Before discussing the basic elements that constitute a law, we consider briefly how the physicist, or the scientist in general, performs his magic of discovering order (laws) in what often appears as a chaotic ensemble of events. This process consists of two complementary aspects, each sterile by itself but the two together constituting the most powerful tool devised by man for probing the universe and revealing its nature.

The first aspect is called ''experimental'' or ''observational physics'' and is performed by the experimental physicist in a laboratory; it consists of collecting data in the form of measurements of various sorts associated with simple events. Groups of such events generally appear as streaks on photographic plates, or they activate instruments from which numerical readings of various sorts can be obtained. In any case, the role of the experimentalist is to obtain numerical data from observed events. The second aspect is called ''theoretical physics'' and is performed by the theoretical physicist with ''pencil and paper.'' The principal purpose of this activity is to discover the laws that account for the events revealed by the experiments performed by the experimentalist in his laboratory.

An important feature of this tandem activity of the experimentalist and the theoretician is that its two components are intimately interrelated in the sense that the experimentalist, in designing his experiments, is guided by the theoretician, and the latter checks the truth of his theory by using the data of the experimentalist. With this distinction between the experimentalist and theoretician in mind, we may call Newton the first and one of the greatest theoretical physicists, although he did remarkable experimental work as well. In any case, both the experimentalist and the theoretician deal with events in nature, with one (the experimentalist) describing these events and the other (the theoretician) formulating the laws, that is, the relationships among measurable quantities that correlate or predict events. Since an event, in its simplest form, is the coincidence of a particle with a given point of space at a given time, the basic physical elements that enter into the description of the experimentalist and the laws of the theoretician are space and time. We must therefore accept these as the elementary undefinables on which the descriptions of the experimentalist and the laws of the theoretician are built.

Accepting this idea, we consider first how the concept of space is to be introduced and then go on to the concept of time. Two aspects of space strike us at once—the extent of space, which is related to the distances between events, and the orientations in space, which are related to the directions of events from a given reference point. Since we cannot define space in terms of any simpler entities, we accept it as one of our basic undefinables and describe how it is measured; thus, this measurement, which replaces the definition, is what we mean when we speak of the distance between two events. The measurement operation consists of laying off a unit of length (for example, centimeter, foot, mile, kilometer) along the straight line connecting the two events. An important feature of this operation is the stipulation of "straightness," which has meaning only if the geometry of space is known. Since only Euclidean geometry was known in Newton's time, we proceed, for the time being, on the assumption that space is Euclidean or flat and that a straight line between two events is simply the shortest distance between the events in the usual sense of that phrase. By Euclidean geometry, we mean the set of Euclid's basic axioms and all the geometrical theorems that can be deduced from these axioms. Of particular importance here is the so-called "parallel axiom," which states that given a straight line and a point outside the line, only one line can be drawn through this point parallel to the given line. It is easy to deduce from this axiom that the sum of the three angles of a triangle must equal 180° and that the circumference of a circle must equal 2π times its radius. But it is just as easy to see from this that one can never demonstrate by measurements of angles of triangles or circumferences of circles that Euclidean geometry correctly describes our space. We return to this point later when we consider the kinds of geometry that may govern space if Euclid's parallel axiom is discarded. Right now, however, we accept Euclidean geometry and picture all events in our universe (for the moment, the Newtonian universe) connected to each other by straight lines, to each of which we can assign a number (its measured length).

Such a geometrically laid-out array of events has meaning only at a particular instant (an instantaneous photograph) during which nothing moves, but all these distances change from moment to moment. Since at any moment these distances range, however, from the extremely tiny (for example, the spacings between neutrons and protons in the atomic nucleus) to the astronomical (distances to the galaxies), one may ask whether one can attach the same geometrical meaning to all the distances in this vast range. This is a legitimate and important question, since the direct operation of laying off a unit to obtain a distance is applicable only to distances within our reach and certainly does not apply to nuclear, atomic, or cosmological distances, which must be obtained indirectly. We treat this point in more detail later, but a brief discussion here of the measurement of atomic and astronomical distances reveals the problems presented when we call these different kinds of operations "measurements" of distances.

We obtain atomic distances by probing matter with high-speed atomic and subatomic particles (electrons, protons, neutrons) or very high-energy radiation (x rays and gamma rays). By observing how the particles are scattered when they interact with atoms or nuclei, and using the accepted laws of interactions among atomic and subatomic particles, we obtain knowledge of the geometry of the atoms and hence their dimensions. In a sense, this operation is comparing the dimensions of the probing particles with those of the atoms or nuclei being probed. We do essentially the same thing when we probe a structure with radiation; we use the wavelength of the radiation (the distance between successive crests of the radiation wave) as our unit of length and picture the wave as being laid off along the dimensions of the structure we are probing.

We measure astronomical distances more or less directly by noting the apparent change in the position of a star, for example, when the earth moves from one side of its orbit around the sun to the other side (the parallax method). Since this procedure can be applied only to nearby stars (within a few hundred light-years), indirect procedures, which are based on the apparent brightnesses of very distant celestial objects, are used to obtain their distances. These brief discussions of the measurements of atomic and astronomical distances show how far these techniques depart from the simple operation of laying off a unit length along a line, which is the basic distance-measuring operation.

In considering the spatial configurations of events, we must take into account the directional aspects of space as well as its extent. It is quite clear that merely measuring the distance of an event from us does not describe its spatial features completely; we must also specify its direction. The reason for this is that space is multidimensional; it is, in fact, as we know, three-dimensional. What precisely does this mean and how do we specify spatial dimensions?

We start by specifying a particular direction, by imagining a line extending from us to a star. We now draw another line through our position at right angles to the first line and note that these two lines define (or span) a plane; any other line (or direction) drawn in this plane can be expressed as lying partly in the direction of the first line (the line toward the star) and partly in the direction of the second line. We therefore say that a plane in space is two-dimensional; it contains only two independent directions in the sense that given a line in the plane and a point on it, only one other line can be drawn in the plane through the point perpendicular to the first line. But we are still free to draw a third line that is perpendicular to this plane and hence perpendicular to the first two lines. This defines a third independent direction in space, so that space is three-dimensional. Any other line drawn in space can be pictured as lying partly in the direction of each of the three independent mutually perpendicular lines.

Owing to the dimensionality of space, we must differentiate between that set of basic physical entities such as time, mass, and temperature that are specified entirely by their magnitudes, without any spatial aspects (no spatial directional features), and that set of entities such as spatial displacements that have directional aspects as well as magnitudes. The entities in the first set are called "scalars" and those in the second set are called "vectors"; a scalar is completely described by its magnitude, whereas a vector is completely specified only if its direction as well as its magnitude is given.

In the preceding discussion of the distance concept, we dealt only with magnitudes and not directions, but as we have already indicated, the complete description of an event means giving its direction as well as its distance from us. Now we have seen that distance between events is given by a number that we obtain by moving to the event along a straight line and counting the number of unit lengths contained in the distance to the event. But how do we measure direction? We introduce the concept of angle, which is a measure of the amount we turn when we change our orientation from one direction to another.

To be precise about this concept, we choose a direction in space (for example, an imaginary line from us to a star, as previously described) and start turning (without otherwise moving from our location) away from this direction; as we turn, the direction we face changes. If we continue turning until we face the original direction again, we say that we have made one complete turn and assign 360 units of turning to the completed turn. Each of these units is called one degree. The angle formed by two intersecting lines pointing in two different directions is a measure of the amount of turning we must do in changing our line of sight from one of the lines to the other. Just as distance gives the number of steps we must take to go from one point of space to another, angle is the amount we must turn to change our orientation from one direction to another. Distance is a magnitude and angle defines a direction, so that to specify a vector completely, we must introduce angle.

Just as distance may be expressed in any units we please (for example, miles, centimeters, feet), so may angle. Thus, the degree is divided into 60 equal parts (minutes) and the minute again into 60 equal parts (seconds). But another important unit of angular measure is used by the physicist, astronomer, and mathematician; it is called the "radian" and may be better understood if you picture yourself as standing at the center of a circle along the circumference of which a particle is moving. Follow the particle with your eye until the distance it has moved along the circumference, from the moment you started following it, exactly equals the radius of the circle. The amount of turning you have done (the angle through which you have turned) is defined as one radian. One complete turn, or 360°, thus equals 2π radians, since the number of radii contained in the circumference of any circle is exactly 2π. One radian is equal to about 57°.

Since turning or rotation occurs about a specific axis, which defines a direction in space, rotation is a vector quantity that is completely specified by a direction (the direction of the rotation axis) and a magnitude (the angle of rotation). Angle is to rotation as distance is to spatial displacement.

We can derive two new quantities from the concept of distance: area and volume. Area is defined as the product of two distances at right angles to each other and is expressed in distance units squared. If one distance is 4 centimeters, and the perpendicular distance is 5 centimeters, the area defined, or spanned, by these two distances is 20 centimeters squared, written as 20 cm^2. Since the planar area can be oriented in any desired direction in space, it is itself a vector, the direction of which is given by the direction of a line perpendicular to it and the magnitude of which is the numerical value of the area (the number of distance units squared it contains). A few important examples of areas are triangle (1/2 its base times its altitude), rectangle (product of two perpendicular sides), parallelogram (base times altitude), circle of radius r (πr^2), sphere of radius r ($4\pi r^2$), where π is the circumference of a circle divided by its diameter and has the approximate constant value 3.14159265.

Volume is defined as the product of three mutually perpendicular directions and is expressed as a distance cubed. One cubic centimeter is the volume of a unit cube (a cube each side of which is one centimeter long). The volume of a room is the product of its length, its width, and its height and is expressed in cubic feet if each of its three dimensions is measured in feet. The volume of a sphere of radius r is $\frac{4}{3}\pi r^3$; since π is very close to 3, the volume of a sphere is approximately 4 times the cube of its radius. The volume of the earth is approximately 256 billion cubic miles. Mankind has explored and exploited less than one ten-thousandth of this volume. The earth's volume is about 64 times that of the moon, Jupiter's volume is some 1000 times that of the earth, and the sun's volume is about 1000 times that of Jupiter.

TIME AS A BASIC ENTITY

To construct the concepts that enter into the laws of nature, we need more than space, since our universe is not static; it is dynamic and constantly evolving. We must therefore introduce time as a basic entity before we can deduce the derived quantities that enter into nature's laws. Just as we are born with a sense of distance and an ability to estimate ordinary distances fairly accurately, so we inherit remarkable time-estimating capabilities, as evidenced by the performances of, for example, great athletes, musicians, and jugglers. Time perception is also developed to a very high degree in animals, insects, and even plants. However, just as we cannot define space, we cannot define time. Instead, we describe a time-measuring operation that gives us a number as the temporal interval between two events.

To do this, we note first that the flow of time has physical meaning only if changes occur that permit us to differentiate the past, the present, and the future. Without such changes, the measurement of time is meaningless. All our experiences indicate that time flows in one direction only and that time reversal never occurs in the macroscopic world, although it may occur, in a very restricted sense, in the atomic or subatomic domain. Second, the time interval between two events must be measured by some kind of periodic or cyclical phenomenon that, in a sense, divides the total interval into a series of small, equal intervals, each of which corresponds to the period (the time of one oscillation) of the cyclical phenomenon. This cyclical phenomenon constitutes a clock, and its period is called the "unit of time." The clock itself may be any device ranging in complexity from a simple pendulum to a very sophisticated atomic clock, the vibrations of which are those of the individual atoms.

The unit of time is the second, which is so defined astronomically that the tropical year (the interval from spring to spring) contains 31.556926 million seconds. Time intervals that physicists deal with range from one billionth of a trillionth of a second for processes inside the nucleus of an atom to billions of years, which are the ages of stars and the age of the universe. With the most sophisticated and accurate clocks, time intervals can be measured with an accuracy of one part in a trillion and time intervals of the order of a billionth of a second can be recorded.

THE CONCEPT OF SPEED

Having discussed the basic entities—space and time—from which we construct the derived quantities that enter into the laws of nature (or in terms of which these laws are expressed), we begin by introducing the speed of a body as the simplest and most basic of these quantities, although, as we shall see, speed is also associated with the motion of energy (for example, radiation). Newton, however, was primarily concerned with moving bodies, and the understanding of the nature of speed is essential to the study of such bodies. Indeed, speed is the simplest derived quantity that we can construct by combining space and time. To that end, we follow a particle from moment to moment and suppose that it is at some point A when we observe it initially, at time t_0, and is at some point B at some later time t_1, and that it has gone from A to B along the straight line connecting these two points. We now measure the distance s from A to B, and divide it by the time interval $t_1 - t_0$ to obtain the quantity v. This symbol v represents the average or mean speed of the body along the straight line from A to B. The formula for v shows us what basic measurable quantities go into the makeup of speed—distance and time. Since the time divides the distance, we specify speed as centimeters per second (or cm/sec, also cm sec^{-1}), or miles per hour, or some other unit of distance per time. We do not have to introduce any new units to measure speed, since speed is a derived quantity and its units are a combination of those that measure distance and those that measure time, as indicated above.

Why do we refer to v as the average or mean speed? Note that the distance from A to B is finite, so that the particle spends a finite length of time traversing that distance. During that time, the speed of the particle may fluctuate from one value to another, just as that of an automobile moving on a highway does. The mere division of the total distance by the total time smoothes out the fluctuations to give us an average value. This average value does not tell us about the speed of the particle at any specific moment (the instantaneous speed), which we must now discuss.

The problem of determining the instantaneous speed of a particle was solved brilliantly by Newton, and its solution led Newton to the discovery of the calculus, which is the single most powerful analytical tool man has yet developed. Hardly any of our modern technology or deep insight into the laws of nature could have been obtained without this remarkable and beautiful mathematical instrument.

The difficulties involved in determining the instantaneous speed of a particle can be seen when we note that the particle must be observed at two distinct points along its path, that is, at two different times, to determine its speed. The introduction of two distinct times in itself contradicts the idea of an instantaneous speed. Despite this contradiction, however, we may still ask whether we can deduce the instantaneous speed from the measurement of the average speed, as defined above. It is clear that we cannot do this if the time interval during which we observe the particle is fairly large, for this leaves room for measurable changes in the speed. The question here is what we mean by "fairly large."

Let us suppose that we could carry out the same measurements on the same particle over and over again, but each time taking a shorter time interval. Another way of looking at this is to picture a large group of observers, each measuring the average speed but using different time intervals from very, very short ones to longer ones. In general, each observer gets a slightly different answer. These differences between the measurements of successive observers grow smaller and smaller with decreasing time intervals, however, until, below a sufficiently small interval, the values for the average speed are the same. This value may then be called the instantaneous speed of the particle at a particular moment. This is really not the actual instantaneous speed, since a finite time interval is still involved, but our procedure tells us that we can always find a time interval that is so short that our available instruments do not reveal any change in the speed of the particle during that interval. More sensitive and accurate instruments that detect a change in the speed in such an interval force us to go to still smaller intervals to obtain the instantaneous speed, but the same general idea applies.

To express this idea algebraically, we introduce the symbol delta (Δ) to represent a small value of some quantity, so that Δt is a small time interval. If this is the small interval of time during which we observe the particle and during which it travels the small distance Δs, we have $v = \Delta s / \Delta t$. To obtain the instantaneous speed, we must take Δt so small that v shows no measurable change for still smaller values of Δt. The actual measured value obtained for the instantaneous speed is thus determined by the accuracy of our instruments.

But we can introduce a theoretical instantaneous speed in the way it was introduced by Newton, who pondered the consequences of allowing Δt to become infinitesimally small. Newton understood that in theory, the instantaneous speed can be found only if Δt is allowed to go to zero. This simple idea, and the mathematical consequences that stem from it, led Newton to the differential calculus.

This procedure of permitting the denominator of a fraction (in this case, the fraction $\Delta s / \Delta t$) to go to zero may appear to be most hazardous (and, indeed, forbidden by the rules of algebra) to the reader who has not studied calculus, but Newton saw that no calamity results if Δs goes to zero at the same time as Δt. It may at first appear that we then obtain the indeterminate result 0/0, but the actual situation is that the ratio $\Delta s / \Delta t$ has a finite value if Δt goes to zero, and this value is the instantaneous speed of the particle. Newton's great contribution was to prove that this fact is indeed so and to show how this value, or "limit," as it is called, can be calculated if one knows how the distance a particle moves depends on the time. The instantaneous speed of a particle is obtained by allowing Δt to become infinitesimal. Newton's procedure is the basis of the differential calculus, and when we apply it, we say we are finding the "derivative" of s, or "differentiating" it with respect to time. Newton represented this process by placing a dot over s and called it a "fluxion," which Leibnitz, who is also credited with the invention of calculus, called a "derivative."

THE CONCEPT OF VELOCITY

We introduced the concept of the average speed of a body above and saw that it is obtained by dividing the distance the body moves in a given time interval by the time it takes the body to move through the distance. But the speed of a body is only one aspect of its motion. It is clear that if we say that a car is traveling at 50 miles/hr, we are giving only part of the information we need to plot its course. To plot a body's course, we must know not only the speed of the body at each moment, but also the direction of its motion. When we have both the direction and the speed of the motion of a body, we have its velocity, which is a vector quantity. The speed is the magnitude of the velocity, and its direction can be represented by an arrow that points in the direction of the body's motion at each moment.

To represent the velocity graphically, consider the path of a particle in some frame of reference (note that without some frame of reference, we cannot assign any meaning to the velocity at all). To each point of the particle's path such as point *A*, we can assign a number that gives the speed of the particle at that point. But how shall we represent the direction of the motion? We best understand how this is to be done by considering another point, *B*, on the path very close to *A*. Suppose that we picture the particle as going directly from *A* to *B* in a straight line, instead of going along the arc of its curved path from *A* to *B*. Then its direction of motion would be along the straight line *AB* instead of along the actual direction. If we take *B* closer to *A*, the direction of the straight line from *A* to *B* gets closer to the actual direction of the motion at *A*. We see that we can find the direction of the motion (or of the velocity) of the particle at *A* by allowing *B* to come infinitesimally close to *A* and then taking the straight line from *A* to *B* as this direction. But we see that when we do this, this line (and hence the direction of the velocity at *A*) is just coincident with the tangent line at point *A*. By the tangent line at *A*, we mean the line that just touches the curve at *A* but does not cut the curve.

We now see that to represent the velocity of a particle at any point of its path as a vector, we draw a line tangent to the curve at a particular point and choose its length (on some scale) to be equal to the speed of the particle. As we move along the path of the particle, this vector changes in magnitude and direction. If we know the velocity vector of a particle at each moment of its motion, we can draw the path of the particle. All we need to do is pass a curve through the foot of its velocity vector (the end of the vector without the arrow) in such a way that the curve just touches each vector without cutting the vector. Each velocity then becomes a tangent of the path of the particle.

THE CONCEPT OF ACCELERATION

Velocity is the first, the simplest, and among the most important dynamic physical entities that we can construct from space and time, and as we shall see, it enters into all our considerations of the laws of nature. But not much can be done with it alone. In comprehending and acting on this idea, Galileo and, in particular, Newton departed from Aristotelian science. Whereas Aristotle did not go beyond velocity in formulating his dynamic theory, Newton saw that a correct dynamics must take into account not just velocity but velocity changes. If the velocities of the bodies that constitute our universe were constant (unchanging), structures (for example, nuclei, atoms, molecules, planets, stars, galaxies) could not exist. But change in velocity by itself does not give us all that we need to formulate the correct dynamic laws—we must use or work with, the rate of change of velocity, which we call ''acceleration.'' The recognition of the crucial role that acceleration plays in dynamics was certainly one of the most momentous discoveries in the history of science. It is therefore desirable and useful to have a fairly clear understanding of the nature of acceleration (any change in speed or direction of motion) as a preliminary to a discussion of Newton's laws of motion.

Whether or not we understand the concept of acceleration consciously, our bodies certainly do, for we are constantly changing our state of motion as we move about. At one moment, we are at rest, then walking leisurely, and some time later, running, and we are aware of these changes from one state of motion to the next. This is true also if we are driving a vehicle, but then we are also aware of how rapidly or slowly these changes occur, and that is the essence of acceleration—the rate of change of our state of motion or our velocity, that is, a rate of change in speed and/or direction.

Our degree of concern with the rate of change of our velocity (our acceleration) depends on our activity. If we are strolling along leisurely, acceleration is not important, but it is extremely consequential if we are playing a fast game of tennis, for we are then constantly changing our velocity and our success depends, to a great extent, on our acceleration. If we are driving an automobile, we are constantly aware of the automobile's acceleration as we slow it down, increase its speed, or change its direction. These experiences show us that acceleration in general involves two typical phenomena: change in the speed of a body without change in the direction of its motion or change in its direction of motion without change in its speed. These two changes may occur together, as they generally do, but it is convenient to discuss them separately, which we now do briefly.

If a body's speed (but not its direction) is constantly changing, as in the case of an automobile moving at a uniformly increasing speed along a straight highway, its acceleration is easily calculated from its initial and final speed in a given time interval. If the automobile's speed has increased steadily from 5 miles per hour to 45 miles per hour in 10 seconds, its acceleration is 4 miles per hour per second. A freely falling body in a vacuum is an important example of this kind of acceleration, which, as we saw, was first accurately measured by Galileo, who also demonstrated that all bodies falling freely in a vacuum at a given point on the earth's surface fall with the same acceleration.

In analyzing the acceleration of a moving body, we simplify the analysis by dividing the body's motion into three independent motions that occur along three mutually perpendicular directions. We may study the motion of a projectile (a body thrown upward in some arbitrary direction) by separating its vertical motion from its horizontal motion. This division simplifies things, because only the vertical motion of the body is accelerated by gravity; the horizontal motion is unaccelerated. The complete motion is then obtained by combining the horizontal and vertical motions, which do not influence each other at all.

Having considered accelerated motion of a body moving in a straight line, in which the acceleration is along the line of motion of the body (either increasing or decreasing the body's speed), we discuss the acceleration of a body moving with constant speed in a circle of fixed size (fixed radius, r). In this case, the direction of the acceleration is always at right angles (perpendicular) to the velocity of the body, which is always around the circumference of the circle. No part of the acceleration can be along the body's velocity, for if it were, the body's speed would increase or decrease steadily. We see that the direction of the acceleration of a particle moving with constant speed in a circle changes constantly, but always points toward the center of the circle.

Though the direction of this acceleration changes constantly, its magnitude is constant. It is not too difficult to figure out the magnitude if we imagine ourselves driving a car around a perfectly circular track at a constant speed because only the direction of the car is changing, so that the car's acceleration (the rate at which the direction of the car's velocity is changing) is determined by the speed of the car and the size of the track. Suppose that the track size is such that we get halfway around the track in 10 minutes at the speed we are going; the direction of our velocity changes by 180° in 10 minutes. If we had been traveling along a track of twice the size (a track with twice the radius), it would have taken us 20 minutes to get halfway around, so that the car's acceleration would have been half as great. The acceleration of a particle in circular motion similarly varies inversely as the radius of the circular orbit (the smaller the circle, the greater the acceleration).

Consider now how the car's acceleration depends on its speed. Clearly, the faster the car is moving (the higher its speed), the faster its direction of motion and hence its velocity is changing, so that its acceleration certainly depends on (increases with) its speed, but this dependence is more drastic than the dependence on the radius of the track. This is a subtle point, but let us see if we can analyze it from the kinematics and obtain the right answer. That the acceleration depends directly on the speed, as already stated, is clear, but another speed factor comes in, because the faster the car moves the more it will depart from the circular track if it is not constrained by its acceleration to move along the track. The faster the car moves, the greater the change in its velocity will have to be in a unit time just to keep it moving along the track. Hence, its acceleration depends on its speed in two different ways, each of which increases the acceleration by the same factor v (the speed). The circular acceleration goes as v^2 (the speed squared). The acceleration of a particle moving in a circle of radius r at constant speed v thus equals v^2/r; this acceleration is perpendicular to the direction of the particle's velocity and is always directed toward the center of the circle.

The importance of circular motion cannot be overemphasized, for it plays a crucial role throughout physics. It is of particular value in astronomy, as illustrated by the rotation of the earth, the revolution of the planets around the sun, and the orbiting of the stars in our galaxy around its center.

Another important kind of acceleration is associated with a particle moving with constant speed in a circle. Imagine illuminating the particle as it moves in the circle so that its shadow is projected onto a given diameter of the circle. The shadow thus oscillates back and forth along the diameter as the particle revolves in the circle; its very special acceleration along the circle's diameter leads to what is called "simple harmonic motion" (SHM). We note that the acceleration in SHM is always toward the center of the diameter and thus always opposite to the velocity of the particle's shadow; its magnitude increases as the shadow's distance from the center increases, and its velocity decreases, reaching a maximum when the shadow is at rest at either end of the diameter. The SHM acceleration is zero when the shadow is at the center of the diameter, moving at maximum speed. Two elementary examples of SHM are (1) the motion of the bob of a pendulum that is swinging through a small arc (only approximate SHM) and (2) the motion of a particle at the end of a vibrating spring (true SHM).

Three important physical (measurable) quantities are associated with SHM: (1) the amplitude of the motion (the size of the diameter along which the shadow is oscillating), (2) the period of the motion (the duration of one complete oscillation or cycle), and (3) the frequency of the motion (the number of vibrations per second). The period is the numerical reciprocal of the frequency. The importance of SHM in the study of physics and in the analysis of the motions of bodies cannot be overestimated, for it was shown by the great French mathematician Baron Jean Baptiste Joseph Fourier in the 19th century that any motion, however complex, can be expressed as, or broken down into, a sum of SHMs. Such a sum, called a Fourier analysis, can often be used to simplify dynamic, kinematical, optical, and thermodynamic problems.

LAWS OF MOTION—THE CONCEPT OF FORCE

The basic concepts of space and time lead to the derived entities velocity and acceleration, but do not give us any insight into the dynamics of motion; for that, we need one other basic concept that is essential for dynamics and that bridges the gap between geometry and matter. In discussing the nature of a law, we introduced the concept of an event, which we defined as the coincidence of a particle with a point of space at a given time so that events have meaning only in relation to matter (particles). Indeed, we could not even conceive of the space and time-measuring operations discussed previously without introducing matter, nor can we conceive of matter without space and time. One might wonder, at this point, whether or not matter can be derived from space and time so that everything reduces to geometry. Since thus far nobody has shown how this task can be done, we must introduce the "space–time–matter bridge" as a new basic entity.

We might, or course, introduce matter itself (mass) as the third basic element and deduce the nature of the "bridge" from Newton's dynamic laws, but this procedure is not satisfactory because it is difficult to see how a unit of mass is to be introduced and used; without introducing additional assumptions, we have no direct operational way of measuring the mass of a body by comparing it with a unit of mass. Mass would have to be introduced quite differently from the way space and time were introduced. We avoid this objectionable procedure by introducing force as our third and final basic quantity.

Since we are as sensitive to and as aware of forces as we are of space and time, the force concept comes naturally to us; it is part of our very psychological and physiological makeup, so that we have no trouble establishing a scale of forces to which we have all been subjected from the time of birth that enables us to perform our daily functions with little effort. This scale, which operates through special sensory nerve endings called "proprioceptors," permits us to estimate forces with remarkable accuracy. Since our muscles can pull and push, we picture forces as being both attractive and repulsive. Finally, we note that force is a vector quality, since we can pull or push a body in any direction we please and with a range of magnitudes (strengths).

Although we estimate forces quite accurately, we do not do so with any unit of force in mind, but if we are to relate forces to dynamics, we must introduce a precise unit of force, which physicists have done; it is the dyne—a very tiny unit. About 450,000 dynes equals a pound of force. Forces are measured by the amount they stretch a given spring.

With the force concept understood, we can now state Newton's three laws of motion, which are the basis of classical mechanics and dynamics and which lead to mass as a derived quantity. The first law—often referred to as the "law of inertia"—states that a body at rest or in a state of uniform motion in a straight line remains at rest or continues its uniform linear motion unless acted on by an unbalanced force. This simply means that the state of motion (rest is included as a state of motion) of a body can be changed only by applying an unbalanced force to the body. The presence of an unbalanced force is thus indicated by changes in the state of motion of a body. It should be noted that forces may be constantly acting on a body, but only if the sum of these forces does not vanish (that is, if an unbalanced force exists) does the body's state of motion change. Since the change in the state of motion is acceleration, we see that an unbalanced force accelerates a body on which it acts.

Newton's second law of motion gives the quantitative relationship between the force applied to a body and the acceleration it imparts. It was clear to Newton that the acceleration must be directly proportional to and in the direction of the force. He expressed this idea by stating that if an unbalanced force is applied to a body, the body is accelerated in the direction of the force and the magnitude of the force divided by the magnitude of the acceleration is the same (a constant) regardless of the magnitude of the force. This concept is stated algebraically as $F = ma$, where F is the applied force, a is the acceleration, and m is the constant that Newton identified with the inertial mass of the body. This simple equation, which is referred to as "Newton's equation of motion" and is one of the most famous statements in the history of science, ushered in modern physics. It is the basis of all Newtonian dynamics, and it dominated physics until the beginning of the 20th century, when the quantum theory and the theory of relativity altered it in a number of remarkable aspects.

A few things about this equation should be noted. First of all, it defines the mass of a body, which is to be treated as a derived rather than a basic quantity, and establishes a procedure (operation) for measuring it: Apply a force of known magnitude to the body and measure its acceleration in the direction of the force while this force is acting; divide the magnitude of the force by the magnitude of the acceleration. The number so obtained is the mass of the body, expressed in grams, if the force is expressed in dynes and the acceleration in centimeters per second per second (cm/sec^2). This establishes a unit of mass in the following sense: If a body acquires an acceleration of 1 cm/sec^2 when a force of 1 dyne (unit force) acts on it, the mass of the body is 1 gram (unit mass).

The second notable feature of Newton's law of motion is its vector character; it relates one vector quantity—force—to another—acceleration—which means that it is three equations rolled into one, since we may picture the force acting on a body as a combination of three mutually perpendicular components acting independently of each other along three mutually perpendicular directions. Thus, we obtain three independent equations of motion, which can be treated more easily than the single equation, since each of the three equations deals with motion that is restricted to only one direction.

Finally, the law is extremely general in the sense that it is the same regardless of the nature of the force. The only forces known to Newton were the force of gravity and the push and pull of one body on another, but he did not distinguish between these forces in stating his second law. All forces acting on a given body produce accelerations of the same kind. The nature of the force does not enter into the law; only its magnitude and direction do. But this law, by itself, important and powerful as it is, does not solve the dynamic problem of finding the path or orbit of a particle acted on by a force. Only if we know the nature of the force (its geometrical and physical features) can we solve the problem by replacing F in Newton's equation by an algebraic expression and then solving the equation for the body's acceleration to obtain an orbit.

Newton's third law is as simple as the other two laws, but it introduces into the action of forces a remarkable symmetry that does not appear in the first two laws. It states, essentially, that forces appear in equal and opposite pairs in nature, which we can best understand by studying the interaction of two bodies that pull upon each other gravitationally, such as the sun and the earth. The third law states that the magnitude of the pull of the sun on the earth exactly equals the magnitude of the earth's pull on the sun, and the two pulls are along the same line but opposite to each other. Newton stated this law in the most general way as follows: If body A exerts a force of any kind on body B, then B exerts an exactly equal and opposite force on A (the law of action and reaction: every action in nature is accompanied by an equal and opposite reaction). We discuss this law in greater detail in Chapter 6.

CHAPTER 6

Newton's Law of Gravity and His Contemporaries

Every great scientific truth goes through three stages. First, people say it conflicts with the Bible. Next they say it had been discovered before. Lastly, they say they always believed it.

—LOUIS AGASSIZ

The laws of nature that scientists study may be divided into distinct categories, which are interrelated but are often developed independently. The basic category consists of those laws, theories, or speculations that deal with space and time. In Newtonian mechanics and Newtonian theory, space and time are assumed to be independent of each other and space geometry is assumed to be Euclidean or flat. Moreover, space is accepted as absolute (distances between events are the same for all observers) and infinite in extent. Time is also assumed to be absolute and flowing continuously from an infinite past to an infinite future. In our discussion of the quantum theory and the theory of relativity, we shall show how these Euclidean and absolute concepts had to be changed.

The second category consists of the laws of motion, which establish the relationship between the motion of a body and the force acting on the body, as illustrated by Newton's three laws of motion. These laws—without which we cannot deduce the nature of the basic structures (for example, atoms, molecules, planets, stars) in the universe—are general statements in which the nature of the forces that produce changes in the states of motion is not specified. Since the description of the state of motion of a body involves both space and time, the laws of motion depend on the geometry we impose on space and our assumption about the relationship between space and time.

The laws that describe the forces in nature constitute the third category, among which is Newton's law of gravity. Today, we recognize or assume the existence of four basic forces: gravity, the electromagnetic force, the weak interaction (a special kind of weak force in the nucleus of an atom that maintains a proper balance between the number of neutrons and protons in a nucleus), and the strong nuclear force. Here, we limit ourselves to a discussion of gravity, which may be defined as the prime force in the universe. It governs such large structures as planets, stars, the solar system, galaxies, and the universe itself. If gravity disappeared, none of these structures could exist.

A law of force generally contains two features that are independent of each other. The first is geometrical in that it describes the dependence of the force (its magnitude and direction) on the geometrical relationship between two interacting bodies; the second feature is intrinsic to the interacting bodies in that it gives the dependence of the magnitude of the force on some physical property of each of the interacting bodies. A force between two interacting bodies may also depend on their states of motion, but Newtonian gravity does not.

Newton was probably directed to his formulation of the law of gravity by Kepler's third law of planetary motion and his own second law of motion. One can imagine Newton contemplating a falling apple while thinking of his own second law of motion, which tells him that acceleration implies force. The apple is accelerated toward the ground; hence, the ground exerts a force on the apple, and so the law of gravity is conceived. Whether or not it actually happened this way is of little importance, but Newton's bold conjecture that the earth and, in fact, all bodies generate gravity was the beginning of a vast development of profound ideas that culminated in Einstein's great achievements.

Newton's conjecture was just the beginning of the formulation of his law of gravity—he still had to show its geometrical and intrinsic aspects discussed above. From Kepler's third law of planetary motion, one can deduce quite easily the dependence of the magnitude of the gravitational force between the sun and a planet on their separation. Kepler did not do this because he did not think of the sun's action on a planet as being along the line to the sun, whereas Newton did. But Newton went beyond thinking of the force of gravity as being associated only with the earth, the sun, and the planets. He generalized the gravity concept and assumed it to be universal in the sense that all particles in the universe pull on each other gravitationally. To show this, he wrote the algebraic expression for the gravitational interaction of two particles of arbitrary mass separated by an arbitrary distance.

To simplify his task, he considered the force between two mass points (particles with no size or shape), thereby eliminating the problems that the shape and size of the interacting bodies might present. Newton confronted the following problem: Given two particles with masses m_1 and m_2 (m standing for mass and the subscripts 1 and 2 identifying the particles) separated by a distance r, how large and in what direction is their gravitational interaction? By "interaction" here, we mean the pull of m_1 on m_2 or of m_2 on m_1, which, according to Newton's third law, are equal and opposite.

From very general symmetry considerations, Newton reasoned that the force must be along the line connecting the two masses; otherwise, the two particles would begin to revolve around each other, and a body allowed to fall freely on the earth would acquire a horizontal as well as a vertical motion, contrary to what actually occurs in the real world.

With the direction of the force settled, Newton turned to the magnitude of the force, which, as we have already noted, has a geometrical component (the distance between the particles) and an intrinsic component (the masses of the particles). To obtain the distance part of the law of gravity, Newton reasoned that since the force of gravity spreads out from a particle uniformly and equally in all directions, the intensity of the force decreases with the square of the distance between the particles. If this distance is doubled, the force is reduced 4-fold; if the distance is tripled, the force is reduced 9-fold; and so on. We may say the force varies inversely as the square of the distance (the inverse-square law). Note that owing to this distance dependence, the force of gravity gets weaker as the distance increases, but it never drops to zero. Therefore, the notion that one body can escape from the gravitational pull of another (for example, from the earth's gravity) is false.

To incorporate into his laws the mass dependence of the force, Newton first assumed that gravity is generated by mass and then that the strength of the force between two particles must depend symmetrically on their masses, since otherwise the action and reaction of the particles would not be equal and opposite. The only algebraic combination that gives the correct symmetry is the product m_1m_2 of the two masses.

Combining this expression with the distance dependence of the force, Newton wrote his law of gravity as follows, where F is the force of gravity:

$$F = G\frac{m_1m_2}{r^2}$$

We have already discussed the mass and distance parts of this formula, but what about the factor *G*? From where does it come? This factor, known as Newton's "universal constant of gravity," is absolutely essential; without it, the formula would be wrong on two counts: It would give much too large a value for the strength of the gravitational force, and it would be incomplete dimensionally in the sense that its space, time, and mass ingredients would be wrong. Note that force is mass times acceleration according to Newton's second law of motion and acceleration is length over time squared, so that force is mass times length over time squared. But without *G*, the formula for force would be mass times mass over length squared, which is not force. To change this last combination of mass and length to the correct one for force, Newton multiplied it by the constant *G*, which has just the right combination of space–time and mass ingredients to give the complete Newtonian formula the correct space–time–mass ingredients, and thereby obtained the correct dimensions in terms of our basic physical entities. The space–time–mass composition of *G* is length cubed divided by the product of mass and time squared. To put it differently, *G* is expressed as $cm^3/g\ sec^2$, and its numerical value (first measured by Henry Cavendish about a century and a half after Newton discovered the inverse-square law of gravity) is $6.668 \times 10^{-8}\ cm^3/g\ sec^2$.

With Newton's law of gravity and his second law of motion, astronomy advanced from a hit-or-miss science to a precise mathematical discipline. This was particularly true of the study of planetary orbits, which developed into what was called "celestial mechanics," one of the most beautiful and greatest achievements of the human mind. The power of Newton's law of gravity, along with his laws of motion, to solve problems that stem from the gravitational interactions of bodies is evident at once when these laws are applied to determining the orbits of a planet around the sun. The combined efforts of Brahe and Kepler, which led to the empirical discovery of the nature of the planetary orbits (Kepler's three laws of planetary motion), represented some 60 man-years of labor, but with Newton's laws, one can reach the same goal in a matter of an hour or more. This difference illustrates the great predictive power of a correct theory combined with correct logic (mathematics).

Newton further demonstrated the power of his laws by correctly explaining the phenomenon of the tides as a consequence of the gravitational pull of the moon and sun on the earth's oceans, lakes, and rivers. He showed that the gravitational force varies inversely as the cube of the distance because the tidal pull is given by the difference between the pull (per unit mass) on the earth's bodies of water and the pull per unit mass on the earth as a whole. Because the moon is about 400 times closer than the sun, its tidal pull is about twice that of the sun, even though the sun's total gravitational pull on the earth as a whole is about 180 times that of the moon.

Although Newton is most famous for his laws of motion and his law of gravity, he made other important scientific discoveries. His work in optics is particularly noteworthy because of its practical and theoretical importance. Although Kepler had developed his dioptrics showing how images are formed with lenses and Galileo had built his telescope, very little was known about the nature of light or how it is propagated. Newton took the first big step in explaining the nature of light by showing that a beam of white light is spread out into a band of colors (a spectrum) ranging from red to violet, after passing through a triangular prism, thus proving that white light is "white" because it is a mixture of light of all the different colors. He supported this conclusion by showing that if one of the colored beams leaving the prism is sent through an identical second prism, the colored beam does not fan out into the spectrum, as the white beam did, but remains the same as it was before entering the second prism. This means that the colors are in the white beam and not in the prism. Newton showed that if the fan of colors from the first prism passes through an identical inverted prism, the beam of white light reappears, showing that the second prism reverses the action of the first prism by recombining all the colors. Newton thus demonstrated a very important property of light propagation—its reversibility. This means that if the direction of a beam of light passing through a given medium or a series of different media (for example, air, water, glass) is reversed at any point, the beam simply retraces its path. This phenomenon of reversibility is very important in nature because, in general, it does not occur. That it does in this instance has raised some very important questions about the laws of nature in general, because they seem to indicate that reversibility should be as common in nature as irreversibility. Newton experimented with lenses and mirrors and showed that a concave spherical mirror (which he ground and polished himself from new glass) forms an image just as a lens does, but one that is free of the undesirable chromatic aberration (a surrounding colored rim) that attends the image formed by a lens. He correctly attributed the chromatic aberration of a lens to its slight prismatic shape, as one passes from its center to its edge, and noted that owing to the absence of chromatic aberration, a telescope with a concave mirror (a reflector) gives sharper images than one with a lens (a refractor). This has made a vast difference in the use of large astronomical telescopes, without which many of the remarkable cosmic discoveries following Newton could not have been made.

From the time of Archimedes, the students of natural phenomena had wondered and speculated about the nature of light and how it is propagated. Galileo had tried, very crudely and unsuccessfully, to measure the speed of light, and Kepler had speculated about its nature, but Newton was the first to propose a precise theory about its propagation. He was led to his theory by noting that when a beam of light strikes an opaque screen with a very sharp edge, the edge of the shadow of the screen cast on a white wall close to the screen appears to be as sharp as that of the screen itself. It appeared to Newton that none of the beam bent around the edge of the screen and into the shadow, as it would if light were propagated as a wave. Newton believed this result meant that light travels in a straight line past a sharp edge (contrary to the behavior of, for example, a wave of sound or water) instead of being diffracted as a wave would be. He therefore concluded that light consists of particles that, according to his first law of motion, must travel in straight lines, since the edge of the screen exerts no force on the light. This conclusion marked the birth of the Newtonian corpuscular theory of light, which was later discarded in favor of the wave theory when more accurate observations showed that the edge of the screen's shadow is not sharp but fuzzy, so that the beam is diffracted around a sharp edge. Moreover, the Newtonian corpuscular theory of light requires that the speed of light in a dense medium such as glass or water be greater than its speed in a rare medium such as air or in a vacuum, which is contrary to the experimental evidence.

Newton, in his *Opticks,* propounded a series of remarkable questions, which showed how deeply he was concerned about light. In the last of these questions, he suggested that under certain circumstances, light and matter might be changed into each other. This does not mean that Newton had the slightest notion about relativity theory, but the theory shows that Newton's conjecture was correct.

Although Newton was the dominant physical scientist from the late 17th to the early 18th century, other brilliant scientists made substantial contributions during that time, some of whom strongly challenged Newton's ideas. Among the most important figures in this group of Newton's contemporaries were Robert Hooke, Robert Boyle, Christian Huygens, Ole Roemer, James Bradley, and Edmond Halley. No story of physics would be complete without at least some discussion of the principal contributions of these remarkable men.

Like Newton, Robert Hooke (1635–1703) was a sickly infant who suffered from headaches throughout much of his childhood. The son of a curate who worked in a small parish on the Isle of Wight, Robert received his early education from his father. Although his poor health made it impossible for him to continue his lessons for several years, Robert did develop an interest in art and was sent to London in 1648 to seek an apprenticeship with the painter Peter Lely. He soon abandoned his efforts to become an artist, however, and enrolled in the Westminster School, where he excelled in the classics and mathematics. He attended Oxford University for two years before his chemistry professor, Thomas Willis, recommended him to Robert Boyle, who had recently arrived at Oxford and was searching for an assistant to aid him in improving his air pump.

Hooke spent nearly a decade with Boyle, honing his experimental and mechanical skills. In 1662, he accepted the position of curator of the Royal Society (the only paid position in it at that time) and assumed the burdensome duties of furnishing three or four significant experiments for each weekly meeting of the society: "It is a monument to Hooke's fertile imagination that he carried out his duties with outstanding skill for many years."[1] In 1664, Hooke was appointed professor of geometry at Gresham College, replacing Isaac Barrow, who had resigned to become the first Lucasian professor of mathematics at Cambridge. Despite his duties as curator of the society, in 1665 Hooke published his *Micrographia,* which "represents not only the results of his original research and observations using the microscope he had constructed, but also contains a number of beautifully drawn illustrations that further emphasize his unusual gifts as a draftsman."[2] The detail of Hooke's illustrations coupled with the book's accessible prose made it a best-seller in England and contributed to Hooke's scientific reputation. Following the Great Fire of London, Hooke used his drafting skills to draw up a plan for rebuilding the destroyed city. Although his proposal was never adopted, Hooke's work attracted the attention of Sir Christopher Wren, and the two became close friends.

Hooke was elected secretary of the Royal Society in 1677 following the death of Henry Oldenberg, a man who, Hooke believed, had tried to steal his spring-driven clock invention. Although Hooke continued to maintain an exhausting regimen of academic duties, his tenure as secretary was undistinguished because his fertile intellect was better suited for analyzing natural phenomena than for mediating disputes among society members. He resigned as secretary in 1683, but retained his position as curator.

Although Hooke was plagued by poor health for much of his life, he was an innovative scientist whose greatest flaw may have been an impatience that prevented him from developing many of his ideas and inventions beyond their crudest forms. Unlike Newton, Hooke did not appear to bear grudges or abhor controversy. However, Hooke had the misfortune to disagree with Newton about fundamental aspects of optics and celestial mechanics. As a result, Newton refused to acknowledge Hooke's own substantial contributions to the study of light and delayed the publication of his *Opticks* for two decades until after Hooke's death in 1703.

Next to Newton, Hooke was one of the most productive and inventive scientists of that era, spending his many daily waking hours (he slept only three or four hours per night) tinkering with all sorts of mechanical devices, performing numerous experiments, and developing fanciful theories of all kinds. He made many refinements of the air pump, which Boyle used in his experiments on gases. He also invented or improved a number of scientific devices including telescopic sights, fine screw adjustments, a clock-driven telescopic mount, spring-driven clocks, the universal joint, and the iris diaphragm, to increase the sharpness of optical images.

In the realm of physical theory, Hooke proposed an imperfect-wave theory of light with the wave vibrations at right angles to the direction of propagation of the light. He argued that the motions of the planets must be treated as mechanical phenomena and suggested an incorrect theory of gravity to account for their motions. He was the first to propose a kinetic theory of matter, insisting that all bodies consist of invisible particles that are in constant vibratory motion, with each different kind of particle capable of vibrating in its own peculiar manner and no other. This is very suggestive of the present picture of electrons having their own states of energy in an atom.

Despite his many inventions and speculations, Hooke is known only for his law of elasticity, which was, in a sense, the beginning of solid-state physics and is the foundation of mechanical engineering. This law, known as "Hooke's law," states that within the elastic limit in a solid, its extension (strain), under the action of a force (stress), is proportional to the applied force ("strain is proportional to stress"). The "elastic limit" in this phrase means the point of stretch beyond which the solid no longer returns to its original state after the stress (force) is removed. This law tells us that we may use a solid to establish a unit of force (the force needed to produce a given stretch) and to measure the magnitude of any force by comparing the stretch produced by this force with the stretch produced by the unit force.

Hooke's scientific partner and lifelong friend, Robert Boyle (1627–1691), was one of 14 children, the son of a Cambridge-educated lawyer who had left England and made his fortune in Ireland. Robert grew up on the family estate at Lismore in the province of Munster. He was educated by tutors until the age of eight, when his father sent him to Eton. Robert spent four years at Eton before moving with his brother to Geneva to be tutored. Although he received an adequate education in Switzerland, he was more profoundly affected by a violent electrical storm there that convinced him he should somehow use his intellectual talents to promote Christianity on earth.

In 1644, Boyle returned to London, where he was influenced by several Utopian philosophers who believed that experiments are the best way to uncover the secrets of nature. This Baconian philosophy of science provided Boyle with the direction he needed; he settled down at the family estate at Stalbridge in Dorsetshire, where he began to devote all his time to the study of science and theology. He was especially interested in chemistry and soon had so thoroughly mastered the subject that in 1654 he was asked by John Wilkins, the warden of Wadham College, to come to Oxford. Boyle found a kindred spirit in Wilkins, for the latter was a confirmed experimentalist who, while deeply religious, recognized the shortcomings of a strictly theological philosophy of nature.

After Boyle came to Oxford in the summer of 1754, he began to conduct experiments on the properties of the atmosphere that later led him to the law that bears his name. His interest in the subject was inspired by von Guericke's experiments in Germany, which showed "that if two hollow metal hemispheres were placed base to base and the spherical interior thus formed was exhausted of air, the pressure of the atmosphere held the halves together against the force exerted by two teams of six horses pulling in opposite directions."[3] Boyle believed that a better pump could be built and, lacking the requisite mechanical skills, decided that he needed an assistant who could actually construct the device. On the recommendation of an associate of Wilkins, Boyle hired Robert Hooke, then an unknown student at Oxford. Boyle and Hooke began a successful partnership that continued until Boyle's death; it was especially fortunate for Hooke because it is doubtful that he would have otherwise been able to sharpen his scientific skills so quickly. In any event, the pair conducted a series of experiments that led Boyle to conclude that air is not chemically

Robert Boyle (1627–1691)

homogeneous, but contains a ''vital quintessence essential to life although most of it serves no purpose,'' and that air has weight. These insights led Boyle to conclude that the atmosphere near the earth is compressed and that its pressure is sufficient to support a 29-inch column of mercury in a vertical tube with its open end in a bath of mercury.

Boyle is most famous for his gas law, which states that in a gas at constant temperature, the pressure in the gas multiplied by its volume remains constant. To understand this simple law, which we discuss in greater detail in Chapter 11, we must express or define pressure in terms of our basic space–time–force concepts. The effect of a force depends on whether it is spread over a large area or concentrated on a small area; the smaller the area, the greater the effect. As a result, we experience sharp pain when a pin is pushed against our skin with a force that we merely find annoying when applied to a blunt object against our skin, and we can easily walk on snowshoes or skis across deep, soft snow that we would sink into if we were wearing ordinary shoes. The effective physical entity in both of these examples is the pressure, which we define as the force per unit area and express as dynes per square centimeter (dynes/cm^2) or pounds per square inch. A solid, owing to its weight (which is a force), exerts a pressure only on the surface that supports it and only on that part of the surface with which it is in direct contact. A liquid (water in a tank or reservoir) exerts pressure on the bottom and on the four walls of its container. The pressure against the walls and within the liquid increases with depth; it is the same at all equally deep points. A gas exerts pressure against all the points of its container, and if the size of the container is small (compared to the height of the atmosphere), the pressure in the gas may be assumed to be the same throughout. In considering a gas, we speak of the pressure at a point within the gas just as for a liquid. An important feature of a gas is that if a pressure is exerted by a force at any point against a gas, the pressure is transmitted equally to all points in the gas. The pressure of a gas at any point on a wall of its container is exerted perpendicular to the wall at that point.

With these features of pressure understood, we can now discuss Boyle's law more effectively. To do that, we consider a gas in a vertical cylinder at the top of which is a weightless movable piston of unit area (1 cm^2) that can be pushed down to increase the pressure in the gas or raised to decrease it. If no weights are placed on the piston, it remains at rest at a definite height such that the pressure in the gas is just the pressure of the atmosphere. If a weight W is placed on the piston, the piston sinks and the pressure in the gas increases until it equals the atmospheric pressure plus W. As the gas pressure increases, the volume of the gas decreases, and if W were increased sufficiently so that the total pressure in the gas increased to twice the atmospheric pressure, the volume of the gas would be half as great as initially (that is, with no weights on the piston). This relationship may be expressed as pressure times volume equals a constant. This statement is true as long as the temperature of the gas is not altered.

Boyle's productive partnership with Hooke ended temporarily when Hooke became the curator of the Royal Society in 1662. Although Boyle continued his atmospheric experiments, he moved in 1668 to London, where he lived with his sister in Pall Mall. He also became very interested in developing an atomic theory of matter. Like Newton, Boyle was interested in the mystical and conducted extensive experiments on the transmutation of metals. In fact, Newton appears to have been more impressed by Boyle's accounts of his attempts to transform base metals into gold than by his work on atmospheric pressure. However, Boyle was thoroughly modern in the way he conducted his research. Instead of performing all the work by himself, he gave instructions to his assistants, who conducted the bulk of the tedious experimental work. Boyle's supervisory position was dictated as much by social custom (as it was thought improper for a landed gentleman to engage in such labors) as it was by his poor health.

Boyle suffered from kidney stones throughout his life, and his condition caused him to become a hypochondriac, consuming a variety of strange remedies that may have done more harm than good: "His regimen also consisted in observing a rigid diet so that he never ate for pleasure but only to keep himself alive."[4] Toward the end of his life, Boyle's thoughts turned more to religious matters, and he composed a number of theological essays. However, he also found time to complete his *Essay on the General History of the Air;* he passed away in bed while correcting the proofs for that work.

Christian Huygens (1629–1695), like Boyle, was a confirmed experimentalist who left a life of wealth and leisure for the more mundane world of the laboratory. His father was Constantin Huygens, the famous diplomat and poet, who was a formidable figure in the history of Dutch literature as well as a distinguished member the Council of State. After the death of Christian's mother in 1637, the family moved to a town near the Hague. Among the many visitors to the Huygens's household was the French philosopher René Descartes, whose geometrical conception of space greatly influenced the development of mathematics and natural philosophy. Christian, whose affinity for mathematics led him to prefer Newton's *Principia* to the works of Aristotle, published several mathematical works that established his reputation in Europe before he reached the age of 30.

During the same period, Huygens made important improvements in the telescope. He studied the phenomena of reflection and refraction and ground lenses with fewer imperfections. With these tools, he discovered the rings of Saturn as well as its sixth moon, Titan. In 1657, Huygens also built a pendulum clock that he hoped could be used to determine longitude at sea, but the ingenuity of its design was offset by its inability to give accurate readings on ships in motion. Although more useful clocks were soon invented, Huygens continued to tinker with his pendulum device for many years, even though he could not correct its basic defects.

In 1661, Huygens visited London to meet many of the founders of the Royal Society. Two years later, Louis XIV of France honored him for his pendulum clock. In 1666, the French Academy of Sciences was organized, and Huygens was requested by the king's minister, Colbert, to direct its activities. Huygens accepted, and, feeling the need for some administrative guidance, began to correspond with Henry Oldenberg, the first secretary of the Royal Society; their communications helped to disseminate throughout Europe the advances then being made in the sciences. Huygens himself published papers on centrifugal force; his

Christian Huygens (1629–1695)

thoughts on what would become known as Newton's second law of motion apparently predated Newton's conception. Huygens's treatise on the subject appeared in 1673, and it "contained, in addition to his studies relating to the pendulum and its period of oscillation, the theory of evolutes, the cycloid as its own evolute, theorems on the composition of forces in circular motion, and the general idea of the conservation of energy."[5]

Huygens proposed a wave theory of light that ultimately displaced Newton's corpuscular theory because it explained all the optical phenomena and properties of light known at the time, which Newton's theory could not. In particular, the wave theory, as Huygens showed, requires that light travel slower in a dense medium than in a rare one (in water as against air), as opposed to Newton's corpuscular theory, according to which light travels faster in a dense than in a rare medium. To explain the rectilinear propagation of light, Huygens introduced the concept of the wave front of the light wave, which he pictured as a spherical surface, advancing at the speed of light from the source of light. As this spherical wave front advances, its size increases and the light intensity at each point on its surface diminishes. In his thinking here, Huygens was probably influenced by the way ripples advance on the surface of a quiet pond into which a small stone or pebble is dropped. One ripple follows another at definite intervals, and it appears as though each advancing ripple is pushed along by the one behind it. Moreover, if an advancing ripple strikes two barriers in the water separated by a small gap, new ripples move outwardly from the gap as though a pebble had been dropped in the gap itself. To explain this phenomenon, Huygens reasoned that every point on a ripple is a source of new, tiny secondary ripples that move away from the old ripple to form a new one, as illustrated by the phenomenon in the gap described above. The two barriers destroy all the ripple except the little piece that enters the gap, and the points on this piece generate new secondary ripples that advance into the water beyond the gap.

Huygens, transferring these ideas to the advance of a spherical wave front of light, pictured each point on such a surface as vibrating and setting up tiny secondary wave fronts that then combined to form a new wave front just ahead of the old front. This construction of Huygens's was and still is very useful in explaining such optical phenomena as reflection, refraction, diffraction, and interference, which are discussed in greater detail in Chapter 9.

The French Academy of Sciences prospered under Huygens's leadership, but his own health was generally poor. He had to spend much of his time and energy defusing the squabbles between the proponents and opponents of Cartesian physics. Huygens himself accepted the Descartes theory of vortices and believed Newton's theory of universal gravitation to be untenable. He was especially reluctant to accept the idea of action-at-a-distance because of the lack of a visible intervening mechanism.

The death of Colbert in 1683 coupled with rising antiforeign sentiment in France caused Huygens to leave the country. He spent several years in England, attending meetings of the Royal Society and meeting its most famous members, including Boyle and Newton. However, his Cartesian view of the world was rendered obsolete by the publication in 1687 of Newton's *Principia,* which established its author as the most influential scientist in Europe. Huygens, believing that he could no longer make valuable contributions to science, gradually lost contact with most of his colleagues and returned to Holland, where he spent the remaining years of his life, isolated from the great changes being wrought by Newtonian mechanics.

The corpuscular and wave theories of light proved to be so intriguing to Ole Roemer (1644–1710), the son of a Danish shipowner, that he decided to study astronomy at the University of Copenhagen. He was hired by the French astronomer Jean Picard to come to Paris to work as his assistant. Roemer spent much of his time observing the motions of the satellites of Jupiter and found that "it was theoretically possible to predict the precise moment at which they would be eclipsed by Jupiter (as viewed from the earth)."[6]

Roemer's great contribution to physics was his very ingenious measurement of the speed of light by timing the intervals between successive eclipses of one of Jupiter's moons. Since the speed of light is one of the most important constants of nature, playing the critical role that it does in the theory of relativity, it is worthwhile to discuss its measurement in some detail. We do not know just when scientists began to think of light as a physical entity that travels at a finite speed; even Galileo, despite his remarkable physical insights, pictured light as racing with infinite speed and moving instantaneously from source to observer. Later, he rejected this erroneous idea and proposed a crude experiment to measure the speed, but his attempt failed. Newton and Huygens knew that light travels at a finite speed, but proposed no method of measuring it, since the instruments to measure it accurately had not yet been invented. These instruments became available in the 19th century when Fizreau, in 1849, using accurate clocks and a toothed wheel rotating at a very high speed, measured the speed of light in water and showed it to be smaller than the speed of light in air. This was the *experimentum crucis* that led to the rejection of Newton's corpuscular theory and the acceptance of Huygens's wave theory.

Roemer was led to his method of measuring the speed of light in 1675, when he noticed that the interval between successive eclipses of any one of Jupiter's visible moons varies from half year to half year. He observed that this interval is longest when the earth is receding from Jupiter and shortest when it is approaching Jupiter. Between these times, when the earth's motion is at right angles to the line to Jupiter, the interval (the true interval) is half the sum of the other two intervals.

To see why this is so, note that if the speed of the earth away from Jupiter is v and the true interval between two successive eclipses is t, the earth moves a distance vt away from Jupiter during that time, so that the light informing the observer on the earth that another eclipse is beginning must travel this additional distance at its true speed *c;* this requires the additional time vt/c (the delay time), which must be added to t to give the observed interval between eclipses while the earth is receding from Jupiter. If n eclipses occur during the time the earth is receding from Jupiter (from its closest to its farthest distance), the total delay for all such eclipses is therefore nvt/c, which Roemer observed to be 1000 seconds. But nvt is the total distance the earth recedes from Jupiter during the time of the n eclipses, which is halfway around the earth's orbit (186 million miles). Hence, nvt/c equals $186{,}000{,}000/c$, which equals 1000 seconds. This means that c equals 186,000 miles per second. This figure is remarkably close to the accurately measured value of 2.9979×10^{10} cm/sec accepted today. Because the velocity of light was then viewed more as a scientific curiosity than as a value of fundamental importance, Roemer's discovery attracted little attention, although it did gain the support of Picard, Huygens, Newton, and Edmond Halley. Despite the indifference of most scientists, however, Roemer's astronomical skills did not go unrecognized. He was appointed in 1681 by King Christian V to be the royal astronomer of Denmark as well as professor of astronomy at the University of Copenhagen.[6] He also found time to dabble in public service, becoming the mayor of Copenhagen in 1705, and helping to reform the Danish system of weights and measures.[6]

The dependence of the observed frequency of a periodic phenomenon on the motion of the observer with respect to the phenomenon became known as the "Doppler effect" some 200 years later when it was applied to light by Christian Doppler. Roemer's measurement of the speed of light was essential for James Bradley's discovery of the aberration of light, a phenomenon that also depends on (and stems from) the motion of an observer relative to a source of light.

Born in 1693 in Sherborne, Gloucestershire, James Bradley was educated at Oxford, where he developed an interest in astronomy. His uncle, James Pound, a clergyman and skilled amateur astronomer, encouraged Bradley's increasing interest in celestial mechanics and introduced him to Edmond Halley. Bradley's friendship with Halley proved to be crucial to his election in 1718 to the Royal Society. Bradley supported himself financially by becoming the vicar of Bridstow in 1719, though he continued to pursue his interest in astronomy. He resigned his office four years later when he was appointed professor of astronomy at Oxford.

Bradley tried to determine the parallax of stars by measuring the semiannual shift in the positions of nearby stars owing to the motion of the

James Bradley (1693–1762)

earth around the sun. If a close ''fixed'' star is viewed from two opposite points on the earth's orbit, six months apart, the apparent position of the star against the very distant stars changes. In 1725, Bradley tried to measure this change (parallax), but failed because it is much too small to be measured by the kinds of instruments Bradley had at the time. This apparent displacement (angular shift) depends on the star's distance and the radius of the earth's orbit (half the separation) between the two positions of the earth six months apart, from which, Bradley knew, he could have found the star's distance if he had succeeded in measuring the parallax. In this task he failed, but he discovered the aberration of light from the stars, which is also an apparent shift in a star's position; it depends on how fast the earth is traveling, not on how far the earth has moved. Aberration results from the combination of the velocity of an observer and the velocity of the light from the observed object. When we are not moving, we can protect ourselves from rain falling vertically downward by holding our umbrellas exactly vertically. If we start to run through the rain, we must tilt our umbrellas forward. The faster we run, the more we must tilt our umbrellas, because the raindrops seem to come from a forward direction rather than straight down. The reason is that when we run, we must subtract our own forward velocity (vectorially) from the downward velocity of the rain to obtain the observed velocity of the raindrops.

If we apply this analysis to the light from a star, the apparent direction of the light and hence of the star from the earth changes as the earth moves around the sun. The observed direction of the star from the earth (that is, the observed direction of the light from the star) is tilted in the forward direction compared to the true direction of the star from the earth. This effect is most pronounced for stars whose directions from the earth are perpendicular to the direction of the earth's motion; the magnitude of the effect, which depends only on the speed of the earth around the sun and not on the distances of the stars or their own motions, is zero for stars whose directions from the earth are parallel to the direction of the earth's motion. Owing to this effect, stars appear to move in their own small elliptical orbits as the earth revolves around the sun.

From his measured value of the aberration and its formula (the speed of the earth divided by the speed of light), Bradley calculated the orbital speed of the earth. To do this, he used Roemer's measured value of the speed of light and obtained 18.5 miles per second. His star measurements also revealed annual changes of declination in some stars, which he concluded were caused, not by aberration, but by the slight and uneven motion of the earth's axis resulting from the changing direction of the moon's gravitational pull.[7] Bradley thus demonstrated for the first time in the history of science that analysis of the behavior and the properties of light from the stars can reveal important aspects of the dynamics of the earth.

Bradley's stellar observations helped to confirm many of the predictions of Newtonian mechanics. Bradley became the astronomer royal in 1742 and was awarded the Copley Medal by the Royal Society in 1748. He occupied a variety of administrative and academic posts at Greenwich until returning to his native Gloucestershire several years before his death in 1762.

Bradley's longtime friend and mentor, Edmond Halley (1656–1742), was educated at St. Paul's School in London and Queen's College, Oxford, where he met John Flamsteed, who was later appointed astronomer royal in 1676. Flamsteed encouraged Halley's interest in astronomy; Flamsteed's efforts to catalog stars in the northern hemisphere motivated Halley to try to compile a similar catalog of stars in the southern hemisphere. He obtained the necessary financial support for his project from his father and King Charles II. In 1676, he sailed to the island of St. Helena in the South Atlantic, where he spent the next 14 months recording the positions of 341 stars and performing a variety of other astronomical observations.[8] Halley's catalog, published in 1678, led to his election to the Royal Society that same year.

Halley is best known for his work on comets, in particular on the famous comet that bears his name, which he observed in 1680 and correctly predicted to return in 1758. He based this prediction on his application of Kepler's laws to the 1680 comet, the orbit of which he was convinced was an ellipse. Although this was but one of his many contributions to science, his greatest service was perhaps his success in persuading Newton to publish the *Principia,* which contained most of Newton's discoveries and was the first great treatise on theoretical physics. This publication may be considered as the beginning of a systematic treatment of natural science in a rigorous mathematical fashion. Although the *Principia* had begun with Halley's question to Newton about the kind of planetary orbits the force of gravity would produce, Halley played a pivotal role in its publication, as he financed the printing costs, "consulted with Newton, tactfully subdued a priority dispute between Newton and Hooke, edited the text of the *Principia,* wrote laudatory verse in Latin for the preface to honor its author, corrected the proofs, and saw it through the press in 1687."[8]

Halley also devised the first meteorological chart that showed the distribution of prevailing winds over the oceans. He also constructed a mortality table for the city of Breslau that greatly influenced the development of actuarial tables because it was one of the earliest attempts to relate age and mortality in a population.[8] Halley was later given command of a war sloop for a three-year scientific expedition to the South Atlantic to measure the longitudes and latitudes of his ports of call. The charts made from Halley's observations were used by navigators long after his death.

Halley returned to England to become professor of geometry at Oxford in 1704. The following year, he published a compendium describing the orbits of 24 comets that had been observed in the previous four centuries; the similarities among three comets that had been observed at roughly 76-year intervals prompted Halley to conclude correctly that there was only one comet (which now bears his name) and that it would return about every 76 years.[9] Halley also made extensive studies of the orbit of Venus and calculated the distance from the sun to the earth using the parallax method.

CHAPTER 7

The Post-Newtonian Era

Dynamic Conservation Principles

The concepts initially formed by abstraction from particular situations or experimental complexes acquire a life of their own.

—WERNER HEISENBERG

During the 18th and 19th centuries, Newton's laws of motion and his law of gravity, as he had stated them, became the basis for the creation of a vast mathematical superstructure that changed Newtonian mechanics from a set of simple equations that had been designed to analyze and follow the motion of a single particle acted on by a well-defined force to a sophisticated complex of partial differential equations designed to study the interactions and motions of many particles. It is unnecessary to present these equations here in all their forbidding formalism to see their significance and understand their importance in the development of physics in the post-Newtonian era; a brief survey emphasizing the highlights will suffice. Such mathematicians and theoretical physicists as Maupertius, Lagrange, Euler, Laplace, D'Alembert, Poisson, Hamilton, Gauss, and Jacobi were associated with these developments, but since presenting their individual contributions would take us too far afield, we consider here the product of their combined efforts.

Two sets of general principles guided these remarkable men in their construction of post-Newtonian classical mechanics: conservation principles and minimal principles. Physicists were led to the conservation principles by their search for constancy in nature, which in the pre-Newtonian era was associated with material things such as the earth, the sun, the stars, and the heavens, in general. These were pictured in Christian theology as having been created by God to be unchanging and eternal, although the pre-Christian Latin poet Lucretius described the universe as ever-changing, with nothing in it abiding. The post-Newtonian physicists, accepting the Lucretian idea of constant change in material things, were looking for constancy in the principles that govern the behavior of the universe. These, of course, are the laws of nature, which are the same everywhere, at all times, and govern all phenomena. But in probing the Newtonian laws, these physicists found that these laws demand a constancy in the quantities of certain measurable entities that are associated with groups of interacting particles. A constancy of this sort is called a "conservation principle," and each such constancy is governed by or expressed in its own conservation principle.

Unlike the conservation principles, which keep track of quantities of measurable entities to make sure that none gets lost, the minimal principles impose the condition that events in nature must occur in such a way that certain measurable aspects of the events change by the smallest possible amount. The conserved entities (which obey the conservation principles) are not the same as those that obey the minimal principles, but both sets are dynamic entities that can be derived from the quantities that we have already introduced and that enter into Newton's laws. We begin by considering the conservation principles and note that these principles, as we shall see, are intimately related to certain space–time symmetries in nature that manifest themselves in the symmetries of the structures all around us. Since these structural symmetries (for example, those of the snowflake, gems, and living organisms) are related to the symmetries of the forces that govern these structures, they ultimately can be traced back to the space–time symmetries of the conservation principles.

CONSERVATION OF MOMENTUM

We follow a particle moving along a well-defined orbit (for example, an object falling along the arc of a parabola on the earth's surface) and note that it has at each point of its path a definite velocity (a vector) that we can represent by an arrow pointing in the direction of the motion, which is just tangent to the path at the given point. We now multiply the velocity v of the particle by its mass m to obtain a new derived quantity mv, which we call the momentum p of the body; it, too, is a vector that points in the same direction as the velocity. As long as we are dealing only with the kinematics of the body (following it from point to point in its orbit), it is sufficient just to know the velocity from moment to moment, but if we are interested in the dynamics of the body's motion, we must take account of its mass as well. A moment's thought concerning our daily experience with moving bodies shows us why this is so: The effect of an object striking us depends not only on how fast it is moving but also on the magnitude of its mass. In other words, in reacting to a moving body (avoiding a collision with it), we respond to its momentum rather than to its velocity alone. The importance of mass in dynamics is indicated by its presence in Newton's second law of motion, which tells us that the force exerted by a body equals its mass multiplied by the rate at which its velocity changes (acceleration). Stated this way, however, the law can be applied only to the motion of a body the mass of which remains constant while the force acts on it. But picture a body losing (or gaining) mass while being pushed or pulled by a force. The simple formula, force equals mass times acceleration, cannot be applied because the mass is constantly changing. This is the situation when a satellite is being launched into orbit while burning the fuel that launches it. Initially, before the satellite is launched, its total mass is the sum of the masses of its fuel and its payload, but as it rises, its mass diminishes because its fuel is decreasing.

Newton was aware of this difficulty with his second law and saw that it is corrected by equating the force acting on a body to the rate of change of its momentum rather than to its mass times the rate of change of its velocity. This is indeed the way Newton's second law of motion is stated: A force acting on a body equals the rate at which the body's momentum is changing and is in the direction along which the momentum is changing. With this point in mind, one can calculate the force acting on a body in two steps: First, treat the body at any moment as though its mass were not changing at that moment and multiply its mass by its acceleration at that moment; second, treat the body as though its velocity were not changing at that moment and multiply that velocity by the rate at which the mass is changing at that moment. The force acting on the body at that moment is then the sum of the two products obtained in these two steps.

Stated in this fashion, the second law already indicates that momentum is a candidate for conservation. We see that the momentum of a body is unchanged (conserved) if no force acts on it. But the constancy of the momentum of a single body when no force acts on it does not reveal the full importance of the principle of conservation of momentum. To establish that momentum is conserved in a collection of any number of bodies interacting with each other in any manner as they move about, we begin by studying the simplest interaction of two bodies and then extend our conclusions to many bodies. Picture two small noninteracting spheres with different small masses and different velocities relative to the ground moving toward each other along the straight line connecting their centers. Since the spheres interact with each other only when they collide, we analyze the effect of the collision on the total momentum of the system to see whether or not the collision changes this important quantity. We obtain the total momentum of the system at any time by adding the momenta of the two spheres, keeping in mind that momentum is a vector and thus has direction. If the spheres are moving along a left–right line, we assign positive momentum to the sphere moving to the right and negative momentum to the sphere moving to the left. The magnitude of the total momentum of the system (the two spheres) is then the difference between the magnitudes of the two individual momenta; the direction of the total momentum is to the right if this difference is positive and to the left if the difference is negative. In this example we assume the difference to be zero.

During the collision, which lasts for a small fraction of a second, each sphere pushes the other one, and these two pushes, according to Newton's third law, are exactly equal and opposite; therefore, their effects on the total momentum of the system (the two spheres) are equal and opposite. Each sphere loses momentum at the same rate as the other one does in the direction it is moving during the collision, transferring its forward momentum to the other sphere until both spheres momentarily come to rest with respect to each other. But the interaction does not stop at that point, since each sphere (owing to its distortion produced by the collision) continues pushing the other sphere in the same direction as before, so that the spheres separate, each sphere moving, with respect to the ground, at a velocity such that the total momentum of the two spheres, as a single system, is still the same as it was before the collision, that is, zero.

This is the essence of the principle of conservation of momentum, which is really a continuation or a consequence of Newton's third law, which tells us that if two bodies interact in any way, they transfer equal but oppositely directed amounts of a momentum to each other so that, as long as no external force (a force other than their interaction) acts on them, their total amount of momentum remains the same. The action and reaction of Newton's third law causes the momenta changes to cancel exactly. This is just a neatly controlled bookkeeping operation (for momentum) in nature. Momentum never gets lost, it just gets transferred from body to body.

Although we analyzed this principle and stated it for just two interacting bodies, it applies to any number of bodies, such as, for example, the molecules in a gas, which move about randomly and constantly collide with each other, the stars in a moving cluster, or the atoms in a solid. We assign to such a collection of bodies (or particles) a total momentum that we obtain by considering all the bodies in the collection simultaneously at any instant. Each body at that instant has its own momentum (a vector) that is the product of its mass and its velocity and may be represented by an arrow of definite length pointing in the direction of the body's velocity. We now add all these arrows (vectorially, to take account of their different directions) to obtain a single arrow that is attached to a single point in the collection, has a definite magnitude, and points in a definite direction. This arrow gives the total momentum of the collection of bodies, and the point to which it is attached is called the "center of mass" of the bodies. As the bodies move about, interacting among themselves, the arrow associated with each body (its instantaneous moment) changes in direction and magnitude, but the arrow attached to the center of mass (the representative of the total momentum of the system) remains constant in direction and magnitude as long as no external force acts on all the bodies together. If the length of this arrow is zero, the total momentum of the collection of bodies is zero. If the length of this arrow is equivalent to a velocity v, the center of mass of the collection, and therefore the collection itself, is moving in the direction of the arrow at velocity v; the total momentum of the system is Mv where M is the total mass of the system (the sum of the masses of the individual bodies). The conservation of momentum may now be stated as follows: if no external force acts on an ensemble of bodies, the center of mass of the ensemble remains at rest if it was initially at rest or continues to move along the same straight line with constant velocity if it was doing so initially.

This simple conservation principle is of great importance in analyzing the motions of interacting bodies and has played a crucial role in the discovery of new particles in high-energy particle physics. It is also related to the space–time symmetry of the laws of nature: The validity of the principle of conservation of momentum means that the laws of nature are symmetrical with respect to a spatial change in an observer's frame of reference; a spatial shift of an observer's frame of reference does not alter his perception of nature's laws. The laws are invariant to a translation of his coordinate system.

The reason for the relationship of this spatial symmetry to the conservation of momentum is that this conservation principle states that to an observer moving along with the center of mass of a system of particles on which no external force acts, the behavior of the particles (and hence the laws that govern this behavior) are the same whether the center of mass is fixed at one point or moving (with constant velocity) from point to point.

THE CONCEPT OF ENERGY

Closely associated with the principle of the conservation of momentum is that of energy. We deduce the energy concept from the basic entities (length, time, force) and the derived quantities (mass, velocity, acceleration), which we can do, of course, by simply combining certain of these quantities algebraically and calling this derived entity energy. Such a procedure does not give us, however, the insight we need to grasp the physical aspects or properties of energy, which entered physics in its present form long after Newton's laws of motion had been promulgated. Although Huygens and Leibnitz correctly perceived the relationship of the motion of a body to its energy, the full development of the concept of mechanical energy, as we now use it, stemmed from the work of Euler, Laplace, and particularly, Lagrange, who made it the foundation of his famous work on analytical mechanics. In that treatise, he attempted to construct Newtonian mechanics as a self-consistent, axiomatic discipline, starting from the energy concept, rather than from Newton's laws of motion.

In introducing or deriving all the mechanical concepts up to now, we have related them to our physical perceptions, wherever possible, rather than presented them as abstract algebraic entities. We proceed similarly with energy, which, as a bare concept, does not appeal to our perception or senses as do length, time, velocity, acceleration, force, and mass. Fortunately, energy is related (in fact, equivalent) to a physical operation or entity that does have that appeal, namely, work. Since all of us understand work (in its physical sense) as the application of a force to a physical task—for example, pushing an object over a given distance—we take this as the technical definition of work and see how it leads us to the concept of energy in its various forms.

If we do work on a body by applying a force to it, the body changes by acquiring energy, which appears in different forms or kinds, depending on the nature of the change in the body. We equate the quantity of the acquired energy exactly to the work done on the body by the force. To make this equation precise, we define work as the force acting on a body times the displacement of the body (or any part of it) in the direction of the applied force. This is a reasonable and acceptable definition of work, for it is precisely what we would be paid for if we accepted the job of pushing a heavy object a given distance. If we accepted a given amount of money to push one such object 100 feet, we would be justified in asking for twice as much money to push it 200 feet (twice the distance) and twice as much again if we were asked to push two such objects (twice the force). In other words, if pay is a true measure of work, then work is the product of force and distance, which we can write algebraically as work = force × distance, or $W = Fd$. The displacement must be in the direction of the force, so that the exact definition of work done on a body is the product of the applied force and only that part of the body's displacement that is in the direction of the force.

To state the quantity of work done by a force and therefore the energy acquired by the body, we introduce the unit of work, which is obtained from the units of the physical entities that define it. Thus, a unit force (1 dyne) that displaces a body a unit distance (1 centimeter) does 1 unit of work, since W (work) is the product of F and d, and therefore must be 1 if both F and d are 1. This unit is called the "erg," which is also the unit of energy in mechanics. The work done on an object displaced a given distance by a force depends only on the magnitudes of the force and the displacement. A force of 1 dyne applied to a body over a distance of 1 centimeter does 1 erg of work, whether the body is a grain of sand or a freight train.

Since the energy a body acquires when work is done on it appears in different forms, which do not immediately reveal themselves as energy, we describe carefully the various forms that may be assumed by the energy of a body. We apply a horizontal force to a particle at rest on a perfectly smooth (frictionless) horizontal surface in a vacuum (no air resistance) and see that the particle is moving at a definite speed after the force stops acting on it. Since the body's energy is contained in the speed imparted to it by the force, we call this acquired energy of motion "kinetic energy"; it is a simple algebraic exercise to show from the definition of work that if m is the mass of the body and v is its final speed, the kinetic energy acquired by the body is exactly $\frac{1}{2} mv^2$. If a body of 10 grams acquires a speed of 100 cm/sec from rest, its acquired kinetic energy is 50,000 ergs. We may now speak of the kinetic energy of a body in general, regardless of how the body acquired it, and say that if a body is moving with a speed v relative to some observer, its kinetic energy relative to that observer is $\frac{1}{2} mv^2$. It is important to qualify the definition of kinetic energy by specifying an observer because the speed of a body has no meaning in any absolute sense. A body at rest on the earth has no kinetic energy with respect to an observer on the earth, but it has an enormous amount of kinetic energy with respect to an observer on the sun. We shall see that, in general, energy cannot be defined in an absolute way, but always involves an arbitrary additive constant that changes with the observer's frame of reference.

We now consider work done on a body that is not free to move but is constrained in some way, and we begin with a particle in a gravitational field (for example, a particle on the surface of an ideal perfectly smooth earth). If we push such a particle horizontally for a moment, it acquires kinetic energy and continues moving on the ideal earth's surface forever in accordance with our discussion of kinetic energy above. What happens if the particle meets a perfectly smooth crater in its path? It rolls down with increasing speed and then rolls back up again with decreasing speed and continues its horizontal motion. We now analyze its energy while it is moving in the crater. Since it gained kinetic energy while rolling down the crater, work equal to its kinetic energy gain must have been done on it, and we see at once that the earth's gravitational force does this work. We examine this idea of work done by or against a field of force (the gravitational field) and see how the energy associated with it (the field energy) is to be defined.

This is best done by considering the work done in lifting a particle of mass m (or weight $w = mg$, where g is the gravitational acceleration) a height h above the ground. Since the vertical force we now exert is the weight w of the particle and we lift the particle vertically a height h, the work we do on the particle is exactly wh (force times distance), so that the particle's energy at the height h is larger by exactly this amount than it was on the earth's surface. The situation here is quite different from that of the kinetic energy case; there, we saw the result of our work (kinetic energy) in the particle's motion, whereas here we see no difference in the particle's behavior at the height h from that on the earth's surface. The difference is not in its behavior but in its position relative to the earth's surface, which stems from its increase in energy. We call this kind of energy, determined by position rather than by motion, "potential energy," which is an apt name, because as soon as the particle is allowed to fall back freely from its height h, the work we did on it reappears as kinetic energy. The only thing that counts in measuring or stating the potential energy of a particle is its height above the ground; the path along which it was transported to bring it to that height is irrelevant.

Potential energy, like kinetic energy, has an arbitrary constant attached to it because it can be defined with respect to any surface, as can be seen from the following discussion: Consider a particle at sea level on the earth's surface, another identical particle at the bottom of a hole (below sea level), and a third identical particle on a mountaintop above sea level. If we define zero potential energy as that of a particle at sea level, the first particle has zero potential energy, but then the particle in the hole has negative potential energy and the particle on the mountaintop has positive potential energy. If we take the mountaintop as the zero potential energy surface, then the particles at sea level and in the hole have negative potential energy. Whereas kinetic energy can be only zero or positive, potential energy can be negative, zero, or positive, depending on our choice of reference surface.

People generally have no trouble with the concept of zero or positive energy, but negative energy puzzles them greatly. To simplify this concept and make it understandable, we return to the particle in the hole with sea level taken as the surface of zero potential energy. We now lift the particle, doing positive work on it and therefore giving it positive energy, to sea level, where its energy is now zero. Clearly, if its energy after we have done work on it is zero, its potential energy in the hole must have been less than zero—that is, negative. Speaking of a body as having negative potential energy merely means that we must do work on the body to bring it to our reference surface of zero potential energy, so the concept is not all that mysterious.

With these points in mind, we choose the reference surface of zero potential energy to be such that the arbitrary constant in the expression for the potential energy of a particle is the same for all points (for all observers). This is so for the surface at infinity, and so we assign zero to the potential energy of any particle infinitely far off. This means that the potential energy of any particle at a finite distance is less than zero, that is, negative, because we must do work on such a particle to move it to infinity.

With this specification of zero potential energy, we can now write the expression for the gravitational potential energy of a particle at any distance from the center of the earth if the earth and the particle are the only objects in the universe. If m is the mass of the particle, M the mass of the earth, r the particle's distance from the earth's center, and G the Newtonian gravitational constant, the work we do to move the particle to infinity can depend only on these four quantities. Recalling that work is force (here, gravity) times distance and that the force depends only on the four quantities listed, we see why the work done depends only on them. We can now deduce the expression for the work done (the negative of the potential energy) from general arguments. Since the gravitational force is large if M and m are large (it depends on their product), the work done must also depend on this product and hence on GMm. But the closer we are to the center of the earth (that is, the smaller r is), the more work we have to do, because gravity is stronger at our starting point if r is decreased. We now combine these two dependencies into a single one to obtain GMm/r for the work done, so that the potential energy of m at the distance r from the earth's center is $-GMm/r$. This holds only if no bodies other than the earth are present; if other bodies are present, we write a similar negative expression for each, one with M replaced by the mass of the body and r replaced by its distance from m in each case. The total potential energy of m is then the sum of all these negative terms. From this, we now see that the mutual gravitational potential energy of two particles of mass m_1 and m_2 separated by the distance r is $-Gm_1m_2/r$.

Combining this quantity with the expression for the kinetic energy, we have for the total energy of a particle of mass m (for example, a planet) in the gravitational field of and at a distance from a particle of mass M (for example, the sun) $T =$ total energy = kinetic energy + potential energy $= \frac{1}{2} mv^2 - GMm/r$. This is called the "total mechanical energy" of a particle. In addition to the potential energy of a particle in a field of force, we have the potential energy of the distortion of a body under the action of a force. If we distort a body when we apply a force to it, without moving it bodily (for example, we stretch a spring), we do work on it without increasing its kinetic or external potential energy. But if energy is work done, the energy of the distorted body must have increased; it is, in fact, an increase in the internal potential energy, which is important in the study of the physical properties of materials and in civil and mechanical engineering, but not of interest in Newtonian mechanics.

Finally, we do work in pushing a heavy object at constant speed across a rough horizontal surface without increasing its potential or kinetic energy. In what way has our work been changed into energy and where is the energy? Here, the energy stemming from the work is not concentrated in the body we push but is dispersed, ultimately flowing into the environment in the form of heat.

CONSERVATION OF ENERGY

Having discussed work and energy, we can now consider the principle of the conservation of energy, limiting ourselves to the kinetic and potential energies of a particle moving in a gravitational field, unhampered by friction. A good example of this principle is the bob of a pendulum. For this discussion, we take the potential energy of the bob to be zero when it is at the lowest point of its swing (closest to the ground); its total mechanical energy (potential plus kinetic) is zero if it is at rest at this point. Starting with the bob in that position, we do work on it and pull it to one side, thus giving it potential energy, since it is now higher above the ground than initially. Its energy is now entirely potential, but it acquires kinetic energy, losing potential, as it starts its downward swing. Owing to its inertia, the bob does not stop at its starting point (the lowest point of its arc) even though its potential energy is zero there. Its potential energy has been transformed completely into kinetic energy. It swings past its lowest point and sweeps along an arc, reaching a highest point on the other side. Thus, the energy of the bob oscillates from potential to kinetic and back again. If the motion were frictionless (no air resistance and no friction of the string of the pendulum against the nail supporting it), the total mechanical energy (potential plus kinetic) of the bob at each point of its motion would be constant. This is the principle of conservation of energy, which the bob does not actually obey because it finally comes to rest owing to friction, which changes the mechanical energy of the bob into heat. But this principle is adequately verified by the motions of the planets around the sun. The earth, for example, has both kinetic and potential energy at each point in its annual solar orbit, and the sum of these two energies (the earth's total mechanical energy) remains constant (and has done so for billions of years) even though the kinetic energy, like the potential energy, changes continuously. At its maximum distance from the sun (the aphelion), the earth's potential energy is a maximum and its kinetic energy is a minimum; at its closest approach to the sun (the perihelion), the earth's kinetic energy is a maximum and its potential energy is a minimum. If the earth's total energy had decreased steadily, the earth would have fallen into the sun long ago; if its total energy had increased continuously, the earth would now be a cold, dead object floating by itself in interstellar space.

Just as the conservation of momentum implies a spatial symmetry of the laws of nature (they are the same regardless of where we place our frame of reference), so the conservation of energy means that the laws of nature are symmetrical in time; the laws are invariant with respect to a shift in time. This means that if we represent the physical state of a system by some quantity that is a function of (that is, depends on) space and time, the momentum of the system is related to how the physical state of the system changes if the system moves from point to point in space and the energy is related to how the state changes from moment to moment. This plays a very important role in quantum mechanics, the theory that has replaced the Newtonian laws of motion.

CONSERVATION OF ANGULAR MOMENTUM

Although Newton's laws of motion apply to the two kinds of motion that a body can possess, translational and rotational, most people rarely think of them except in their translational application, and yet pure translational motion rarely occurs in nature; rotation (or revolution) is the rule, rather than the exception. In translation, a straight line connecting any two points of a body remains parallel to itself (its spatial orientation does not change); in pure rotation, no straight line, except one, remains parallel to itself. The line that does, called the body's "axis of rotation," consists of all the points in the body that do not move as the body rotates. Note that whereas translation means changing the body's position without changing its orientation, rotation means changing its orientation without changing its position.

Since Newton's second law of motion cannot be applied directly to a rotating (spinning) body (it has no bodily acceleration), we must enlarge (or extend) the law to encompass such motion. We do so by beginning with the translation of a particle of mass m (a point mass) with constant speed v around the circumference of a circle of radius r. The magnitude mv of the particle's momentum is constant, but its direction changes continuously, so that its momentum is not conserved. This means that a force directed toward the center of the circle acts on the particle continuously, to keep it moving in the circle. But although the particle's momentum is not conserved, there is conserved some other physical property of the motion, which leads us to the concept of angular momentum. To find this dynamic property, we note that the quantities m, v, and r remain constant during the particle's motion, so that the product mvr of these three quantities also remains constant. But the direction of the axis (the line through the center of the circle) around which the particle is revolving also remains constant. We therefore assign to the particle's motion a new vector the magnitude of which is mvr and the direction of which is along the axis of revolution of the particle. This constant vector is called the "angular momentum" of the particle, and we show that it obeys a conservation principle just as the momentum and energy do.

Before doing this, we relate this to the rotation of a body by picturing the particle described above as attached to the axis by a rigid weightless (massless) rod. This rod and the particle together constitute a rotating body with angular momentum *mvr*. We note that this quantity and the direction of the axis of rotation remain constant even though a constant force acts on the particle; hence, the angular momentum of a body can remain constant even though a force acts on it, so that conservation of angular momentum is associated with the absence of something other than a mere force. We must have something in addition to a force, and to see what this is, we picture spinning an unattached wheel. If we simply pull along one of the spokes of the wheel, the wheel moves bodily toward us without spinning; to spin it, we must pull (or push) either the circumference or at right angles to one of the spokes at a finite distance from the axle. Following Newton's law of action and reaction, this push or pull generates an opposite and equal reaction (the inertial reaction) of the wheel at its center (the point on the wheel's disk through which the axle passes). Thus, two equal and opposite forces, separated by a distance, act on the wheel. Such a combination of opposite forces acting on a body—which are not in the same line, so that the effect of one of them on the body is not completely compensated by that of the other—is a "torque." A torque must be applied to a body either to spin it, to stop it from spinning or, in general, to change its angular momentum.

The principle of conservation of angular momentum states that if no external torque is applied to a system, its angular momentum remains constant; if a torque is applied, the system's angular momentum changes in the direction of the torque at a rate equal to the magnitude of the torque, which is the product of the perpendicular distance between the two forces that define the torque and the magnitude of either of the forces. The direction of the torque is along the perpendicular to the plane of the two forces.

A few simple points about torque and angular momentum in general are noteworthy. In our daily activities, we exert torques far more often than simple forces, and in doing so we are aware of and apply the torque magnitude formula to simplify daily tasks such as turning keys in locks, opening doors, unscrewing tops of jars, steering our cars, lifting objects, and engaging in all kinds of physical activities. The torque magnitude formula tells us that it is easier to unscrew a nut with a long wrench than with our thumb and fingers or with a short wrench, a fact that we know instinctively because the longer the wrench handle, the greater the torque (the lever principle).

Though we introduced the angular momentum concept by using a particle moving in a circle, a particle has angular momentum even if it is moving in a straight line with constant speed, but there is an important difference between the angular momentum for these two types of motion. The angular momentum of the particle moving in a circle has the same average value mvr for all observers no matter where they are, but the angular momentum of a particle moving in a straight line depends on the observer's position with respect to the straight line. If the observer is at a perpendicular distance y from the line and the particle is moving at constant speed (along the line), its angular momentum with respect to this observer is just mvy, which diminishes as the observer approaches the line, becoming zero when the observer is on the line. When the observer goes to the other side of the line, the angular momentum changes sign, with ever-increasing magnitude as the observer recedes from the line. Thus, the particle in the circle has intrinsic angular momentum, whereas the other does not.

If a line is drawn from the observer to the particle moving at constant speed v along the straight line, the line from the observer sweeps out the area of a triangle in a unit time, which just equals $\frac{1}{2}vy$, since v is the base and y is the altitude of the triangle. But this is exactly the particle's angular momentum divided by twice its mass. We see, then, that the line from the observer to the particle sweeps out equal areas in equal times, which is equivalent to the conservation of angular momentum. Kepler's second law of planetary motion, which states that the line from the sun to a planet sweeps out equal areas in equal times, therefore states that the angular momentum of a planet is conserved under the gravitational action of the sun. This is so because the sun's gravitational force is along the line from sun to planet, so that no torque acts on the sun–planet system.

Rotational symmetry is associated with the conservation of angular momentum; the laws of nature are the same no matter how we orient or rotate our frame of reference. This means that space is isotropic; the geometry of space is the same in all directions.

The importance of angular momentum in understanding the laws of nature and the structure of space, time, and matter can be only broadly indicated here. Its full significance will be apparent when we discuss the quantum theory, atoms and nuclear structure, and elementary particles.

We began this chapter by introducing two general kinds of principles that have evolved from Newton's laws—conservation principles and minimal principles—but we have discussed only the former, which have taken us far beyond the simple laws of motion in our understanding of physical phenomena. But this was only the beginning of the transformation of Newtonian physics from its meager origin to the elegant sophisticated mathematical structure that is now known as Lagrangian—also Hamiltonian—mechanics (classical Newtonian mechanics garbed in its fanciest mathematics). This transformation required combining the conservation principles with the minimal principles, which we discuss in the next chapter.

CHAPTER 8

The Post-Newtonian Era

Minimal Principles and Lagrangian and Hamiltonian Mechanics

In questions of science the authority of a thousand is not worth the humble reasoning of a single individual.

—GALILEO GALILEI

THE CONCEPT OF ACTION

We saw in the previous chapter that the Newtonian laws of motion take on a deeper meaning if we introduce dynamic conservation principles that stem from and are equivalent to the laws of motion. The conserved quantities, constructed algebraically from the more elementary physical entities that enter Newton's laws directly, may be vectors (momentum, angular momentum) or scalars (energy). The minimal principles deal only with scalars and therefore have the very desirable property that if they are valid in one frame of reference, they are valid in all frames of reference.

Minimal principles in physics were first introduced in the 17th century by the French mathematician Pierre de Fermat in his investigation of the propagation of light. The hypothesis of René Descartes that light travels faster in a dense medium (for example, water) than in a rare medium (air) was challenged by Fermat, who argued that such a behavior of light would violate the "principle of economy," according to which events in nature unfold in the shortest possible time. From his studies of the reflection and refraction of light, he concluded that light moves along a path that it traverses in the shortest possible time. This is known as Fermat's "principle of least time"; from it, the laws of reflection and refraction of light can be deduced. This general idea of the economy of nature was first introduced into Newtonian mechanics in the 18th century by the French mathematician P. Maupertuis, who argued that the principle of economy is best fulfilled not by time of transit, but by a quantity he called "action" and defined (incorrectly) as the distance a body travels multiplied by its speed.

Since the modern action concept and the principle of least action have played an extremely important role in the development of physics, we define action very carefully and precisely at this point. Two measurable entities of a particle, its position and its momentum, are associated with the particle at each point in its orbit, and if these entities are known for any point, the particle's orbit can be deduced from Newton's second law of motion and knowledge of the law of force acting on the particle. Since over a very tiny stretch of the particle's path its momentum is practically constant, it is meaningful to speak of the product of this momentum and the tiny stretch of path; this product is called the "increase in the particle's action." Action is a scalar quantity that the particle carries with it and that increases as the particle moves from point to point along its orbit. This definition of action is quite different from Maupertuis's, since he used speed rather than momentum.

Applying the "principle of economy" to action, Maupertuis introduced the principle of least or minimum action for a particle, stating that a particle chooses that path, from its starting to its final point, along which the increase in its action is a minimum. This statement is known as the "principle of least action." This principle eliminated the difficult concepts of action at a distance, which is inherent in Newton's law of gravity and which Newton himself abhorred, for it shifted the focus of Newton's laws from the way one body affects another body over a distance to how each of these bodies responds to some property (action) of its own neighborhood or path. It is as though each body or particle sees all possible paths laid out before it and chooses the one along which its action changes by the smallest amount. This conclusion was the beginning of the concept of the force field, which has been so useful in the development of the electromagnetic theory, general relativity, high-energy particle physics, and cosmology.

The action concept and the principle of least action were enlarged, extended, and generalized by the great Irish mathematician and mathematical physicist William Rowan Hamilton, so that they are applicable not only to the trajectories of particles but also to the propagation of light, to the behavior of complex systems of particles, and to fields (for example, electromagnetic, gravitational). Before we discuss Hamilton's great contributions, however, we point out a very important deduction that flows from the action concept if a certain restriction is placed on action. Newtonian mechanics assumes, implicitly, that natural phenomena unfold continuously in space and time and that processes such as the motion of a particle along its path can be followed in infinite detail and with infinite precision. This assumption would be true if one could simultaneously measure or observe the exact position and the exact momentum of a particle at any point in its orbit. If this were possible, the motion of the particle would not have to be followed in detail to obtain its orbit, for Newton's equations of motion, combined with our knowledge of the particle's position and momentum at a single point of its orbit, give us the complete orbit forever after.

This conclusion would be true if all physical entities were infinitely divisible, so that no matter how small they were they could still be observed or measured. However, this would not be so if this kind of divisibility were limited. If action during a process changed in small but finite amounts, h (never less than h), rather than continuously, the precise orbit of a particle could never be determined for the following reason: To determine the orbit precisely, we must know the particle's precise momentum and position at the same point, but action is the particle's momentum multiplied by a measured spatial interval that must be taken as infinitesimal if we want to know the particle's position. But if we do that, the product that defines the action becomes infinitesimal and hence smaller than the permitted minimum h unless the momentum becomes infinite and we lose all knowledge of it. In other words, if action comes in lumps (that is, if it is quantized), we cannot know the particle's position and momentum simultaneously. We shall see the full significance of this result for physics when we discuss the quantum theory.

HAMILTON'S PRINCIPLE OF LEAST ACTION

Sir William Rowan Hamilton (1805–1865) was born in Dublin, Ireland. At an early age, he showed unusual intellectual powers that prompted his father to turn the responsibility for his son's education over to the former's brother, the Reverend James Hamilton, who was a member of the Royal Irish Academy. Before the age of three, William was sent off to the town of Trim near Dublin to live with his uncle, who worked as a schoolmaster there at a Church of England school; William remained with his uncle until he entered Trinity College, Dublin, in 1823.

Soon after Hamilton arrived at his uncle's home, he learned to read English and perform complex arithmetic. By the age of five, he could translate Latin, Greek, and Hebrew and recite long passages from works by authors ranging from Homer to Milton.[1] In the next five years, he became intimately acquainted with Sanskrit and taught himself Arabic, Chaldee, and several Indian dialects and mastered Italian and French. Before he turned 12, he compiled a grammar of Syriac and, two years later, had become sufficiently advanced in his studies of the Persian language to compose a welcome note to a visiting dignitary from Persia.[1]

Hamilton's interest in mathematics began in 1820 when he met an American named Zorah Colburn who could calculate mentally the solutions to problems involving extremely large numbers.[1] Impressed with the utility of mathematics, Hamilton immersed himself in the classical scientific treatises such as Newton's *Principia* and Laplace's *Traité de mécanique céleste;* his detection of a logical flaw in Laplace's monumental treatise brought him to the attention of John Brinkley, a professor of astronomy at Trinity College, Dublin.[1] Hamilton's growing mastery of classical mechanics led him to consider the subject of optics. He completed his first paper on geometrical optics in 1824 while an undergraduate at Trinity College and submitted it for publication to the Royal Irish Academy. Because the paper was so mathematically abstract, however, the academy members did not really understand it, so they asked Hamilton to demonstrate his findings. Hamilton followed their advice and completed the project while still an undergraduate student. He formally presented his paper, "An Account of a Theory of Systems of Rays," to the academy in 1827: "It is one of the great classics in theoretical phsycis and has remained the basis of most treatises on geometrical optics. Moreover, it contains the germ of the idea which was to lead Hamilton to his famous formulation of dynamics. In this paper Hamilton introduces the characteristic function of a system of rays from which all the properties of the system can be deduced by simple mathematical operations such as differentiation. The importance of this function is that it is directly related to the action integral, which plays the dominant role in Hamiltonian dynamics and which was the starting point of Schroedinger's derivations of his wave equations."[2]

Hamilton's paper established his reputation as a mathematical physicist because it anchored the subject of geometrical optics in mathematics by showing that all problems in that subject could be solved with the single uniform method offered by Hamilton.[3] John Brinkley resigned from his post of professor of astronomy at Trinity College while Hamilton was still an undergraduate. Because of the brilliance of his paper, Hamilton was offered the vacant position even though he was only 21 years old. Hamilton accepted the job and devoted himself to his researches into mathematics, physics, and astronomy as well as his teaching duties and the more mundane daily responsibilities of running the observatory.[4]

Hamilton's discovery of conical functions led to his being awarded the Cunningham medal of the Royal Irish Academy and the Royal medal of the Royal Society in 1834. Three years later, he became the president of the Royal Irish Academy, a post he held for eight years. During his tenure, Hamilton did his most important work in mathematical physics, the results of which were published in 1835 in a famous paper entitled "General Methods in Dynamics": "In it he applied his idea of the characteristic function to the motion of systems of bodies and expressed the equations of motion in a form that revealed the duality between the components of momentum of a dynamical system and the coordinates determining its position."[5] These equations were developed by using the so-called "Hamiltonian function," which has contributed so strongly to the development of quantum mechanics. This function is explained in the following manner: "Suppose we have a system, say a particle, moving in a field of force. We may then write down the total energy of this system (kinetic plus potential) in terms of the components of momentum and the coordinates. This expression is called the Hamiltonian of the system and the equation describing the motion of the system can be obtained from this quantity by certain mathematical operations. The advantage of such a formulation of dynamics is that it can be applied to very complex systems by a judicious choice of coordinates."[6]

Hamilton's studies of algebra also led him to the discovery of quaternions, which made possible an entirely new method of computation. The reason is that quaternions behave like numbers, but are not numbers, because they violate the commutative law $a \times b = b \times a$ for ordinary numbers. By freeing algebra from the commutative postulate of multiplication, Hamilton provided a powerful tool for studying "quantities in magnitude and direction in three-dimensional space."[7] This discovery in 1843 ended 15 years of labor by Hamilton to find a way to multiply vectors, a concept that had resulted from the efforts of Möbius to add points and forces. The answer to the problem came to Hamilton in a flash of insight as he was strolling with his wife along the Royal Canal on his way to an academy meeting in Dublin: "He could not resist the impulse to cut with a knife, on the stone of Brougham Bridge, as they passed by it, the fundamental formula $i^2 = j^2 = k^2 = ijk = -1$, indicative of the

Quaternions which gave the solution to the problem."[8] Although Hamilton's quaternions had a significant impact on modern algebra, they were not greatly used in applied mathematics for long because of the development by Josiah Willard Gibbs of his simpler vector analysis method.

Hamilton spent the last two decades of his life writing mathematical papers that addressed problems in algebra, calculus of probability, theory of equations, and theory of functions. Although his work was characterized by a profound intuition, his papers were mathematically detailed and his logic of reasoning was always clear and precise. At the same time, he was a deeply religious man who tried to establish a metaphysical basis for his mathematical discoveries, as when he attempted to "derive algebra from the elements of space and time and establish it as a science rather than as a branch of mathematics."[9] Hamilton was familiar with the works of George Berkeley and Immanuel Kant, and his physicalistic approach toward science had been greatly influenced by René Descartes and Isaac Newton. Like Newton, however, he was so devoutly religious that he elevated the importance of the spiritual life above that of science and mathematics. Neither man viewed such an ordering as inconsistent, for mathematics was seen by both as the only tool that could demonstrate the remarkable dynamic beauty of a universe that each believed to be the handiwork of a divine creator.

Hamilton began his investigations into the dynamics of moving bodies by analyzing the propagation of light as expressed in Pierre de Fermat's principle of least time and noting its similarity to the Maupertuis principle of least action for the paths of particles. In doing this work, he noted and developed a remarkable analogy between optics and dynamics that Erwin Schrödinger took many years later as his starting point for the construction of the wave mechanics of particles. This is one of the most remarkable examples of the interrelationships of different branches of physics in its history, as will become more evident when we discuss the quantum mechanics.

Hamilton pointed out that "the science of optics" had developed along two different paths, one based on the propagation of rays along straight lines, which he called the "theory of systems of rays" (geometrical optics), and the other based on wave propagation (physical optics). In considering ray optics, he was struck by the similarity between Fermat's principle of least time for ray optics and Maupertuis's principle of least action for particles, and he was convinced that a kind of geometrical optical theory of dynamics could be developed, so that the laws of optics and dynamics could be combined or represented by the same minimal principle of a universal action. He was driven to develop these ideas by his great admiration for Lagrange's work on dynamics as developed in Lagrange's *Méchanique Analytique*.

Hamilton described these ideas in a famous paper, "On a General Method of Expressing the Paths of Light, and of Planets by the Coefficients of a Characteristic Function," which appeared in the *Dublin University Review* in 1833. He prefaced his work there by stating:

> Those who have meditated on the beauty and utility, in theoretical mechanics, of the general method of Lagrange—who have felt the power and dignity of that central dynamical theorem from which he deduced, in the *Mécanique Analytique,* from a combination of the principle of virtual velocities with the principle of D'Alembert—and who have appreciated the simplicity and harmony which he introduced into the research of the planetary perturbations, by the idea of the variation of parameters, and the differentials of the disturbing function, must feel that mathematical optics can only then attain a coordinate rank with mathematical mechanics, or with dynamical astronomy, in beauty, power, and harmony, when it shall possess an appropriate method, and become the unfolding of a central idea.
>
> This fundamental want forced itself long ago on my attention; and I have long been in possession of a method, by which it seems to me to be removed. But in thinking so, I am conscious of the danger of a bias. It may happen to me, as to others, that a meditation which has long been dwelt on shall assume an unreal importance; and that a method which has for a long time been practised shall acquire an only seeming facility. It must remain for others to judge how far my attempts have been successful, and how far they require to be completed, or set aside, in the future progress of the science.
>
> Meanwhile it appears that if a general method in deductive optics can be attained at all, it must flow from some law or principle, itself of the highest generality, and among the highest results of induction. What, then, may we consider as the highest and most general axiom (in the Baconian sense) to which optical induction has attained, respecting the rules and conditions of the lines of visual and luminous communication? The answer, I think, must be, the principle or law, called usually the Law of Least Action; suggested by questionable views, but established on the widest induction, and embracing every known combination of media, and every straight, or bent, or curved line, ordinary or extraordinary, along which light (whatever light may be) extends its influence successively in space and time: namely, that this linear path of light, from one point to another, is always found to be such, that if it be compared with the other infinitely various lines by which in thought and in geometry the same two points might be connected, a certain integral or sum, called often Action, and depending by fixed rules on the length, and shape, and position of the path, and on the media which are traversed by it, is less than all the similar integrals for the other

> neighbouring lines, or, at least, possesses, with respect to them, a certain stationary property. From this Law, then, which may, perhaps, be named the Law of Stationary Action, it seems that we may most fitly and with best hope set out, in the synthetic or deductive process, and in search of a mathematical method.
>
> Accordingly, from this known law of least or stationary action, I deduced (long since) another connected and coextensive principle, which may be called, by analogy, the Law of Varying Action, and which seems to offer naturally a method such as we are seeking. . . .[10]

To Hamilton, then, the principle of least action, in which he included Fermat's principle of least time, stood at the very peak of natural laws and was the door to the unification to physics. To obtain a simple principle of least action that covers both optical and mechanical phenomena, Hamilton started from Fermat's principle and showed that it can be replaced by one that closely resembles Maupertuis's principle of least action. He did this by replacing the time in Fermat's principle by the length of the path the ray travels between two points in any medium divided by the speed of the ray in that medium. This is equivalent to multiplying the distance the ray would move in a vacuum by the index of refraction of the medium at each point of the ray's path. By this procedure, Hamilton showed that Fermat's principle of least time is formally similar to Maupertuis's principle of least action.

In doing this work, Hamilton obtained a remarkable synthesis of the laws of optics and Newton's laws of motion. The index of refraction at a point in a medium determines the speed of light at that point, so that it behaves for light somewhat the way a force field behaves toward a particle moving through it. From considerations of the similarity between the principle of Fermat for geometrical optics and the Maupertuis principle of least action for dynamical systems, Hamilton was thus led to the idea that the behavior of particles might be described by a kind of wave mechanics. He showed that the trajectories of particles having the same total amount of energy are identical with the paths of rays of light in a medium having the proper index of refraction. In other words, it is possible, according to Hamilton, to find an index of refraction such that the trajectory of any particle can be described by the path of some ray of light in a medium having the given index of refraction. Since, however, rays are only an approximation (which becomes more accurate as the wavelength gets smaller) to the correct wave description of light, it follows that the Newtonian trajectories are only an approximation to a wave description of the motion of particles. Just as in optics the rays of light are orthogonal to the wave fronts (which are surfaces of equal phase), so in mechanics the trajectories of particles are orthogonal to another kind of wave front (the surfaces of equal action). In other words, in particle wave mechanics, the action would play the role of the phase in optics. This Hamiltonian formalism is just what was needed to go from the classical mechanics of a particle to the quantum and wave mechanics; it was taken over bodily by Schrödinger.

We may now summarize Hamilton's great contribution to classical dynamics: Suppose we have a particle that is moving in a force field. We may now, in accordance with Hamilton's procedure, describe the path of the particle as though the particle were a ray of light moving through an optical medium having an index of refraction that is related in a definite way to the force field through which the particle is moving. One may therefore surmise from this idea (which Hamilton did not, but which Schrödinger did) that just as ray optics, that is, geometrical optics, gives only an approximate description of light, so classical dynamics gives only an approximate description of the motion of a particle. And just as ray optics has its wave–optical complement, so particle dynamics has its wave–dynamic aspect.

If one compares Newtonian dynamics with classical optics, it appears that the dynamics describes only half a picture as compared to the optics; whereas the latter appeared in two different forms, the Newtonian corpuscular form and the Huygens wave form, the latter had no wave aspect at all. To one like Hamilton, who passionately believed in the unity of nature, this was a flaw in Newtonian physics that had to be eliminated, and he took the first step in this direction by extending the action concept to include the propagation of light.

In his search for a universal action principle, Hamilton enlarged the Maupertuis principle of action by extending it to take account of the energy as well as the momentum of a particle in its motion. Instead of defining the action over a short stretch of the particle's path as the product of the particle's momentum and the small stretch of the path, he considered the particle's motion during a very small time interval and defined its action as the product of this tiny time interval and an entity called the "Lagrangian" that Lagrange had introduced into generalized Newtonian mechanics. The Lagrangian of a particle is merely its kinetic energy minus its potential energy, so that Hamilton's definition of action can be used to describe particles moving in a force field in terms of their potential energies. For a single particle, the Hamiltonian action is the sum of two products: the product of the particle's momentum and a tiny stretch of its path (the Maupertuis action) and the negative of the product of the particle's total energy and the tiny time it spends in the small stretch of path. This equation may be expressed as follows: Hamiltonian action = (momentum × distance) minus (energy × time) = Maupertuis action minus (energy × time).

That Hamilton's definition of action links momentum with space and energy with time is remarkable in two respects: It shows that the measurement of the momentum of a particle in its path is in some way related to the measurement of its position, which anticipates quantum mechanics. It also indicates that momentum must be combined with space and energy with time in setting up quantities such as action to describe physical systems, thus anticipating the theory of relativity, which combines space and time into a single space–time continuum.

THE CONTRIBUTIONS OF LAGRANGE

The French-Italian mathematician Joseph-Louis Comte de Lagrange (1736–1813) was born into a wealthy family with some ties to the Italian nobility in the city of Turin. His father was the treasurer to the King of Sardinia, but soon lost the family fortune in speculative investments. Forced to rely on his own resources, Joseph concentrated on his studies in school, especially the classical Greek and Roman poets such as Homer and Virgil, but only after he happened to read a memoir by Edmond Halley did Lagrange discover that his true calling lay in mathematics. Lagrange read every mathematical treatise he could lay hands on, and he soon mastered the subject, so that by the age of 19 he was teaching mathematics at the royal artillery school at Turin. His command of the subject was so thorough that despite an uninspiring speaking manner, he could hold the attention of even the most skeptical senior professors. Lagrange's unassuming personality and his enthusiasm for mathematics won the respect of his colleagues; this collection of scholars became the earliest members of the Turin Academy of Science, the founding of which was due in large part to the efforts of Lagrange. However, his greatest talent lay in his mathematical essays: "With a pen in his hand Lagrange was transfigured; and from the first, his writings were elegance itself. He set to mathematics all the little themes on physical inquiries which his friends brought him, much as Schubert set to music any stray rhyme that took his fancy."[11]

Lagrange first came to the attention of mathematicians throughout Europe when he solved the so-called "isoperimetrical problem," which

Joseph-Louis Comte de Lagrange (1736–1813)

had teased mathematicians for over half a century. Lagrange sent his solution to Leonard Euler, then Europe's most famous mathematician, who, despite having arrived at a similar conclusion, graciously allowed Lagrange to receive the full credit for the discovery. Lagrange's solution to the problem had required that he invent a "calculus of variations," the development of which became the focus of his career and was of crucial importance to the concept of economy in nature (the principle of least action). This minimal principle later influenced the work of Hamilton, James Clerk Maxwell, and Albert Einstein; it continues to be relevant in all areas of modern physics.

Lagrange continued to make important mathematical discoveries, and he became recognized as one of the most talented mathematicians in Europe. Unlike Laplace, he graciously acknowledged the contributions of others, but he could also detect the subtlest of errors in the work of others, including several mistakes committed by Newton himself. His contemporaries freely acknowledged Lagrange's abilities; he was later widely regarded as the greatest living mathematician in Europe.

In 1764, he received a prize from the Paris Academy of Sciences for his essay "on the libration of the moon, the apparent oscillation that causes the slight changes in position of lunar features on the face that the moon presents to the earth."[12] The solution to the problem was aided by Lagrange's invention of the equations that bear his name. Two years later, he received another prize from the academy for his essay on the theory of the motions of the moons of Jupiter. In the following decade, he received three more prizes for his lucid and impeccably reasoned essays in mathematics. "In 1776, on the recommendation of Euler and the French mathematician Jean d'Alembert, Lagrange went to Berlin to fill a post at the academy vacated by Euler, at the invitation of Frederick the Great, who expressed the wish of 'the greatest king in Europe' to have 'the greatest mathematician in Europe' at his court."[12]

Lagrange's unassuming personality was not greatly changed by the court appointment, and he continued an exhausting regimen of academic study. After Lagrange contracted several illnesses owing to lack of rest, Frederick himself gave his mathematician a lecture on the need for him to cut back his grinding work schedule. Lagrange appears to have taken his benefactor's advice, for he "changed his habits and made a program every night of what was to be read the next day, never exceeding the ration."[13] He spent 20 years in Prussia and wrote a vast number of outstanding mathematical papers that were later incorporated into his *Mécanique Analytique*. Among his works were "papers on the three-body problem, which concerns the evolution of the orbits of three particles mutually attracted according to Newton's law of gravity; differential equations; prime number theory; the fundamentally important number-theoretic equation that has been identified (incorrectly by Euler) with John Pell's name; probability; mechanics; and the stability of the solar system."[14]

After the death of Frederick, Lagrange left Prussia to accept Louis XVI's invitation to come to Paris. Although he was given an apartment and many honors, Lagrange's first two years in France were totally unproductive because his passion for mathematics apparently left him soon after his arrival. He appeared to friends as distracted and indifferent; it may be that the many years of unceasing labors in mathematics had finally worn out his mind. Like Newton before him, Lagrange turned to other subjects, including metaphysics, philosophy, and chemistry; he showed little interest when his *Mécanique Analytique* was finally published in 1788 and did not even open a first edition copy for two years. In any event, Lagrange's book was a remarkable "synthesis of the hundred years of research in mechanics since Newton, based on his own calculus of variations, in which certain properties of a mechanistic system are inferred by considering the changes in a sum (or integral) that are due to conceptually possible (or virtual) displacements from the path that describes the actual history of the system."[14] This work included the use of independent coordinates (generalized coordinates), which are necessary for specifying the positions of particles in a complex system, and the Lagrangian equations (instead of Newton's equations), which relate the kinetic energy minus the potential energy of a classical mechanical system to the generalized coordinates, the corresponding generalized forces, and the time.[14]

The outbreak of the Revolution in France did not affect Lagrange directly, even though many of his friends fled the country; he continued to be treated well by the government throughout that turbulent period, even though the beheading of the famous chemist Antoine-Laurent Lavoisier caused Lagrange to wonder whether his own days were numbered. Despite the danger to his life, however, Lagrange remained in Paris and devoted his energies to a committee that had been assembled to reform the metric system so that all denominations of money, weights, and measurements would henceforth be strictly based on multiples of the number ten. By 1791, Lagrange had shaken off his mental lethargy about mathematics and again began to produce many papers on a variety of topics and problems.

The contributions Lagrange made in almost every area of mathematics in his 76 years have seldom been equaled. Few scholars could match the originality and grandeur of Lagrange's writings or even the sheer quantity of his output. Lagrange's interests ranged from Newton's classical mechanics to Fermat's theory of numbers, and his work inspired the accomplishments of many distinguished mathematicians, including Laplace, Fourier, Monge, Legendre, and Cauchy.[15] Laplace sketched the broad design of the canvas that became modern mathematics, leaving the details to be filled in by his contemporaries and successors; in the same way, Newton laid the foundation of classical physics with his three laws of motion and his theory of gravitation, thereby providing the "bricks and mortar" for the subsequent construction of a monumental intellectual edifice. As Lagrange found his greatest inspiration in the formidable mathematical conception that had been formulated by Newton, it is valuable to consider how Lagrange's own contributions developed from the work of Newton.

Newton's equations of motion deal explicitly with the motions of individual particles subject to external forces and their mutual interactions, so that the complete Newtonian description of the motions of a group of such particles (an ensemble of particles) becomes very complicated and in fact impossible to handle for large numbers of particles. To rid Newtonian mechanics of this difficulty, Lagrange developed procedures to simplify the Newtonian equations by reducing the number of such equations required to describe an ensemble. This procedure stems from Lagrange's concept of the degrees of freedom of a system of particles, which we can best understand by first considering the motion of a single particle. It is free to move in any one of three mutually perpendicular directions if it is subject to no forces at all. We represent three such directions by three mutually perpendicular lines that meet at a point. Thus, a particle on the earth's surface (for example, an automobile on a winding hilly road) moves north, south, east, west, and up and down. We say that such a particle has three degrees of freedom, and to describe its motion when a force acts on it, we need three Newtonian equations of motion. Since three equations are required for each such particle in an ensemble, the total number of equations becomes forbiddingly large as the number of particles increases. Lagrange showed how this number can be reduced to a mathematically manageable set of equations by considering the constraints (generally of a geometrical nature) to which particles in an ensemble are subjected; these constraints permit one, with the aid of the Lagrangian of the system, to replace Newton's large number of equations of motion with a few equations. The equations are simplified still more by replacing the usual coordinates of the particles by so-called "generalized coordinates" (certain algebraic geometrical combinations of the old coordinates) and "generalized velocities," which are the time rates of change of the generalized coordinates. A few examples illustrate this important simplification of Newtonian dynamics.

The first example is the simple pendulum: a bob of mass m at the end of a weightless, inextensible string of length l. The bob is constrained by the string to move along the arc of a circle the radius of which is the length of the string, so that its three degrees of freedom are reduced by the string to one, which we may take as the angle (designated by the Greek letter θ) between the string and the vertical when the bob is at any point in its circular path. This angle is the generalized coordinate of the bob, and the time rate at which it changes is its generalized velocity $\dot{\theta}$ (the dot means rate of change). The pendulum's Lagrangian is the bob's kinetic energy ($\frac{1}{2} m l^2 \dot{\theta}^2$) minus its potential energy [$mgl(1 - \cos\theta)$], where g is the acceleration of gravity and $l(1 - \cos \theta)$ is the height of the bob above its resting position. From this expression, one obtains by certain mathematical operations the Lagrangian equation of motion for the angle θ, which replaces Newton's three equations of motion for the spatial coordinates of the bob. All the characteristics of the dynamics of the oscillating pendulum are deducible from this single equation for θ.

Another example is a system of two spheres connected by a rigid weightless rod. If the spheres were free (no connecting rod) and moved independently, each would have three degrees of freedom, and the total number of degrees of freedom would be six. But the rod is a constraint that reduces the number of degrees of freedom to five: the three degrees of freedom of the translational motion of the center of mass of the system (a point on the rod at a distance from each sphere that is inversely proportional to the mass of the sphere) and the two rotational degrees of freedom the system has in being free to rotate about either one of two perpendicular axes at right angles to the rod at the center of mass.

The Lagrangian concept of generalized coordinates and generalized velocities and Hamilton's extension of the principle of least action, with action constructed from a Lagrangian as described by Hamilton, permit physicists to treat dynamically fields of force such as the gravitational field and the electromagnetic field. As a result, one can introduce generalized coordinates and velocities of a field and construct a field Lagrangian from which the action of the field can be constructed. The Hamilton principle of action then leads to the equations of motion of the field.

All this may appear very abstract without the explicit presence of any of the "real" measurable quantities, such as space, time, force and mass, on which the Galilean–Newtonian physics was built, but the importance of the concepts of action, the Lagrangian, and the principle of least action for theoretical physics today cannot be overemphasized. The peak of these developments is contained in the famous Hamilton–Jacobi equation for the action of a system, discovered by Hamilton and an outstanding 19th-century German mathematical physicist, Karl Gustav Jacobi. From the action, obtained as a solution of this equation, one can deduce all the observable dynamic properties of the system such as momentum, energy, and angular momentum, so that in a sense, it represents the ultimate synthesis of Newtonian mechanics. But it went beyond that and was applied to classical electrodynamics (the interactions of electrical charges with electromagnetic fields) and to relativity dynamics. Although the Hamilton–Jacobi equation for the action of a system was developed for classical (Newtonian) mechanics, it is closely related to the famous Schrödinger wave equation for a particle in quantum mechanics; indeed, the Schrödinger wave function for a particle can be constructed from the many values of the classical action that can be assigned.

Finally, one can obtain the orbit of a planet around the sun quite easily by using the conservation principles, rather than by solving Newton's equations of motion. Here, we sketch the procedure without going through the algebra, which is no more demanding than elementary high school algebra. We consider a planet of mass m (the earth, for example) and the sun of mass M, interacting gravitationally—that is, pulling on each other along the imaginary line connecting them. Since the earth and the sun are spheres, and spheres behave gravitationally as though their masses were concentrated at their centers, we may in this analysis replace the earth and the sun by mass points (masses concentrated in a single point) separated by a distance equal to the distance between the center of the sun and the center of the earth. Newton's third law of motion tells us that the earth pulls on the sun with exactly the same force as the sun pulls on the earth, but the sun's response to this force is smaller than the earth's response by a factor equal to the proportion of the mass of the earth to the mass of the sun. In short, the earth's response is much greater than the sun's response because the earth's mass is much smaller. Keeping this point in mind, we can now see how the relevant geometrical, kinematical, and dynamic elements determine the orbits of the two particles.

We begin with the particle m (the earth) and the particle M (the sun) held at rest (kept from moving by some force) at the distance r from each other. We then apply the conservation principles to see how they move when the force that holds them in place is removed. Before this is done, the total momentum of the system (the two particles, earth and sun) is zero, its total angular momentum (no rotation) is zero, and its total energy is the mutual potential energy (negative) of the two particles; the kinetic energy of the system is zero. The conservation principles now tell us exactly how the two particles must move to obey these principles. Since they can move only along the line connecting them, they must do so in such a way that their total momentum remains zero, which means that m (the earth) moves along the line at a speed that is greater than the speed of M (the sun) by the factor M/m. This simply means that the product of m and its speed equals the product of M and its speed. If the earth and the sun were doing this, the earth's speed along the line would always be 340,000 times the sun's speed.

Since the speeds of m and M increase as they move toward each other, their kinetic energies increase, but these increases are exactly compensated by decreases in their potential energies; the total energy of the system thus remains constant, as demanded by the conservation of energy principle. After colliding at their center of mass, they do not stay attached to each other because energy is then not conserved, so they must separate, and they do so, each moving away from the center of mass at exactly the same speed of separation at each point as its speed of approach at that point. The two bodies come to rest when the distance between them exactly equals their initial separation, and they then start moving toward each other again to repeat the cycle. As long as no mechanical energy is lost in any such cycle, the two bodies simply keep oscillating along the straight line forever.

This is a very special orbit of the system that certainly will not do for a planet. To obtain a planetary orbit, we must apply a torque to the system to give it angular momentum (rotation). We can do this very easily by simply applying force to m for a very short time, exactly at right angles to the line connecting m and M. The magnitude of the torque is just this lateral force times the distance of m from the center of mass; the angular momentum imparted to the system is this torque multiplied by the time during which the force is applied to m.

To obtain the orbit of m around the center of mass (the shape of M's orbit around the center of mass is exactly the same as that of m, but is smaller by the mass factor m/M), we note that when m is pushed laterally (at right angles to the connecting line) for a moment, it acquires a lateral velocity that makes m move in an arc of a shape determined by how fast it was moving laterally after being pushed. At each stage of m's motion, one can write its total energy (kinetic plus potential), which cannot change, and its total angular momentum, which also remains constant as m moves along its path. From these two constant dynamic properties, two algebraic equations are obtained that contain (depend on) the speed of m and its distance from the center of mass at each point of its path. The solutions of these equations show that m's orbit is an ellipse and that the line from M to m sweeps out equal areas in equal times. These are just the first two of Kepler's three laws of planetary motion; the third law also follows directly from the equations.

We have gone to some lengths to develop these ideas and to show how Newton's laws of motion and his law of gravity are applied to this very important case of two mutually interacting bodies because this case represents the height of Newtonian mechanics. Moreover, it is the simplest gravitationally bound system of particles (two mass particles), the motions (orbits) of which can be obtained exactly. The orbits of three or more interacting particles (the famous n-body problem where n is greater than 2) cannot be written in explicit mathematical forms, except for very special cases such as the restricted three-body problem that was solved by Lagrange. The general many-body gravitational problem can be solved computationally only by the method of perturbations, which uses successive numerical approximations to get as close as may be desired to the correct solution (the observed orbits).

CHAPTER 9

The Growth of Optics, Electricity, and Magnetism

The cause is hidden, but the effect is known.

—OVID

THE END OF THE NEWTONIAN ERA

Newtonian mechanics and gravity dominated physics into the mid-19th century, when it reached the peak of its mathematical developments through the work of the mathematical physicists of that post-Newtonian period. The other branches of physics, such as optics, electricity, magnetism, and heat (thermodynamics), grew and evolved independently of Newtonian mechanics and quite slowly. This slow growth was partly due to strong intellectual competition offered by Newtonian mechanics, which was developing into a very beautiful and aesthetically satisfying mathematical structure. Moreover, its successful applications to planetary theory, with their immediate rewards and recognition, were much more tempting than the relative obscurity of working in a hidden laboratory on some abstruse, difficult problem concerning the nature of matter.

The dramatic success of Newtonian mechanics, applied to astronomy, and its payoff are indicated in the independent calculations by John Couch Adams in Great Britain and Urbain Jean Joseph Leverrier in France in 1846, which showed that the departure of the observed orbit of Uranus from its calculated orbit can be accounted for by an unobserved planet (named Neptune when observed later within a degree of where Adams and Leverrier had predicted it to be) that orbits the sun in a path far beyond that of Uranus. That the existence of a planet that had never been thought of could be deduced by pure thought (mathematics) from an abstract scientific theory created tremendous excitement and hoisted science to a level in people's thinking as high, if not as exalted, as that of religion.

Moreover, it showed that in the hands of experts, a combination of science and mathematics is a very powerful intellectual tool.

The preference for working in Newtonian mechanics rather than in other branches of physics during that post-Newtonian period also stemmed from its ease of use; the physical theory, the mathematics, and the problems to which the theoretical machinery were applied were all at hand, and these problems went far beyond the purely theoretical ones of idealized particles moving under ideal conditions. Such problems as the rise and fall of the tides, the flow of rivers, the orbits of projectiles (ballistics), the motions of wheels and parts of machinery, and the conditions for the stability of structures were all successfully tackled. Thus, mechanical and civil engineering were established on a sound scientific basis.

Compared to the ease with which such scientific work could be done with Newtonian theory, discovering the properties of light and probing the structure of matter were difficult, and the discoveries in these areas came slowly and, in many instances, accidentally, because, unlike Newtonian mechanics, these branches had no mathematical basis from which to develop. Much of the work was trial and error, done with very crude laboratory experiments. Even so, the wealthy amateur scientist Henry Cavendish, in 1794, measured the gravitational force between two lead spheres of known mass, with centers separated by a known distance, and thus obtained the first numerical value of the Newtonian gravitational constant with remarkable accuracy.

POST-NEWTONIAN OPTICS

Many facts about the propagation and nature of light had been discovered, and with Fizeau's and Foucault's measurements in the 1850s showing that the speed of light is smaller in a dense medium than in a rare one (in water as against air or a vacuum), the Newtonian corpuscular theory was discarded in favor of the Huygens wave theory. This choice was strongly supported by the experimental work of Thomas Young and Augustin-Jean Fresnel, who, using the wave theory, explained such things as the polarization, the interference, and the diffraction (bending around corners) of light, which were completely incomprehensible phenomena from the Newtonian corpuscular point of view. Though the wave theory was fully accepted in time, nothing was known about the physical nature of the wave or how it is transmitted over such vast distances as those of the stars from the earth. Though it was known that waves are transmitted through a medium (for example, sound waves through air, water waves over a water surface), there was no obvious medium for the transmission of light waves; therefore, such a medium, the "luminiferous ether," was invented.

The nature of the light wave was not discovered until James Clerk Maxwell, in 1862, working with Michael Faraday's experimental discoveries in electricity and magnetism, wrote his famous equations of the electromagnetic field, which show that light is an electromagnetic phenomenon. But even though the nature of light waves was not known in the first half of the 19th century, some of their important properties were known and were used in a practical way. Thus, the basic wave formula—that the wavelength of the wave multiplied by its frequency equals its velocity—was fully understood. The wavelength is the distance between two successive crests of a wave; the frequency is the number of crests that pass a given point in 1 second. The unit of frequency (1 per second) is the hertz, which was named after the great German experimentalist Heinrich Rudolf Hertz, who experimentally verified Maxwell's electromagnetic theory of light. It was also shown that the wavelength of red light is about twice as long as that of violet light and that all wavelengths (colors) travel equally fast in a vacuum.

Newton's discovery of the optical spectrum (that white light is a mixture of all the visible colors) ultimately led to the discovery of the spectroscope, which, despite its simplicity, is probably the single most important scientific instrument ever devised. Since its invention, it has led to more great scientific discoveries, ranging from the nuclear to the cosmological domain in physics and astronomy, and incorporating all branches of geology, chemistry, and medicine, than any other instrument or combination of instruments.

The design of lenses of all kinds was the most important practical application of optical theory, and to this end a knowledge of wave or physical optics was unnecessary; ray or geometrical optics, which neglects the wave aspects of light, is all that is needed to design the most complex optical systems. Thus, all kinds of telescopes, microscopes, and camera lenses were designed and constructed, all of which were based on Snell's law, which states how the path of a ray of light is bent as the ray goes from one medium to another. The difference between two media, as far as a ray of light is concerned, depends on the speed of the ray in the medium, and this speed is expressed by a number called the "index of refraction" of the medium; the larger this number, the smaller the speed of light in the medium and the more the ray is bent on entering the medium at some angle to the surface other than 90°. Thus, practical optics, based on pure geometry, enjoyed a very healthy growth during this period. Today, optical systems are designed with more attention to the wave properties of light than at that time, consideration of the wave properties being absolutely essential if high precision is desired.

ELECTRICITY AND MAGNETISM

Although electricity and magnetism (the lodestone) were known to the early Greeks, and William Gilbert, the court physician to Queen Elizabeth, had done extensive experiments with magnets, the great surge in the development and theories of electricity and magnetism occurred after the discovery of electric currents and, following that, the discovery of the magnetic properties of such currents. Until then, electricity (electrostatics) and magnetism (magnetostatics) were often treated as amusing curiosities to entertain people in a drawing room, rather than as natural phenomena worthy of scientific study.

During the late 17th and early 18th centuries, electrostatics was seriously studied by a few devoted amateur scientists, but no body of theories such as those that followed Newton's work was developed. Electrostatic machines (to concentrate large amounts of electrical charge on spheres) were invented, and devices such as the Leyden jar (essentially an electric condenser) to store electric charge were constructed, but the step from such devices to the concept of electric conductors and the idea of electric circuits was not taken. Benjamin Franklin was closest to taking such a step, in his investigation of lightning discharges, when he showed that such discharges produce electric sparks at the ends of wires near the ground if the other ends are high enough above the ground to be hit by the discharges. From these observations, Franklin, in 1752, proposed the idea of an imponderable electric fluid that permeates all space and is present within all matter. If the amount of this fluid within a body exceeds that outside it, a net outward flow occurs and the body becomes negatively charged; if the flow occurs in the other direction, the body becomes positively charged. Thus, charge of one sign or the other was associated with an excess or a deficiency of a universal electric fluid.

Although the early Greeks studied electrical and magnetic phenomena, the science of electrostatics and magnetostatics developed slowly because no quantitative law of the electrical force between electrical charges had been discovered. It was known, beginning with the experiments of Dufay in France in 1730, that electrically charged bodies attract one another (if oppositely charged) or repel one another (if charged the same), but this knowledge was qualitative; it became quantitative when Charles-Augustin de Coulomb, using a delicate torsion balance, measured the repulsion between two small spheres carrying equal amounts of positive charge and separated by a carefully measured distance. He discovered that the force, like gravity according to Newton's law, varies inversely as the square of the distance between the centers of the two spheres and directly as the product of the charges. This is known as Coulomb's "inverse-square law of electric force"; it also applies to magnetic charges (magnetic poles), which attract each other (if unlike) or repel each other (if like). Since this law applies to electrically charged bodies whether they carry equal or unequal amounts of charge, the Coulomb law of force is stated algebraically as $F = q_1 q_2 / r^2$, where q_1 and q_2 are the quantities of charge on the two spheres and r is the separation of their centers. From this formula, one defines the unit of electric charge (1 electrostatic unit, or esu) as the amount of charge placed on a sphere such that the force between it and a sphere carrying an equal amount of charge is exactly 1 dyne if the centers of the spheres are separated by 1 cm. This is true only if the electric charge on each sphere is distributed uniformly over its surface so that it behaves as though all the charge were concentrated at the sphere's center. This is automatically true if the sphere is an electrical conductor, since the charges are then free to move, and they repel each other to form a uniform surface layer of charge. Coulomb's law of force for electric charges is the basis of electrostatics and the basis of magnetostatics for magnetic poles; whereas positive and negative charges can exist independently of each other, individual magnetic poles cannot. This difference produces a remarkable asymmetry between electricity and magnetism even though the two phenomena are intimately related, as we shall see.

Even though many of the observed electric phenomena during this period implied the flow of electric charges (electric currents), no one then thought of an electric current in the sense of a continuous flow of charge that could be set up and controlled at will. Such an electric current was first discovered accidentally in the 1780s by Luigi Galvani, a professor of anatomy at Bologna, who noted that the muscles of a frog's legs in contact with two unlike metals (zinc and copper) contracted spasmodically. His colleague Alessandro Volta, a professor of physics, rejected Galvani's suggestion that the phenomenon was biological and proposed the idea that the current was produced by the two different metals when in contact with the nerve endings of the frog's leg. The nerve did not produce the current, as Galvani thought, but was merely the conductor of the current. Volta then showed that if the two dry ends of a copper and a zinc rod immersed in sulfuric acid are connected by a wire, a strong current flows. This device was the first voltaic cell and the beginning of electrical technology. Volta showed that the strength of the electric current so produced depends on the kinds of metallic rods and the kind of acid used. With that discovery, a rapid shift occurred in electrical research from electrostatics to the study of electric current. It was therefore inevitable that electromagnetism would be discovered, as it was not many years afterward, by Hans Christian Oersted, in 1820.

Oersted (1777–1851) was born in Langeland, Denmark, the son of an apothecary. Because there was little money to support the many members of the family, Christian and a younger brother, Anders, were sent while they were still very young children to live with some friends of the family, a wigmaker and his wife. The couple took an interest in the boys' education and taught them German with the family Bible, as well as some Latin and mathematics. Both boys learned quickly and, despite little formal education, showed a degree of inquisitiveness uncommon in children of such tender years. For Christian, his scientific education at this time in his life came not from primary school classes but from helping his father out in the pharmacy shop, which he did while living away from home.

Despite his limited educational background, Christian passed the entrance examinations at the University of Copenhagen and began studying there in 1794. Although he was greatly interested in philosophy, particularly the works of Immanuel Kant, he studied astronomy, physics, mathematics, chemistry, and pharmacy, receiving a degree in the last subject in 1797. After a short stint as an editor of a philosophy journal, Oersted began writing his doctoral dissertation, which examined the importance of Kant's philosophy in the sciences. In 1801, Oersted took a series of trips around Germany to study the work that was being done on the relationship between electricity and chemistry, work that had made possible the invention of the voltaic pile.

He returned to Copenhagen in 1804 and tried unsuccessfully to secure a position teaching physics. Pressed for money, Oersted began to give a series of public lectures on scientific and philosophical subjects. The lectures were so well received that the warden of the University of Copenhagen initiated proceedings that resulted in the creation of a special teaching position for Oersted. Now that he was finally situated in a university position, Oersted began to write a number of scientific papers that helped to establish his professional reputation.

From the accounts of Oersted's discovery given by those attending his lecture demonstration where it happened, it seems that Oersted was puzzled at what had occurred, as though he had not expected it. He had sent a strong current through a wire connected to the positive and negative ends of the electrodes of a voltaic cell and was surprised to see a magnetic compass needle, originally set parallel to the wire along the north–south direction, rotate 90° and remain fixed in the east–west direction at right angles to the wire. When he reversed the direction of the current in the wire, the needle immediately rotated 180°, and regardless of the direction of the wire, the needle always aligned itself perpendicular to the wire with the north pole of the needle on one side or the other of the wire depending on the direction of the flow of current in the wire. This is certainly one of the greatest scientific discoveries of all time, for it opened the doors to the vast scientific and technological domain of electromagnetism that revolutionized all phases of life and society. The discovery itself had an immediate influence on scientific research, for it indicated a force acting on the magnetic needle that is quite different in its behavior from the electrostatic, the magnetostatic, or the gravitational force. Indeed, that the needle rotated indicated that it was acted on by a torque emanating from the electric current and not just a force of attraction or repulsion, as in the case of two masses, two electric charges, or two magnetic poles acting on each other. That the rotation was a magnetic effect was clear from the fact that a compass needle is not affected by an electric charge, and so Oersted correctly concluded that an electric current generates magnetism. To explain this fully, we must introduce the concept of the electric and magnetic fields.

ELECTRIC AND MAGNETIC FIELDS

We saw that we can replace the concept of gravitational action at a distance by the interaction of a mass with a gravitational field produced in its neighborhood by another mass; thus, the force between two separated bodies becomes the interaction of each body with an entity (the field) in the space it occupies. The field itself is an all-pervasive physical object or structure that is defined at each point of space by its intensity (or magnitude) and direction. Thus, the gravitational field at a point produced by some distribution of masses equals the acceleration of a unit mass placed at that point. All masses in a gravitational field tend to move from the weaker to the stronger part of the field. We proceed in the same way with the electrostatic and magnetostatic forces between electric charges and magnetic poles. These fields, however, are somewhat more complicated, because electric charges and magnetic poles repel as well as attract each other, but the general field idea is the same.

The strength of the electric field produced at a distance r by a point of electric charge of magnitude q equals the magnitude of the charge divided by the square of the distance, that is, q/r^2. The direction of the field is the direction in which a positive charge placed at that point moves. In other words, the electric field at a given point is given by the force acting on a unit positive charge placed at that point. Thus the electric field at a point may be positive or negative depending on whether the charge that produces the field is positive or negative; a positive and a negative charge placed at the same point of an electric field move in opposite directions. These same ideas apply to magnetic fields produced by magnetic poles.

Determining the electric or magnetic field of a single electric charge or magnetic pole is easy enough, but the problem becomes forbiddingly complex as the number of charges or poles is increased. We can simplify the problem here, just as we did for gravity, by defining the field at each point by its potential at that point. The concept of the potential at a point in a field stems from the concept of potential energy. Just as a particle (which is best illustrated by the bob of an oscillating pendulum) can have kinetic and potential energies in a gravitational field, a moving electric charge can have kinetic and potential energies in an electric field. The electric potential at a point in an electric field is defined as the work needed to take a unit electric charge from infinity to that point. If this point is at a distance r from the charge of magnitude q producing the field, the potential at the point is q divided by r, that is, q/r. Thus, the potential is zero at infinity and is positive or negative at a point depending on whether the charge producing the field is positive or negative. The potential energy of a charge at a point in an electric field is the magnitude of the charge times the potential of the field at that point; it is positive or negative depending on whether the charge placed at the point of the field is positive or negative.

The advantage of working with the potentials rather than the field intensities is that the latter are vector quantities, whereas the former are scalars. If the electric field arises from many different electric charges, we obtain the potential at any point in the field produced by all of them by adding the individual potentials numerically. The electric field intensity at any point in the field can be obtained from the potential at that point by a simple mathematical operation. The motions of interacting electric charges may now be described by saying that they always move from points of higher potential to those of lower potential in the electrostatic field. This fact explains the flow of electric charge from one pole to the other in the voltaic cell.

Before we return to the discovery by Oersted, we describe a graphical way of representing fields by lines of force that is much more useful in discussing electromagnetism than in describing the gravitational field. The gravitational lines of force produced by a mass particle are just straight lines diverging uniformly in all directions from the particle, and this is also true of the electric lines of force emanating from a point charge. But differences between gravitational and electric lines of force arise when two mass particles interact gravitationally and two electric charges interact electrically. The gravitational lines of force between the particles are always the same, but the electrostatic lines of force between the two charges depend on whether the signs of the two charges are the same or opposite. If they are the same, the lines of force repel each other, but if the signs are opposite, the lines emanating from one charge converge onto the other charge just as the gravitational lines of force from one mass converge onto one or many different masses in the mass's neighborhood. Since the total electric charge of the universe is zero (as many positive as negative charges exist), no stray lines of force are present in the universe. Magnetic lines of force are similar, geometrically, to electric lines of force and may be described in the same way. The strength of a field of force in a given direction at any point (electric, gravitational, or magnetic) is defined as the number of lines of force in a given direction that pass through 1 cm^2 of area placed at right angles to the direction of the lines of force at that point. Lines of force in opposite directions cancel each other; those in the same direction reinforce each other.

With these simple ideas of lines of force, we can now describe Oersted's discovery somewhat more physically than simply in terms of the rotating compass needle. To begin with, the Greeks' discovery of the magnetic properties of a lodestone stemming from their observations of the behavior of the lodestone in the earth's magnetic field pointed to the concept of the lines of force of the magnetic field. To see this concept, consider a magnetic field (such as the earth's magnetic field in a small region of space or the magnetic field between the parallel flat surfaces of the north and south poles of two magnets). We may picture such a field as pointing in a definite direction and as described by parallel lines of force, all pointing in that direction, which is given by the direction in which the north pole of a magnet moves when placed in the field. If a bar magnet (equivalent to a lodestone) is now placed at some angle to the lines of force in such a field, it experiences a torque because the field pulls the north pole along the lines of force and pushes the south pole with an equal force in the opposite direction. Since these two opposite forces are equal in strength but are not in the same line, they exert a torque on the magnet that causes it to rotate and line up along the field. This simple explanation of how a magnet behaves in a magnetic field shows the usefulness of the lines of force concept.

Returning now to Oersted's discovery of the magnetism produced by the electric current, it is easy to understand the initial confusion it generated. Up to the time of this discovery, magnetic effects had always been associated with magnets or magnetic poles, but here was magnetism without magnets. The field concept made it easier to accept this apparent paradox by simply introducing the dualism that magnetic fields can be produced by magnets and by currents. Later, the distinction between these two phenomena was eliminated by the unifying principle (now an accepted fact) that magnetic fields have no independent existence of their own, but are produced by electric currents, and the magnetic fields of magnets are explained by molecular currents within the magnet itself. A direct consequence of this concept is that one should be able to construct a magnet by arranging a current-carrying wire in a certain way. This can be done by winding the wire into a coil, which then behaves exactly like a magnet with a north and a south magnetic pole. Such a current-carrying coil, a solenoid, is called an electromagnet, it is one of the most important devices ever produced in the history of technology.

Using the concept of magnetic lines of force, we can now deduce certain important features of the magnetic field produced by a current in a long, straight wire. The direction of the field at any point outside the wire is perpendicular (at right angles) to the wire. We therefore conclude that the lines of force of this field are concentric circles. The magnetic field strength at a point outside the wire depends on the strength of the electric current and the radius of the line of force (circle) passing through the point. The stronger the current, the stronger the field, and the larger the radius of the line of force (the distance of the point from the wire), the weaker the field. The direction in which a north magnetic pole moves when placed at a point in the magnetic field of the current is the direction in which the fingers of the right hand, grasping the wire, curl around the wire if the thumb points in the direction of the current; a south pole moves in the opposite direction. Using the thumb and fingers of the right hand, we can also determine the polarity of an electromagnet; the thumb points toward the north pole if the fingers are wrapped around the coil in the direction of the flow of current in the coil.

DYNAMICS OF ELECTRIC CURRENTS

Oersted's great discovery was followed by a period of very intense research into the interactions between electric currents, led by the brilliant work of André-Marie Ampère, after whom the unit of electric current (the ampere) is named. He noted that two wires carrying currents in the same direction attract each other, but repel each other if the currents are in opposite directions. We can explain this result using lines of force. The lines of force produced by each current are concentric circles pointing in the direction of the fingers of the right hand when the thumb points in the direction of the current in each case. This means that the lines of force produced by the two currents in the region between them cancel each other (exactly so if the current strengths are equal), but the lines of force on the other side of each wire are intact and behave like stretched rubber bands around both wires, thus pulling them together. This is known as the "pinch" effect, which is used in the large modern devices that try to produce nuclear energy by fusing light nuclei (protons, deuterons, and the hydrogen-3 nucleus) to form the helium nucleus (thermonuclear fusion).

The importance to electromagnetic technology of the discovery of these dynamic properties cannot be overemphasized, for they are the basis of the electric motor, which revolutionized all industry and freed power-intensive industrial processes from the need to be close to their sources of power. That Oersted's discovery leads immediately to the electric motor is clear from Newton's third law of action and reaction, for if a magnet placed near an electric current experiences a force, the wire carrying the current experiences an equal and opposite force. Thus, the wire moves, under the action of this force, in a direction opposite to that of the magnet's motion. If a coil of wire carrying a current is placed near the north (or south) pole of a fixed magnet, the coil experiences a torque and begins to rotate; the motor action of the current is thus immediately obvious.

The connection between the magnetic force (or field) produced by a current and the electric force (or field) produced by an electric charge was quite puzzling at first, since a stationary electric charge exhibits no magnetic properties; if an electric charge and a magnetic pole are placed next to each other, neither one detects the other's presence. The electric charge does not know that it is in a magnetic field produced by the pole, nor does the pole know that it is in the charge's electric field. But the physicists following Oersted saw the connecting link—the relative motion of charge and pole. An electric current is a flow of electric charges, and the flow is the crucial connection between the presence and the response of the magnetic pole and the electric charge to each other. As noted above, if charge and pole are at rest with respect to each other, no recognition exists between them, but as soon as the charge moves, it creates a magnetic field to which the magnet responds. This is the essential explanation of Oersted's discovery, which was not fully accepted until Henry Rowland showed that individual electric charges produce magnetic fields exactly as electric currents do.

This explanation of Oersted's discovery, as the product of electrical charge and its motion, points to a remarkable symmetry between electricity and magnetism that was not fully discovered and understood until some ten years later when Faraday produced electromagnetic induction. That such symmetry exists is indicated by Newton's third law of action and reaction. Since a moving charge exerts a force on a magnet (by producing a magnetic field in the magnet's neighborhood), the magnet must exert an equal and opposite force on the moving charge. But an electric charge responds only to an electric field, which means that the moving magnet produces an electric field. In all of this, only the relative motion of the magnet and the charge counts. If the two moved together, they would be ignorant of each other. But relative motion means that the same effects are produced at the magnet and the charge whether we hold the magnet fixed and move the charge or hold the charge fixed and move the magnet. In either case, both experience a force; this force is discussed in greater detail in the next chapter, in which we describe Faraday's and Maxwell's electromagnetic researches.

The electrostatic potential, which was introduced to simplify the theory of the electrostatic field, also simplifies the study of electric currents. The electrostatic potential is so defined that electric charges move from points of higher potential in an electric field to points of lower potential, which means that electric charges flow from states of higher energy to states of lower energy. Since an electric current is a flow of electric charges, we therefore represent this flow in terms of a difference of potential between any two points in the circuit; one then speaks of a "potential drop" between the two points in the circuit.

The application of the potential concept to electric circuits follows naturally from Volta's production of an electric current with his chemical cell, the essential features of which are two different metal electrodes (for example, zinc and copper) immersed in an acid medium (for example, sulfuric acid). Each electrode acquires an electric charge because the acid leaches ions (positively charged atoms) from the electrode's surface, leaving it with an excess of negative charge. The zinc ions go into solution much more readily than the copper ions, so that when equilibrium exists between the electrodes and the acid, the zinc electrode has a higher negative charge than the copper electrode. Thus, the copper electrode has a positive potential (smaller negative charge) relative to the zinc, and so an electric current is said to flow from the copper (called the positive pole of the cell) to the zinc when they are connected by a conductor (metal wire). Actually, the electric charges flow from the zinc to the copper, but for historical reasons, the current is described as flowing from the copper to the zinc (from the higher potential to the lower potential). Since such a flow of charge represents an energy drop, the current can be maintained only if energy is supplied continuously to the circuit, and this work is done by the acid in the form of chemical energy. Thus the electric cell is a device that transforms chemical energy (the work done by the acid on each electrode in forcing zinc and copper ions to go into solution and thus to establish a potential difference between the two electrodes) into electromagnetic energy. Energy is thus associated with an electric current that depends on the strength of the current; the stronger it is, the higher is its energy content.

This energy manifests itself in various ways; part of it is in the magnetic field that accompanies the current, which is demonstrated by the motion (kinetic energy) of a magnet placed near the current, and part in the kinetic energy of the electric charges flowing in the current. This kinetic energy produces heat in the electric circuit because the light, mobile charges that account for the current constantly collide with the heavy atoms that are more or less fixed in the circuit and thus resist the motions of the charges. This resistance accounts for the heat produced by the current. With the introduction of the concept of the potential difference between two points in an electric current, the development of electric circuit theory and technology progressed very rapidly. Here, we present only the basic ideas that are associated with Georg Simon Ohm (Ohm's law) and Gustav Robert Kirchhoff (Kirchhoff's law of circuits).

Ohm's law, which is essential for understanding the flow of electric currents through various substances, relates the magnitude of the current to the potential difference between the point where the current begins and the point where it ends and to the resistance of the substance through which it flows. The simple formula that states Ohm's law is the basis of the electrician's practice and guides him in all his work; without it, the vast electrification of the world, which has changed life on earth so drastically, would not have been possible. We saw that when two different metals (electrodes) are immersed in an acid, they acquire different amounts of negative charge; the potential difference thus established drives a current through any metal wire connecting the two electrodes. The magnitude of this current depends on this potential difference and the resistance of the connecting wire to the flow of electric charge; the larger the potential difference, the stronger the current, and the larger the resistance, the weaker the current. This relationship forms the basis for Ohm's law, which is generally stated as follows: The potential difference required to drive a given current through a given wire equals the product of the current strength and the resistance of the wire. This law may be expressed algebraically as $V = IR$, where V is the potential difference, or voltage, I is the current strength, and R is the magnitude of the resistance. In this formula, the quantities V, I, and R are expressed in certain practical units that we do not relate directly to our basic units of length, time, and mass (or force) on which Newtonian mechanics is structured. The reason for this incongruity is that current and potential difference involve electric charge (current is electric charge per second and potential is energy per electric charge), and we have not expressed charge in terms of space, time, and mass. However, casting charge in such terms is possible, as shown in the following example: Force, which is mass times acceleration, equals electric charge squared over distance squared (a force), so that electric charge is a distance times the square root of a force, or a distance times the square root of mass over time multiplied by the square root of a length.

Owing to this complicated dimensional formula for charge expressed in our basic space–time–mass units, the dimensional formulas for potential and current expressed in basic units are too complex for practical purposes, so that a different set of basic electrical units is used: the coulomb for charge, the volt for potential difference, the ampere for current, and the ohm for resistance. With these units, Ohm's law is as follows: volts = amperes × ohms. This formula means that if the potential difference between two points on a current-carrying wire is 1 volt and the resistance of that stretch of wire is 1 ohm, the current in the wire is 1 ampere. If one knows the voltage between the two ends of a wire and the resistance of the wire, one can calculate the current using the formula $I =$

V/R. A current of 1 ampere means a flow of 1 coulomb of charge per second through any cross section of the circuit. Since $V = IR$, the electrician speaks of the IR drop from point to point along a current, rather than of the voltage drop.

The electrical resistance of a wire depends on its chemical nature (for example, gold, silver, copper), its temperature (in general, the lower the temperature, the lower the resistance), and its length and diameter (the longer and thinner a wire, the greater its resistance). The unit of resistance, the ohm, is defined as the resistance of a uniform column of mercury 106.300 centimeters long and having a cross section of 1/100th of a square centimeter at 0°C.

Since a current is a flow of electric charges, it carries energy because the charges have kinetic energy, and if the wire were resistanceless, the current, once started, would go on forever, since the charges would not lose their kinetic energy. The constant friction produced by the resistance robs the charges of their kinetic energy, however, so that a constant voltage difference must be maintained along the wire to keep the current flowing. The kinetic energy lost by the moving charges, owing to the resistance, appears as heat in the circuit, and the rate at which such heat is produced is given by I^2R, that is, the square of the current multiplied by the resistance.

The energy carried by an electric current can produce chemical reactions, drive motors, amplify electronic processes, and do many other useful things. The rate at which an electric current produces energy, measured in watts, equals IV, that is, the product of the current and the voltage; a current of 1 ampere flowing between a potential difference of 1 volt is a 1-watt current, which means it produces 1 joule (10 million ergs), or slightly less than $\frac{1}{4}$ of a calorie per second. To define the latter units, 1 erg is the kinetic energy of a 2-gram mass moving at 1 centimeter per second, and 1 calorie is the heat needed to raise the temperature of 1 gram of water 1°C.

Kirchhoff's laws of circuits are the basis for all circuit theory, which is indispensable for the analysis of complex circuits such as are used in computers and communications equipment of all kinds. Kirchhoff's first law states that in any branch point in a circuit, the sum of the currents approaching the point equals the sum leaving that point. This means that electric charge cannot accumulate at any point in a circuit. Kirchhoff's second law states that the IR drop around any closed section or loop of the circuit must be zero. This statement says that the potential drop between any two points in a circuit is the same no matter whether the two points are connected by a single conductor or by a number of different conductors.

Though the discovery of the electric current and its associated magnetic field was the beginning of electromagnetic technology, it revealed only half of nature's remarkable electromagnetic dualism. The second half of this dualism is described in the next chapter.

CHAPTER 10

The Faraday–Maxwell Era

It is the customary fate of new truths to begin as heresies and to end as superstitions.

—T. H. HUXLEY

Michael Faraday (1791–1867) was one of the most remarkable scientists of all time, a man who, with no formal education, taught himself enough science to become the outstanding experimental physicist of his time. He had little mathematical training and depended more on his intuition and inspiration than on theoretical knowledge. Like a man groping in the dark, Faraday painstakingly devised and performed a number of experiments that enabled James Clerk Maxwell to deduce his mathematical equations to unite the electric and magnetic fields into a single entity.

Faraday was born in a grimy section of London, the son of a blacksmith. At that time, it was uncommon for children in the working class to be literate, let alone to have a formal education, so not until Michael was apprenticed to a bookbinder at the age of 13 did he have an opportunity to read many books. Despite the family poverty, however, he had always shown a keen interest in learning. However, Michael never enjoyed mathematics, perhaps finding its detailed proofs and rigorous logic to be unsuited to his tastes.

Although Michael was a diligent bookshop worker, he spent most of his spare time reading the books that came into the shop to be bound. He was particularly influenced by an accessible book on chemistry that provided him with a rudimentary knowledge of that subject and acquainted him with some of the methods and terminology of science. Because he had never cared for the business of bookbinding even though he had two boys working for him before he reached the age of 20, Faraday began to search for some other career that would satisfy him intellectually. The answer became clear to Faraday after he attended a series of science lectures at the Royal Institute that convinced him that his true calling was

Michael Faraday (1791–1867)

science. After Faraday completed his bookbinding apprenticeship in 1812, he went to work at another bookshop for several months, but by that time he had convinced himself that he would never be happy as a bookbinder. His employer was not very sympathetic toward Faraday's growing interest in science and tried to discourage him from leaving his job. But the hostility of his employer only strengthened Faraday's determination to follow what he believed was his true calling.

He began his scientific career as an assistant to the British chemist Sir Humphrey Davy, whom Faraday had heard speak at several institute lectures. Since Faraday had no academic credentials to recommend himself, he had sent Davy the volume of notes he had taken while at the lectures. Faraday's repeated requests were finally answered when Davy hired him as a laboratory assistant at the Royal Institute in 1813. Davy had recognized the flashes of brilliance in the notebook sent by Faraday and, refusing the advice of an acquaintance to begin Faraday as a bottle washer, extended the offer of employment.

Faraday's work as an assistant was characterized by his outstanding care and attention to detail. He also listened carefully to the many lecturers at the institute and began to think about how he would set up the experiments he wanted to do. Faraday was reappointed to the institute three years later in 1816. He also published that year his first scientific paper, which attracted the attention of the British scientific community and led to his election to the Royal Society in 1823. This period in Faraday's life marked the beginning of his work in electrical and chemical experiments that culminated in his discovery of electromagnetic induction.

Although Faraday was devoted to his work and was doubtless aware of his own considerable abilities, he was always tolerant of other opinions. He was especially fond of children and gave a series of basic science lectures that he had written especially for youngsters. He was an extremely religious man who belonged to a sect that believed the words of Christ were a sufficient guide to live a Christian life. Consequently, Faraday had little use for organized religion and rarely ventured inside a church. His religious beliefs did help him to persevere in his work even though he was often tormented by headaches and loss of memory: "From 1831 to 1840 these symptoms became quite severe and he himself, believing that no doctor understood his ailment, was convinced he was suffering from a decay of his physicomental faculties and that there was no way to stop this deterioration. Nevertheless, in spite of this progressive disease, which at times left him momentarily paralyzed, his creative faculties remained as great as ever, and his productivity continued until he died on the afternoon of August 25, 1867, just one month short of his seventy-sixth birthday."[1]

Faraday's discovery of electromagnetic induction was originally prompted by Hans Christian Oersted's experiments, which showed that an electric current deflects a magnetic needle. However, Faraday was one of the few to proceed in the opposite direction and wonder what effect, if any, a magnetic force would have on an electric current.[1] His experiments on magnetism and electricity led him to formulate the concept of the field as a way of explaining the interaction between the magnet and the electric current. The field concept, which has continued to dominate much of physics to the present, was prompted by Faraday's desire to avoid the Newtonian notion of action at a distance and his "conviction that all forms of physical action are basically one, that is, the concept of the unified field."[3]

Faraday discovered the phenomenon of electromagnetic induction in 1854 and wrote that "I have long held an opinion, almost amounting to conviction . . . that the various forms under which the forms of matter are made manifest have one common origin: in other words, are so directly related and naturally dependent that they are convertible as it were into one another and possess equivalents of power in their action."[2] After his discovery of eletromagnetic induction, Faraday began a series of electrochemical experiments that led him to the laws of electrolysis and provided the first clear indication as to the nature of the forces that bind the atoms inside molecules together. From these experiments, Faraday concluded that "electrical forces are at work in a molecule, for he reasoned that if one can obtain an electric current from a voltaic cell, which operates by the chemical reaction of the electrodes with a chemical solution, then the atoms in solution must have electrical charges in them."[2] Faraday found that the electric currents of the same strength he sent flowing through solutions always decomposed the same quantity of material or, alternatively, the same number of ions of a given chemical compound. He later concluded that "all charges on ions are integral multiples of a single fundamental unit of charge and that one never finds fractions of this charge."[2] His work revealed that a large amount of electricity can be obtained from the combinations of very small quantities of atoms, showing that the electrical forces between ions are very large.[3]

Faraday also discovered benzene in 1824 and the chemical and physical properties of organic compounds such as butylene and ethylene, but he was drawn to the study of electricity, for it offered the opportunity of research in physics as well as in chemistry, to which he remained devoted all his life. That he was a very skillful chemist is indicated by his electrochemical studies, which laid the foundation of electrochemistry and demonstrated the atomicity of electric charge. Before his great discovery of electromagnetic induction, which revealed the complete electromagnetic symmetry first indicated by Oersted's work, Faraday performed numerous electrostatic experiments. Since he had only a very meager mathematical background and was basically suspicious of mathematical models as correct descriptions of physical phenomena, he developed his own physical models to explain his experimental results. He was thus naturally led to the concept of the electric and magnetic fields, which he pictured as real, physical entities. To describe the interactions of electric charges (and magnetic charges), he introduced the concept of "tubes of force" that extend from one charge to another and thus transmit the interaction between the charges. By introducing tubes of force, Faraday did not have to contend with the repugnant notion of action at a distance, which was as unacceptable to him as it had been to Newton, who had dismissed it as "so great an absurdity that I believe that no man who has in philosophical matters a competent faculty of thinking can ever fall into it.''[1]

Before discussing Faraday's greatest scientific achievement, his discovery of electromagnetic induction, we mention briefly some of his other accomplishments, particularly his electrochemical experiments. He began his electromagnetic work with experiments in electrostatics and magnetostatics and demonstrated experimentally a number of basic facts that he deduced from his concept of lines or tubes of force. By assuming that the number of lines of force emanating from an electric charge is proportional to the magnitude of the charge, he showed that the field strength at any point in the field (electric or magnetic) is proportional to the magnitude of the charge (pole) divided by the square of the distance of the point from the charge (or pole). This relation gives Coulomb's inverse-square law of force for electric charge.

Faraday proved another important feature of electrostatics, namely, that electric charges are produced in equal and opposite pairs (every positive charge is accompanied by an equal negative charge). He did this by introducing a positive electric charge into an ice pail and showing that an equal positive charge is produced on the outer surface of the pail, which remains unaltered as the positive charge is moved about inside without touching the pail. Faraday explained this result by picturing the lines of force from the positive inner charge attaching themselves to negative electric charges on the inner surface of the pail, thus causing the positive charges in the metal of the pail itself to move to the outer surface owing to their mutual repulsion. Thus, for every negative charge produced by the lines of force on the pail's inner surface, an equal positive charge was produced on the outer surface. This result led Faraday to his concept of the atom as a structure held together by electrical forces among its constituent particles.

He stated this important notion in one of his papers on his experimental researches in electricity as follows: ''Although we know nothing of what an atom is, yet we cannot resist forming some idea of a small particle which represents it to the mind; and though we are in equal, if not greater, ignorance of electricity, so as to be unable to say whether it is a particular matter or matters, or mere motion of ordinary matter, or some third kind of power or agent, yet there is an immensity of facts which justify us in believing that the atoms in matter are in some way endowed or associated with electrical powers, to which they owe their most striking quality and among them their mutual chemical affinity.''[3] Faraday's electrochemical experiments proved these surmises and were the first indications that electrical charges come in multiples (positive and negative) of a basic unit of charge, which, many years later, was demonstrated to be the charge on the electron, which is now called the ''element'' (the elementary unit) of electric charge, although today's elementary particles (the so-called ''quarks'') are assumed to exist with fractional values of this charge. Quarks are discussed in Chapter 19.

In his electrochemical investigations, Faraday discovered the basic law that the amount of a substance liberated at an electrode in a given time when a current is passed through an electrolyte (a salt solution) depends on the total amount of electric charge carried through the solution by the electric current. To Faraday, this meant that salts break up into their constituent electrically charged atoms (ions), each of which carries one, two, three, or more units of the elementary charge; all ions of the same kind carry the same (positive or negative) charge.

This basic law is best illustrated by a simple example. In a solution of sodium chloride (salt), the sodium chloride molecule breaks up into a positive sodium ion and a negative chlorine ion, and each such ion carries one element of charge, so that the sodium ions are attracted to the negative electrode (the cathode) immersed in the solution and the chlorine ions are attracted to the positive electrode (the anode). An electric current of a given strength (amperage) flowing through the salt solution for a given time deposits the same number of sodium ions on the cathode as it does chlorine ions on the anode, but the quantities of sodium and chlorine deposited are not the same, since the mass of a sodium atom is not the same as that of a chlorine atom; their masses are in the ratio of 23 to 35 (their atomic weights), so that for every 23 grams of sodium deposited in a given time, 35 grams of chlorine are also deposited. Faraday found that the mass of the ion of any atom deposited by a current on an electrode is proportional to its atomic weight if each ion carries one unit of charge. The ions of copper, for example, carry two units of charge each, so that the same current that deposits 23 grams of sodium deposits 32 grams of copper (half the numerical value of copper's atomic weight) in a given time. Since each ion of copper carries two units of charge, it transports twice as much electric charge in the current as an ion of sodium, so that half as many copper as sodium ions are deposited from solution in a given time. The number of unit electric charges carried by the ion of an atom in solution, called the ''valence'' of the atom, plays a very important role in chemistry.

Faraday's electrochemical studies convinced him that matter consists of different kinds of atoms, each of which is an electrically balanced structure with equal numbers of positive and negative units of electric charge, and that such charges are multiples of a basic unit of charge. He could not measure the elementary unit of charge, but he did measure the ratio of the unit of charge to the mass of a single atom and discovered that this number is larger for the hydrogen ion than for any other ion, so that the mass of the hydrogen atom is smaller than that of any other atom.

THE DISCOVERY OF ELECTROMAGNETIC INDUCTION

Although Faraday's electrostatic and electrochemical research mark him as a giant among experimental physicists, he is most famous for his discovery of electromagnetic induction. Believing passionately in the unity of the laws of nature, Faraday was sure that Oersted's discovery that an electric current produces a magnetic field is only half the story of the relationship between electricity and magnetism. He was certain that if electricity (the electric current) produces magnetism, then magnets should somehow produce electric fields. But in pursuing this idea, he stumbled initially by trying to produce electric fields with stationary magnets. In doing this, he had overlooked the important contribution of motion in the relationship between electricity and magnetism. An electric current is a flow of electric charges (electric charges in motion), so that motion is the crucial element in the production of a magnetic field by electric charges.

Whether Faraday made his discovery accidentally or by analyzing the Oersted discovery consciously is not important; his law of electromagnetic induction is one of the great breakthroughs in pure science as well as in electromagnetic and, in fact, in all other technology. Electromagnetic induction completed the circle of discoveries begun by Oersted, which were the base on which Maxwell built his magnificent electromagnetic theory, so that Faraday's discovery was a watershed in the history of science.

Accidentally or not, Faraday discovered that if a magnet is at rest near a loop of wire, nothing happens, but if the magnet is moved, however slightly, a current is generated in the loop. If the loop and magnet are moved together in exactly the same way, nothing happens, but if the magnet is moved while the loop is held fixed or the loop is moved while the magnet is held fixed, current flows through the loop; only the relative motion of the loop and magnet counts. The direction of the current depends on the direction of the relative motion; if the direction of the relative motion is reversed, so is the direction of the current.

Faraday explained this phenomenon in terms of his lines or tubes of force by picturing such lines or tubes as threading or filling the area of the circle formed by the loop and stating that when a wire cuts tubes or lines of force, a current is generated. He formulated this idea as a law that states exactly how large a current is produced by cutting lines of force of a magnetic field with a wire. He expressed this law in terms of the potential difference between the two ends of such a wire. This can best be understood by picturing a wire (conductor) in the form of a rectangle, one side of which can be moved at right angles to its length along the other two sides with which it is in contact. If this wire rectangle is placed between the north and south poles of two vertical magnets so that the lines of force of their magnetic field are perpendicular to the area of the rectangle, a current flows in the rectangular circuit as the movable wire is drawn horizontally through the magnetic field, cutting the lines of force. Faraday's law of electromagnetic induction states that the magnitude of the potential difference between the two ends of the movable wire is proportional to the rate at which the wire cuts the magnetic lines of force. This law is often stated somewhat differently by picturing the magnetic field as a flux through the rectangular area that changes as the wire is moved because the area of the rectangle is reduced. Faraday's law then states that the induced potential across the moving wire is proportional to the rate of change of the magnetic flux. This remarkably simple law is the basis of the electric generator, which is essentially a huge coil of wire rotated in a strong magnetic field so that each strand in the coil cuts the magnetic lines of force. The more rapidly the coil is rotated and the stronger the magnetic field, the larger the induced current in the coil.

Considering the electric charges in the wire that produce the current, we see that their velocity along the wire is perpendicular to the magnetic flux and to the velocity of the wire as it cuts the lines of force. This means that the magnetic field exerts a force on a moving charge that is proportional to the velocity of the charge multiplied by the product of the magnitude of the charge and the magnitude of the magnetic field. This formula is known as the "Lorentz force" after the great Dutch physicist Hendrik Antoon Lorentz, who discovered it. The direction of the force acting on the moving charge is perpendicular to the velocity of the charge and the direction of the magnetic lines of force.

We see that Oersted's and Faraday's discoveries are two aspects of the same basic phenomenon: An electric charge and a magnetic pole are not aware of each other's presence when at rest with respect to each other, but if they are moving with respect to each other, they are aware of this motion because each experiences a force (electric for the charge and magnetic for the pole) generated by the motion.

MAXWELL'S ELECTROMAGNETIC THEORY

Faraday discovered many other electromagnetic phenomena, but did not have the mathematical skills to develop a single all-encompassing theory of the electromagnetic field; this task was left for the British physicist James Clerk Maxwell. It is not always easy to rank scientists, particularly physicists, since their work generally covers many areas, no one of which is revolutionary or a notable synthesis of previous discoveries, but a few are so outstanding that their preeminence is immediately recognized. Maxwell was one such physicist who ranks with Newton and Einstein as a unifier of scientific principles. An expert mathematician, he saw that he could unify the electric and the magnetic fields into a single electromagnetic field by means of a set of equations that relate the variations of either of the fields (electric or magnetic) from point to point in space with the variation of the other field (magnetic or electric) with time. In other words, his equations show that if one of the fields at a point (for example, the magnetic) changes from moment to moment, the electric field changes from point to point in space in a definite way, and vice versa.

James Clerk Maxwell (1831–1879) was born at the family estate in Dumfrieshire, the son of a lawyer who divided his time between his practice in Edinburgh and his hobby of tinkering with inventions and mechanical devices. James seemed to have inherited his father's curiosity and usually wanted to know what made a particular device or invention work. His idyllic life-style came to an end when he was eight, however, with the death of his mother. Although his father tried to keep the family together, James needed to begin his formal education, so his father sent the boy to Edinburgh to live with an aunt. For the next decade, James spent his winters in Edinburgh studying at the Edinburgh Academy and his summers at Dumfrieshire. Although his father had to devote most of his time to running the estate and maintaining his legal practice, he always tried to be available for his son.

Although his academic record was undistinguished at first, James soon found that he had a knack for geometry. He became so skilled in solid figure geometry in particular and mathematics in general that he was awarded the academy's annual mathematics medal at the age of 13. The

James Clerk Maxwell (1831–1879)

following year, Maxwell's father began taking him to meetings of the Edinburgh Royal Society. James seems to have benefited from these meetings, as they encouraged his interest in ovals and prompted him to write a paper on the subject that was read to the society in March 1846.[4] He also conducted a number of experiments with prisms and made several detailed colored drawings to record his findings.[4]

In 1847, Maxwell graduated from the academy, first in English and mathematics. The following autumn, he enrolled at Edinburgh University, where he spent three years developing his skills in mathematics and physics. During his summers at the family estate, he carried out a number of experiments in a homemade laboratory. He also found time to compose two more papers on rolling curves and the equilibrium of elastic solids that were read to the society in 1849 and 1850, respectively.[4]

Maxwell entered Peterhouse College, Cambridge, in 1850, but soon transferred to Trinity College, where he believed he had a better chance of winning a fellowship. His brilliance was recognized early in his college career, and the distinguished Cambridge tutor Hopkins believed Maxwell to be the most extraordinary man he had ever met.[4] Maxwell was elected a scholar at Trinity College at the end of his second year.

In June 1853, Maxwell suffered a nervous breakdown while preparing for the grueling tripos (a special examination for awarding highest mathematical honors at Cambridge) examinations. He was still suffering from the effects of his illness when he sat for the examination in January 1854. He finished second behind Edward Routh, who later became a distinguished mathematician, though the two tied for first place in the more advanced competition for the Smith prize.[5] Finding the academic environment to be particularly well suited for his own somewhat idiosyncratic personality, Maxwell continued his studies at Cambridge after receiving his undergraduate degree. He was elected a fellow of Trinity College at the age of 24, and as part of his duties began to give lectures and conduct experiments on electricity and magnetism. However, he soon left Cambridge to assume the chair of natural philosophy at Marischal College in Aberdeen.

In 1857, Maxwell submitted a paper on the structure of Saturn's rings that won the Adams prize by demonstrating that the only stable structure would be one consisting of minute particles.[6] This paper not only established Maxwell's reputation but also encouraged his interest in the motions of large groups of particles, which is the basic concern of the kinetic theory of gases: "This interest soon led to his brilliant deduction of the distribution of molecular speeds in a gas at equilibrium at any temperature. This great step forward in the understanding of the behavior of the elementary particles of gases represents one of the major advances in the progress of the atomic theory of matter."[6] Maxwell published his results in 1860, the same year that Marischal College was absorbed into the University of Aberdeen. The consolidation eliminated Maxwell's position, but he immediately joined the faculty at King's College in London, where he spent the next five years formulating his theory of the electromagnetic field. During this time, he performed many of his experiments at home, and his wife, whom he had married in 1858, served as an able assistant. Maxwell also gave a number of lectures on scientific topics and worked on his books on electricity and magnetism, and heat.

Maxwell resigned his position in 1865 and joined the staff at Cambridge. He worked as an examiner in the tripos examinations in mathematics; his questions on thermodynamics and electricity and magnetism prompted the formation of a university committee to recommend revisions in the curriculum. The committee concluded that courses should be given in these subjects and that a physics laboratory for conducting experiments should be established. These recommendations might never have been acted on owing to lack of funding had it not been for the generous financial contributions by the Chancellor of the University, the Duke of Devonshire, who was independently wealthy and possessed outstanding academic qualifications in his own right, having been awarded second wrangler (one who wins highest honors at Cambridge) in the mathematical tripos examination and first place in the Smith prize competition.[7] In 1870, the duke provided the necessary funding to build and stock a laboratory that was later named after Henry Cavendish. It became and remains the site of much of the significant work in atomic physics in Great Britain.

Because a professor knowledgeable in the subjects of electricity, magnetism, and thermodynamics was needed to guide the construction of the laboratory, the position was offered to Maxwell after being declined by Lord Kelvin. Maxwell assumed his official duties in the autumn of 1871. Unfortunately, he also accepted the arduous task of going through the unpublished manuscripts of Henry Cavendish, a distant relative of the duke, and editing them for publication, a task that took up most of the next five years and, aside from his official administrative duties, consumed most of his working hours.[8] His wife was also taken ill and remained bedridden for many months; Maxwell devoted most of his spare hours to her care. However, the constant strain took its toll, and in 1877, Maxwell began to suffer from stomach pains. He suffered in silence for two years until he finally consulted a doctor in early 1879. He became progressively weaker during the summer and finally succumbed to the disease on November 5, 1879, the victim of stomach cancer at the relatively young age of 48.

To see how Maxwell achieved his electromagnetic synthesis, we consider the electric field associated with an electrical condenser, which consists of two parallel metal plates, one of which is grounded, separated by a small distance. If a negative electric charge is place on one of the plates, the other plate immediately acquires an equal positive charge, and an electric field, with its lines of force perpendicular to the plates, is set up between the plates. Maxwell carefully analyzed this simple device with particular attention to the events in the space between the electric plates (occupied by an electric field) when the two plates are connected by a conductor. The negative charges from one of the plates flow through the conductor, thus establishing a current. While this flow occurs, the electric field between the plates changes. These changes greatly interested Maxwell, for, if not properly interpreted, they seem to introduce an impermissible discontinuity in the behavior of the condenser's electromagnetic field.

To understand these changes, we note that when the electric current flows through the conductor connecting the two plates, it is surrounded by circular magnetic lines of force that grow in intensity from zero to a maximum when all the negative charges from the plate have passed through the conductor onto the other plate and have canceled the charges on the other plate. At this point, no current is flowing through the conductor, the condenser plates are chargeless, and no electric field exists between the plates, but the magnetic field surrounding the conductor, produced by the current, is still present. To Maxwell, this situation lacked the symmetry of Oersted's and Faraday's discoveries. Moreover, the difference between the conditions in the space surrounding the conductor (a magnetic field) and those in the condenser space (no field) represented an intolerable discontinuity. To eliminate this objectionable feature and to achieve the unification he desired, he suggested that the decreasing electric field between the condenser plates accompanying the current that was building up in the conductor was itself a current, not in a conductor but in the vacuum. He called this current the "displacement current" and argued, correctly, that its flow generates a magnetic field with its lines of force circling the lines of force of the electric field.

The situation in the condenser and in the conductor, at the moment that the plates were chargeless and the current was zero, was, then, a very strange one that had never been conceived of before: a magnetic field in empty space with its lines of force circling a wire in which no current was flowing and circling the empty space between the condenser plates with no charges on them. What happened after that? The magnetic lines of force, with no current to support them, began to collapse, contracting like rubber bands, back to the conductor again and within the condenser. But Faraday's discovery was that such collapsing magnetic fields produce electric fields that cause electric charges in conductors to move. Thus, a current is generated in the conductor again, but in the opposite direction from that of the initial current, so that the condenser plates become charged again, but with opposite polarity. The whole phenomenon repeats itself, and this goes on periodically over and over again, so that we have an electromagnetic oscillator somewhat like a mechanical oscillator (a pendulum or a spring).

To see how close this analogy of the mechanical and the electromagnetic oscillator is, we consider it from a slightly different point of view, taking the electromagnetic energy of the oscillating condenser into account. When the condenser is not discharging (no current in the conductor), all the energy is in the electric field between the plates, and this is the potential energy of the charges on the plates. The energy is thus all potential, as it is in the pendulum when the bob is lifted above its equilibrium position (the lowest point of the swing). Just as the potential energy of the pendulum is transformed into kinetic energy, so, too, is the potential energy of the condenser transformed into the kinetic energy of the current. The kinetic energy is stored in the magnetic field created by the current, which again becomes the potential energy of the condenser plates with reversed polarity. We extend the analogy a bit further by noting that the bob of the pendulum does not stop when it swings to its lowest point, but, owing to its inertia, passes right through the lowest point to reach the other side of the swing. This is also true of the current in the condenser's circuit. The inertia of the moving charges that create the current causes these charges to continue moving until the condenser is charged again with the charges on the two plates reversed, so that one cycle of the oscillation is completed and another is ready to begin.

An important question concerning these oscillations is how quickly they occur (that is, the number of oscillations per second—their frequency—or the duration of a single oscillation—their period, which is the reciprocal of the frequency). Again, we use the analogy with the pendulum, the period of which is determined by the length of the pendulum and the acceleration of gravity—the longer the pendulum, the more slowly it oscillates, and the larger its acceleration (gravity), the more rapidly it oscillates. In a general way, this rule also holds for the condenser; the smaller it is, the more rapidly it oscillates, and the larger the magnetic field its current has to build up, the more slowly it oscillates. The self-magnetic field the current builds up acts like a retardant and slows the oscillations. An exact analysis of the discharge of the condenser leads to a simple formula for this oscillation period that has great practical applications.

Before we return to Maxwell's electromagnetic theory, we note a very important feature of the electric oscillations of the condenser that reveals the essence of Maxwell's theory—namely, his discovery of the wave character of electromagnetism. The electric oscillations of the condenser do not continue forever, but cease after a certain time, which means that the condenser has lost all its energy, just as the pendulum does. A very important difference marks these two energy losers, however; the pendulum loses all its energy through friction (heat), but this is not so for the condenser. Even if the friction in the discharging circuit of the condenser is reduced to zero, the condenser loses all its electrical energy very quickly—it radiates the energy away. To see the physics of this phenomenon, we again consider the displacement current produced by the changing electric field between the condenser plates and note that this current is encircled by a changing magnetic field, which in turn is linked to a changing electric field (owing to induction), and so on, so that a chain of alternating electric and magnetic fields is built up, and at each new cycle the old chain is pushed out into space by a new chain. Thus, an oscillating chain of electric and magnetic fields is pushed out into space in surges.

MAXWELL'S ELECTROMAGNETIC THEORY OF LIGHT

This crude physical description of the radiation from an oscillating condenser conveys the context of Maxwell's electromagnetic theory only superficially, but it is sufficient to show that the theory predicts the existence of electromagnetic waves, which can be derived rigorously from Maxwell's six equations of the electromagnetic field. Three of these equations describe how the variations in time of the three components (one component for each dimension of space) of the electric field determine the variations of the magnetic field from point to point, and the other three describe how the variations in time of the three spatial components of the magnetic field determine how the electric field varies from point to point. From these field equations, which intermix the components of the magnetic and electric fields, Maxwell derived, mathematically, a simple equation for the components of the electric field and a similar equation for the components of the magnetic field. These two equations are called "wave equations" because they show that the electric field and the magnetic field are propagated together as periodic oscillations at right angles to each other and both at right angles to the direction of propagation of the waves. Moreover, the electric and magnetic field oscillations are not in step with each other, but are 90° (one quarter of a cycle) out of phase; that is, when the electric field is a maximum, the magnetic field is zero and vice versa. This type of phenomenon is called a "transverse wave" in analogy with the propagation of a periodic disturbance across the surface of water when a pebble is dropped in the water. The surface oscillations of the water are perpendicular to the direction of propagation of the water wave.

The importance of Maxwell's electromagnetic theory was not fully understood until the velocity of his electromagnetic wave was found to be that of light. A wave equation relates the spatial variations of the amplitude of a field (its intensity) to the time variations. Now, the spatial variations are essentially the wave amplitude divided by the square of a small distance, and the time variations are the same amplitude divided by the square of a small time (a kind of acceleration). But these two variations cannot stand as two terms in a single equation without some additional factor attached to one or the other to make the space–time dimensions of both the same; the two terms are dimensionally different because the first has in its denominator the square of a length and the second the square of a time, which means the second denominator has the square of a speed as a factor. This is just the speed of the wave, the numerical value of which is, as Maxwell showed, the speed of light. This clearly established the correctness of the electromagnetic theory of light and, in fact, of all radiation. Maxwell had thus unified not only electricity and magnetism but also light into a single theory. Ultimately, this work became the electromagnetic theory of radiation, which includes the entire electromagnetic spectrum from the longest radio waves to the shortest gamma rays emitted by atomic nuclei.

The presence of the speed of light in Maxwell's equations is remarkable because its value in these equations does not depend on the frame of reference of the observer, since it is given as the ratio of the two different sets of units (the electrostatic and the electromagnetic) in which electric charge can be expressed. This independence is startling, since in Newtonian mechanics, the observed speed of an object depends on the motion of the observer. The full implications of this property of the speed of light (its nondependence on the motion of the observer) were first delineated by Einstein in his special theory of relativity. Another noteworthy feature of the speed of Maxwell's electromagnetic waves is that it is the same for all wavelengths in a vacuum, but varies from wavelength to wavelength in a material medium. In such a medium, the speed, as Maxwell's theory shows, equals the speed in the vacuum (the maximum speed) divided by the index of refraction of the medium (a number larger than 1), if the medium is nonmagnetic. Since the index of refraction in a dense medium like glass is greater for the short wavelengths of light (the blue and violet) than for the longer wavelengths (red and orange), red light travels faster in a dense medium than violet light, so that the paths of red rays are bent less than those of violet rays when light passes obliquely from the vacuum into a dense medium. This relationship had already been observed by experimental physicists before Maxwell promulgated his electromagnetic theory of light, so that physicists, in general, were predisposed to accept it, but with some skepticism. But all reservations as to its correctness were swept away by Heinrich Rudolf Hertz's famous experiments on the production of electromagnetic waves by electric oscillators and their reception by oscillators of the same kind placed at various distances from the source of the waves. Hertz produced his waves by discharging a condenser that incorporated, instead of plates, two small spheres separated by a small distance. At a high enough potential difference between the two spheres, the discharge occurred in the form of a spark in which very rapid electromagnetic oscillations were generated. The waves thus produced caused a spark to jump between two similar spheres connected by a similar conductor placed at a distance of some yards from the first set of spheres. Hertz thus demonstrated that Maxwellian electromagnetic waves can be produced and are propagated exactly as Maxwell's equations predicted. This discovery was the beginning of radio technology.

Hertz went beyond merely proving that electromagnetic waves are produced by an electromagnetic oscillator (discharging condenser); he demonstrated that they are reflected, refracted, and diffracted (bent around corners) just the way light is, so that all doubts about Maxwell's electromagnetic theory of light were dispelled. In proving Maxwell's theory experimentally, Hertz greatly strengthened the belief that matter consists of electrically charged particles, for light is emitted by hot solids, liquids, and gases. This can only mean that electrical charges are oscillating in the matter when it emits light, which also supports Faraday's surmises about the electrical nature of matter.

The relationship between light and electromagnetism was further clarified by the experiments of the Dutch physicist Pieter Zeeman, who showed that the various wavelengths of light emitted by a glowing gas are altered when the gas is placed in a strong magnetic field. Hertz discovered another important effect that indicates the electrical structure of matter: When either of the two small spheres in his electrical oscillator was irradiated by a beam of light, the discharge occurred more quickly than otherwise. The reason for this is that the light knocks electric charges (electrons) out of the irradiated sphere, and these charges act as conductors for the discharge spark between the two spheres. This phenomenon became known as the photoelectric effect. It could not be explained by the wave theory of light and remained as one of the mysteries in physics until Einstein explained it with his photon concept of light, which stemmed directly from Planck's quantum theory.

As we see in the next chapter, Maxwell contributed to the molecular theory of matter and played an important role in the development of statistical physics (statistical mechanics), but his very premature death cut short his incredible scientific creativity. Despite his short life, however, the last 40 years of the 19th century may correctly be called the Maxwellian Years.

CHAPTER 11

The Broadest Laws of Physics

Thermodynamics, Kinetic Theory, and Statistical Mechanics

The most important discoveries of the laws, methods and progress of Nature have nearly always sprung from the examination of the smallest objects which she contains.

—JEAN BAPTISTE DE LAMARCK

There are in the history of physics many examples of developments that initially proceeded independently of each other and yet were later found to be intimately related. It can easily be seen that this is particularly true of the mathematical formulations of various theories. Thus, the wave equation (the partial differential equation that describes the propagation of a wave), which was first developed for acoustical waves, was carried over bodily, so to speak, to describe electromagnetic waves—that is, to describe the propagation of radiation. The potential theory, which was introduced to simplify gravitational problems, was applied, without change, to electrostatics and magnetostatics. Much later, Erwin Schrödinger applied the same classical wave equation to describe the wave characteristics of the motion of the electron. Another very interesting example is the successful application of the mathematics that describes the interference and diffraction patterns of a light wave passing through a series of slits in a screen to the distribution of the diffraction pattern of an electron passing through such slits. That similar mathematical schemes can be used to describe physical phenomena that appear to be unrelated does not, in general, mean that the phenomena are just different manifestations of the same basic reality. There is always a strong feeling, however, that such a one-to-one relationship of the mathematics must point to an underlying unity of the physics of such phenomena.

Just as it does not follow that similar mathematical formulations imply physical unity, neither does it follow that different mathematical representations mean different physical phenomena. Indeed, the mathematical treatment may at times obscure the underlying unity of the physics, as is very well illustrated by the three remarkable branches of physics called thermodynamics, kinetic theory, and statistical mechanics, which appear in different mathematical guises but describe essentially the same physical phenomena. Physicists have always been attracted to these three disciplines and held them in high regard because they are based on a minimum of assumptions, they are governed by the most general principles known in physics, and all problems within their framework can be solved. In this chapter, we discuss the development of these three branches of physics from the classical (Newtonian) point of view, but later we shall see how and to what extent the quantum theory and the theory of relativity altered them. In particular, we shall see that the simple classical statistical mechanics had to be replaced by two different kinds of statistical mechanics owing to the quantum mechanics, and that a new branch of thermodynamics, the thermodynamics of black holes, was forced on physicists by the general theory of relativity. But the general principles of thermodynamics, kinetic theory, and statistical mechanics have remained unaltered from their inception to the present.

THERMODYNAMICS

Thermodynamics as we now know it began with Julius Robert von Mayer's discovery in 1842 of the equivalence of heat and mechanical energy (that is, work, or kinetic energy and potential energy) and his statement of what is now called the first law of thermodynamics (the principle of the conservation of energy, with heat included as a form of energy). Mayer (1814–1878), the son of a pharmacist in Heilbronn, Germany, decided not to follow his two older brothers into the family business, but instead enrolled as a medical student in 1832 at the University of Tübingen. Although Mayer was arrested and expelled in 1837 for his activities in a secret student society, he returned the following year to finish his studies and take the state medical examination. Mayer spent a year as a physician aboard a Dutch merchant ship voyaging to the East Indies. He formulated his theory that motion and heat are interconvertible manifestations of the same entity in nature and that this entity (energy) is conserved in any such conversion.[1] Mayer's speculation was prompted by

Julius Robert von Mayer (1814–1878)

the surprising redness of the blood of newly arrived sailors he examined. He reasoned that this redness was due to the heat of the tropical climate because a lower metabolic rate could maintain the same body temperature in a hotter climate, so that less oxygen would have to be extracted from the red arterial blood.[1] Mayer saw the oxidation of food as the only possible source of animal heat and concluded that the chemical energy latent in food can be expressed quantitatively as the heat obtained from the oxidation of food.[1] He believed that muscle force and body heat were derived from the chemical energy latent in food and that if the animal's intake and expenditure of energy are in balance, there must be a conservation of this energy.[1] An 1845 paper by Mayer extended this conservation principle to magnetic, electrical, and chemical energy and described the basic energy conversions in the animate world beginning with the conversion by plants of sun energy into latent chemical energy, the consumption by animals of the source (in the form of food) of this energy, and the subsequent conversion by animals of this energy to body heat and the mechanical muscle energy in their life processes.[2]

Despite the originality of his thought, Mayer's ideas were not readily accepted by the society of physicists. The frustration and depression Mayer felt were compounded by the lasting estrangement from his brother Fritz, owing to their differing political stands in the 1848 revolution, and by the deaths of five of his seven children in infancy. Mayer attempted suicide in 1850. He suffered fits of insanity over the next few years and was committed to a succession of asylums.

The German physicist Hermann Helmholtz publicized the importance of Mayer's work after he had read some of Mayer's earlier papers. Helmholtz argued for Mayer's priority in the discovery of the conservation principle; his cause was also taken up by Rudolf Clausius and, later, the English physicist John Tyndall. This long-delayed scientific recognition seems to have had a dramatic therapeutic effect on the health of the long-forgotten Mayer. He began extensive correspondences with his supporters and saw many of his earlier papers translated into English. In 1870, Mayer was voted a corresponding member of the Paris Academy of Sciences; the following year, Mayer was awarded the Copley Medal by the Royal Society. Despite his newfound status, however, it was Mayer's fate that his works would exercise little direct influence on science, since by the time they had become widely known in scientific circles, their principles had already been independently formulated and entrenched in physics. Moreover, Mayer did not use mathematics extensively, which limited the usefulness of his papers to other physicists. In fact, it had been the precise experimental work of the British amateur scientist James P. Joule on the mechanical equivalent of heat—that is, his accurate measurements showing that the amount of heat produced by a given amount of work (for example, in the form of friction) is always the same—that had made possible the recognition of Mayer's genius.

Although Mayer's formulation of the first law of thermodynamics was the first step in the development of the science of thermodynamics, the work of the French engineer Nicolas Léonard Sadi Carnot on the efficiencies of heat engines and his discovery of what is now called the Carnot cycle laid the basis for the second law of thermodynamics. Carnot (1796–1832) was the eldest son of Lazare Carnot, a distinguished engineer who made valuable contributions to engineering mechanics.

Nicholas Léonard Sadi Carnot (1796–1832)

Lazare, young Sadi's first teacher, instilled in his son a wide-ranging interest in mathematics and the sciences. Sadi left home to study chemistry, geometry, and mechanics at the Ecole Polytechnique. His studies were interrupted by several months of duty in Napoleon's army, during which time he saw action against the invading allies at Vincennes before returning to the Polytechnique to graduate in 1814. He then spent two years studying engineering at the Ecole du Génie before being commissioned as a second lieutenant in an engineering regiment. Most of his time was taken up by bureaucratic paperwork, so he eventually transferred to an army staff position, where he was freer to pursue his scientific interests. He took various engineering courses at Paris institutions including the Sorbonne and began to examine critically the principles of the steam engine. In 1823, Carnot began work on his classic book *Réflexions sur la puissance du feu et sur les machines propres à développer cette puissance,* which outlined his thoughts about possible improvements that could be made in the operating efficiencies of steam engines.

Although Carnot's work was published in 1824 and received favorable reviews, most scientists paid little attention to Carnot's own contributions to the study of heat. In his book, Carnot outlined three premises that set universal standards for judging the operating efficiencies of steam engines on the basis of accepted scientific principles. First, Carnot asserted that perpetual motion was impossible, even though it had figured in the study of mechanics, including some of the work done by his father.[3] Second, Carnot used the caloric theory of heat to assert that the quantity of heat absorbed or emitted by a physical system can be measured by examining the initial and final states of the system.[3] Third, Carnot assumed that useful work can be produced whenever a temperature difference exists.[3] His famous waterwheel analogy suggested that the so-called "motive power" of heat depended on both the quantity of caloric [heat] and the size of the temperature difference through which it fell; it also implied that "the expenditure of motive power can return caloric from the cold body to the warm body."[3] Carnot also developed the principles of the ideal heat engine that bears his name and introduced the concepts of completeness and reversibility.

Disheartened by the lack of attention paid by scientists to his work despite a favorable review by the Académie des Sciences, Carnot continued to work on the theory of heat and concentrated on making further improvements in steam engine design. He also investigated the relationship between temperature and pressure in gases for a brief time until his death from cholera. Although scattered references were made to Carnot's work over the next several decades, it was only after the British physicist William Thomson (later Lord Kelvin) published a series of papers that relied heavily on Carnot's *Réflexions* that Carnot's pioneering work in the study of heat was widely recognized. Coupled with Rudolf Clausius's modifications, which assumed, contrary to Carnot's formulation, that some heat was lost in the engine and some heat was transferred to the colder body, Carnot's theorem was formally unveiled as the second law of thermodynamics. Thus, a good deal of experimental work prepared the stage on which the late-19th-century theoreticians played their important roles in the thermodynamics drama.

As we have already indicated, thermodynamics rests on two basic laws, the first and second laws of thermodynamics (the law of energy and the law of entropy), which apply under all conditions to all forms of matter and energy and to their interactions (absorption, emission, and scattering of energy by matter). Although the laws of thermodynamics are universally true and apply to all forms of matter (solid, liquid, and gaseous), they are most easily formulated and understood when applied to gases. All matter (with the possible exception of helium), under the appropriate thermal conditions, can exist in a solid, liquid, or gaseous state. Of these states, the gaseous state is the simplest and most thoroughly understood because the constituent particles (molecules, atoms, nuclei, or electrons), in general, move about independently of each other. In its simplest state, when the molecules or atoms of a gas are completely independent of each other—that is, when they exert no forces on each other as they move about randomly—the gas is said to be a perfect or ideal gas. Such an ideal gas does not exist in nature, but it is a useful theoretical concept from which many correct deductions can be drawn.

The behavior of such an ensemble of particles (a perfect gas) is described by the famous gas laws of Boyle and of Charles and Gay-Lussac. We recall that Boyle's law states that if the temperature of a gas is kept constant, then the volume and the pressure of the gas cannot change independently of each other; they must change together in such a way that the pressure multiplied by the volume is always the same (see Chapter 6 for a discussion of pressure and volume). This very special case of the gas law (Boyle's law) does not give us an insight into the thermal properties of a gas deep enough to lead us to the two laws of thermodynamics, so we must go to the general gas law, which governs the behavior of a gas when its pressure, volume, and temperature all change together. In this case, any two of these three quantities may be changed independently of each other, but then the way the third one changes is set, as already discussed. Charles and Gay-Lussac independently discovered that no matter what one does to a gas (compress it, raise or lower its temperature, or alter its volume in any way), the pressure, volume, and temperature of a perfect gas must change together in such a way that the product of its pressure and volume, divided by its temperature, must remain constant. With this thought in mind, we are prepared to state the first law of thermodynamics and see what it means. To that end, we go back to the concept of energy that we discussed in detail in Chapter 7.

We saw there that the energy concept can best be understood by considering the work done on an object, which is defined as the product of the force applied to the object and the displacement of the object in the direction of the force (that is, the total distance it moves) while the force is acting. Thus, we do work *on an object* whenever we push or pull it over a given distance. The body on which the work is done is not the same, of course, as it was before we did work on it; it has acquired something—we call it ''energy''—that it did not have before we did work on it. This energy exactly equals the work we do on the body if it is completely free to move. If the work done on the body merely sets it in motion without altering its height above the ground, the energy is entirely kinetic energy (energy of motion); if, however, the work merely raises the body to a greater height, where it remains at rest, the energy is entirely potential (energy of position). A change in the speed of the body indicates a change in its kinetic energy, and a change in its height means a change in its potential energy. The bob of an oscillating pendulum is a very good example of a body with both kinetic and potential energies and with a continuous interchange occurring between these two kinds of energies.

With the concept of energy introduced into physics as a direct outgrowth of Newton's laws of motion and his law of gravity, the question as to the permanence or conservation of energy arose. The nature of this problem is again nicely illustrated by the bob of the pendulum as it oscillates: At the highest point of its swing, the bob has only potential energy; at the lowest point, it has only kinetic energy. This conclusion prompted the question of whether the pendulum's total energy, that is, the sum of its kinetic and potential energies, is always the same. The answer, as we know from experience, is no! The pendulum ultimately stops, so that its mechanical energy (kinetic plus potential) disappears. The great classical physicists immediately following Newton knew this and understood that if no friction were present between the bob of the pendulum and the surrounding atmosphere and between the pendulum's string and its support (a nail or a screw), the pendulum would swing forever—its mechanical energy would be conserved. The motion of the earth (or of any other planet) around the sun illustrates the conservation of mechanical energy very well. As the earth revolves around the sun, its distance from the sun and hence its potential energy change continuously; the same is true of the earth's speed and hence of its kinetic energy. But these two energies change in such a way that their sum remains constant. Mechanical energy is thus conserved, fortunately for life on the earth. If the earth lost its mechanical energy continuously (like the pendulum or like an artificial satellite in the earth's atmosphere), it would ultimately fall into the sun.

Although the late-18th- and early-19th-century physicists knew that friction robs a mechanical system, such as a pendulum, of its energy, they did not know what happens to the energy and simply assumed that it disappears. That energy is never diminished or increased but assumes a different form never occurred to them, although if they had noted that the temperature of the air surrounding the pendulum increases slightly as the pendulum slows down, they might have suspected that something that heats the air flows into it from the pendulum. Mayer recognized this "something" (heat) as another form of energy, and so insisted that in any system of bodies interacting in any way whatsoever, kinetic energy plus potential energy plus heat always remains constant. This conclusion is essentially the content of the first law of thermodynamics.

To express the first law in its most obvious form, we consider a gas in a vertical cylinder capped by a freely movable, weightless piston on top and with a thermometer inserted in the side of the cylinder so that the temperature of the gas can be measured. The piston–cylinder concept is familiar to all of us from our experiences with automobile engines, the power capacities of which are rated by the number of cylinders they contain. Here, we assume that the gas in our cylinder is just the same as the air outside it, so that the pressure inside the cylinder is atmospheric pressure. We now put some weights on the piston and note that the piston sinks a certain amount (the heavier the weights, the more it sinks), finally coming to rest again when the pressure of the air in the cylinder has increased sufficiently above atmospheric pressure to support the weights on the piston.

If we now heat the cylinder with a flame, allowing a definite amount of heat to flow into the gas (air), we immediately note two things: The piston rises and the temperature of the gas increases. These events show that the heat entering the cylinder does work (it raised the weights on the piston) and it increases something in the gas that has to do with its temperature. This "something" is the internal energy of the gas, which we cannot observe directly: The higher the temperature, the greater the internal energy. The first law now states that the heat supplied to the gas exactly equals the work done on the weights plus the increase in the internal energy of the gas. One must marvel at the great simplicity of this statement and yet its profound significance for mankind, for it represents the difference between freedom and slavery. For the centuries before the first law of thermodynamics was known, man, water, wind, and beast were the only sources of work, and so man and beast became commodities to be bought and sold by those who controlled the economy and politics of society. Slavery of one form or another was the typical result.

The first law of thermodynamics undermined the slave economy (although not all slavery was immediately abolished with its discovery), for it shows that work can be obtained from heat. In other words, one can construct an engine (essentially a piston and a cylinder containing a gas) that can change heat into work.

Here, indeed, was a law of nature that promised mankind a paradise on earth, for the backbreaking physical drudgery that had enslaved people, whether they were legally free or not, could now be eliminated forever with the proper use of heat. Of course, the remarkable fruits that were to be borne by this law of optimism (as we may call it) were not immediately discerned, for the step between the pure science that reveals a law and the technology that gives birth to its fruits is often a very long one. In the development of the technology that was generated by the first law, the first and most important step was the development of a fairly efficient heat engine, the first example of which is the steam engine; the construction of such an engine, as we know, led to the Industrial Revolution, which sounded the death knell for slavery. The second step was, of course, finding cheap and abundant sources of heat—namely, fuel—and so vast enterprises were organized to mine coal and drill for oil.

With the development of heat (thermal) engines of all kinds, it soon became clear that whereas nature places no restrictions on changing work into heat, it imposes certain severe restrictions on the transformation of heat into work. Nature favors one direction over the other in the work⇄heat exchange; the work→heat process, which proceeds spontaneously, is not spontaneously reversible. Friction, which leads to a spontaneous transformation of work into heat, imposes a severe restriction on the fraction of a given amount of heat that an engine can transform into work. The moving parts of an engine rub against other parts, and the friction between such parts thus changes some of the mechanical energy of the moving parts (for example, pistons), which would otherwise become work, into heat again. In principle, one can reduce friction to any desired degree; in practice, there is a limit to how far this can be done. Friction in itself limits the efficiency of a heat engine. But even if mechanical friction were completely eliminated, the efficiency of a heat engine would always remain less than 1 (1 being complete or perfect efficiency) because of the second law of thermodynamics, which we may dub the law of pessimism, since it places certain restraints on the amount of work that can be obtained from heat under the best circumstances. The first law is, in a sense, just a bookkeeping law that says the energy account book must always balance, whether we change heat into work or work into heat; it says nothing about whether nature prefers one of these processes to the other. The second law does, however, and it is the essence of the whole matter.

To understand the second law fully, we must return to the swinging pendulum, but this time we place it inside our gas-filled cylinder, attached to the movable piston; the bob of the pendulum now takes the place of the weights on the piston to keep the pressure in the gas constant. We now set the pendulum swinging (that is, we do work on it and thus supply it with mechanical energy) and observe the thermometer in the cylinder and the height of the piston. Both the temperature of the gas and the piston rise as the pendulum slows down, losing its energy, and finally stops. Clearly, all the pendulum's mechanical energy now reappears in two forms, as work done on the piston (shown as the increasing potential energy of the rising piston) and as internal energy of the gas, as indicated by the higher temperature of the gas. It is not necessary for us, at this point, to know the nature of the internal energy of a gas, which the kinetic theory of gases will reveal later in this chapter, but simply to note that the initial potential energy of the bob of the pendulum exactly equals the sum of the increase in the potential energy of the piston and the increase in the internal energy of the gas.

This simple pendulum experiment neatly illustrates the essential feature of the second law of thermodynamics: Energy flows spontaneously in only one direction—from mechanical energy to heat and not from heat to work. To see this point more clearly, we note that energy flows spontaneously from the bob of the pendulum in the form of heat and raises the temperature of the gas, which then, as it expands, lifts the piston. But the work it does in lifting the pendulum is less than the energy it received in the form of heat from the pendulum. Although all the mechanical energy of the pendulum was changed into heat, only part of this heat was available to do work (to lift the piston). Natural (or spontaneous) processes in nature are thus irreversible. We shall see why this is so when we discuss the kinetic theory. If we now slowly push the raised piston down to its original height, the temperature of the gas rises, showing that the work we are doing is changing to heat, but none of this heat makes the bob of the pendulum oscillate; heat does not change itself into mechanical energy again.

The second law arose in the last quarter of the 19th century from the recognition by scientists of irreversible processes in nature. Since this law is one of the most profound statements about nature and its behavior, we consider the nature of reversible and irreversible processes in some detail. Physicists define a reversible process in a system as a very minute change that occurs very slowly so that the system is always in equilibrium and, depending on external conditions, may go one way or the other; that is, if the conditions that are changing the system slowly are removed, the system returns to its initial state. If the piston in our cylinder is pushed down very slowly by a small amount, the temperature and pressure of the gas thus increase by a very small amount. But if the push on the piston is slowly reduced to zero, the piston slowly returns to its initial height and the temperature and pressure return to their initial values. This is an example of a reversible process that is never realized in nature because natural processes do not proceed at infinitesimal rates.

The following are examples of reversible processes: In a mixture of ice and water at 0°C and at 1 atmosphere of pressure, a slight increase in the pressure causes some of the ice to melt, which refreezes when the pressure drops to 1 atmosphere again. In a saturated salt solution at a given temperature, some of the salt precipitates when the temperature is reduced slightly, but it is then redissolved when the temperature returns to its initial value.

Since the second law applies basically to irreversible processes, we now consider various examples of such processes in search of some common characteristic or property among them that may direct us to the second law. The simplest example is again found in the behavior of a perfect gas in a cylinder, but now with the gas confined to half the cylinder by a partition that separates it from the empty half. If we now remove the partition, the gas expands spontaneously to fill the entire cylinder. This is an irreversible process, since the gas will never recontract by itself to fill one half the cylinder and leave the other half empty. We can use a piston to recompress the gas into its initial half volume of the cylinder, but to do so we must expend energy (do work). The diffusion of one gas into another is another irreversible process that we can actually observe if the two gases are colored. Thus, if a blue and a red gas are separated by a partition that is then removed, the two different colors give way to purple as the gases mix. Another important example is the spontaneous flow of heat from a hot body to a cold body when they are in contact; the hot body grows cooler and the cold body warmer until they are both at the same temperature. The reverse process, a net flow of heat from a cold to a hot body, never occurs spontaneously, although we can reverse the process by applying energy, as is done in a refrigerator or an air conditioner.

Other phenomena are associated with irreversible processes that are the earmarks of such processes. We thus discover that irreversible processes are associated with disorder. In other words, the total amount of disorder in the universe increases as irreversible processes proceed; this tendency occurs at the expense of order, so that the total amount of order decreases. Further, the total amount of energy available for doing work decreases with each irreversible process—or, to put it differently, irreversible processes change available to unavailable energy, so that such processes result in a net increase in the unavailable energy. Again, irreversible processes lead to a loss of information about a system in which such processes occur. If irreversible processes are the rule and not the exception in the universe, the second law, as we shall see, tells us that the universe and all systems in it strive toward complete disorder, which, when it is achieved, will mean that the universe will be in a state of complete equilibrium so that all processes will cease. Complete equilibrium means death. Associated with all processes in the universe is the flow of time, which is always in the direction from the past to the future. We immediately recognize the relationship of the irreversibility of daily events or phenomena with the irreversibility of the flow of time. We can progress only from the past to the future because such a flow of time is defined by irreversible processes all around us, such as the progress of living beings from infancy to old age. A question that naturally arose and is still unanswered in this context is whether or not the unidirectional flow of time stems from the second law of thermodynamics or whether the flow of time and the second law are related at all.

That disorder occurs spontaneously even under highly ordered conditions while orderly phenomena are occurring at the same time is obvious. The desk of even the most orderly person becomes disordered quickly during the course of a work day, and the highly ordered living being builds up its own order (from the food, water, and oxygen it takes in) while creating disorder (for example, heat, waste products). The flow of heat also leads to disorder because as it flows it spreads out (disperses) and becomes a more disorganized form of energy.

That information is lost during an irreversible process is best illustrated by a gas confined by a partition to half the volume of a container. We then know in which half of the container every molecule in the gas is located. After the partition has been removed and the gas has expanded to fill the entire container, we no longer know which half of the container any particular molecule is in at any given moment. In other words, we know half as much about the locations of the molecules as we did before.

We thus see that the second law of thermodynamics, irreversibility, the flow of heat, disorder, loss of information, and the direction of the flow of time are related. But merely stating that such a relationship exists is not enough for the physicist; he must have a precise formulation of the second law, in terms of measurable quantities, which enables him to evaluate the degree of irreversibility of disorder associated with any pro-

Rudolf Julius Emanuel Clausius (1822–1888)

cess. Such a law was formulated independently by Rudolf Clausius in Germany and Lord Kelvin in Great Britain.

Rudolf Julius Emanuel Clausius (1822–1888) was the son of a preacher who received his early education at a parochial school taught by his father. Rudolf studied at Stettin Gymnasium before entering the University of Berlin in 1840, where, despite an initial attraction to history, he studied physics and mathematics. Clausius received his doctorate at Halle in 1847. His appointment as professor at Zurich in 1855 was prompted by his 1850 paper, which set out his theory of heat and established the basis for the modern study of thermodynamics.

This paper contained Clausius's discovery that the ratio of the heat content of a system to its temperature always increases in any process in a closed system. In an ideal system operating at perfect efficiency, this ratio does not change. Clausius referred to this ratio as a measure of a system's "entropy," which he derived from the Greek. He defined entropy as a measure of the extent to which the energy in a system could be converted into work. The higher the entropy, the less energy available for doing work. Clausius argued that the entropy of a system irreversibly increases, which led to the speculation that since the universe is defined as the only completely closed system, its entropy continues to rise and the amount of energy available for conversion into work continues to fall until the entropy reaches a maximum state and thermal equilibrium (equal temperature) is reached everywhere. Physical changes of any sort are then impossible, since heat flows can no longer occur.

All of Clausius's work was characterized by a strong grasp of fundamental facts, a detailed knowledge of relevant phenomena and real-world analogies, and sustained efforts to correlate the two with mathematics.[4] He ventured into kinetic theory and revamped the prevailing billard-ball model to include rotary and vibratory as well as translational motion. In short, Clausius showed that collisions between molecules could transform one type of motion to another, thereby disproving the idea that all molecules move at constant equal velocities.[5] Clausius's kinetic model also provided the first mechanical argument for Avagadro's theory that equal volumes of gases at equal temperatures and pressures have the same number of molecules because all gases are governed by the same relationship among temperature, pressure, and volume and all gas molecules have the same average translational energy at the same temperature.[6]

Despite his talents, Clausius was notoriously prone to ignore advances made by others, even in the area of thermodynamics. He seems never to have noticed Boltzmann's work, nor did he search for a mechanical explanation for the tendency of entropy to move irreversibly toward a maximum.[7] He also neglected to follow Gibbs's work on chemical equilibria. Yet he did devote a great effort to developing an electromagnetic theory based on the conservation of energy.[7]

Clausius joined the faculty at Würzburg in 1867 and then took his final professorial position at the University of Bonn, where he remained for the rest of his life. He organized a volunteer ambulance corps of students during the 1870–1871 Franco-Prussian War and was himself wounded. He spent much of his remaining years engaged in bitter disputes with British scientists such as Peter Guthrie Tait over whether the German Mayer or the Briton Joule discovered the equivalence of work and heat.

William Thomson Lord Kelvin (1824–1907)

Perhaps the most distinguished representative of the British scientific community at that time was William Thomson (1824–1907), who was later knighted Lord Kelvin. Thomson was the son of an engineering professor who taught mathematics at the University of Glasgow. Thomson was so much the prodigious son at eight years of age that he attended and enjoyed his father's lectures. Young Thomson matriculated at Glasgow in 1834, where he finished second in his class in mathematics. He wrote his first paper in mathematics at the age of 15; it was read to the Royal Society of Edinburgh by an elderly professor, since it was thought a young schoolboy would detract from the dignity of the society's proceedings. Thomson enrolled at Cambridge in 1841 and excelled in mathematics and the sciences and won several medals for single-shell sculling as well. Much of his time at Cambridge was spent preparing for the grueling House–Senate examinations; he was ultimately rewarded by a first-place finish, but Thomson found to his regret that his education was somewhat specialized and that his study of natural philosophy was totally derived from Newton's *Mathematical Principles of Natural Philosophy*.

After graduating in 1845, Thomson journeyed to Paris to do postgraduate work in Regnault's laboratory. While in France, Thomson became interested in Carnot's theory of the motive power of heat and formulated a methodology that later was of invaluable assistance to Maxwell's efforts to describe the electromagnetic field mathematically. Thomson also met prominent French mathematicians and scientists, including Cauchy and Dumas, and became proficient in French laboratory techniques.

When Thomson returned to Scotland in 1846, he assumed the vacant post of professor of natural philosophy at Glasgow, where he remained for over 50 years. Soon after taking the post, he announced that his calculations of the age of the earth showed it to be about 100 million years based on his assumption that the earth was originally cast off from the sun and had been steadily cooling off since that time. Thomson's conclusion provoked controversy with many geologists who thought that Thomson's estimate was far too modest. With the discovery of radioactive disintegration, Thomson's theory was shown to be erroneous because the earth was found to have an independent radioactive source that did not appear to be cooling off. However, Thomson's theory did cause biologists to consider ways in which the time estimated to be necessary for the development of life might be reduced, and it ultimately led to De Vries's theory of mutation and gave impetus to Darwin's own theory of evolution.

Thomson also performed an invaluable service for experimental science in Britain when he obtained a small room from the university and set up the first teaching laboratory in Britain. Thomson's interest in thermodynamics led him to use Carnot's principles to formulate an absolute temperature scale based on natural law. Thomson's absolute scale stemmed from the discovery by the French physicist Jacques Charles that gases lose 1/273rd of their volume at 0°C with each 1°C drop in temperature. Thomson proposed that it was not the volume but the energy of the molecules that reached zero at −273°C. Indeed, Thomson found this to be so for all matter, prompting him to conclude that −273°C should be considered absolute zero, the lowest possible temperature that can exist in the universe. Thomson then proposed a new temperature scale (eventually to be named after him) having a zero mark equal to −273°C; it was quickly adopted by physicists owing to its usefulness for studying thermodynamics and its precise designation of the amount of work obtainable from an engine operating between two different temperatures.

Thomson also played a major role in encouraging the professional acceptance of James Joule's theory of the interconvertibility of heat and kinetic energy. Joule's theory was revolutionary because it discarded the caloric theory of heat, proclaiming heat to be a form of motion and not a substance. Although Thomson had been reluctant to accept Joule's theory when he first heard about it in 1874, he saw how Joule's argument meshed with his own views of the nature of energy. He eventually embraced Joule's view in his "On the Dynamical Theory of Heat," which outlined Thomson's own version of the degradation of energy associated with the second law of thermodynamics.

Thomson's mathematical skills enabled him to explain, using a few basic equations, diverse phenomena ranging from thermodynamics and mechanics to magnetism and electricity. He tried to express all manifestations of energy in mathematical equations and was unsuccessful in the attempt, but his wide-ranging interests enabled him to play the major role in the synthesis of 19th-century theories about energy. Thomson believed that all energy forms are somehow interrelated; this belief motivated him to search for a grand theory uniting both matter and energy. Thomson was also a kind of mentor for James Clerk Maxwell, whose unification of electricity and magnetism owed a great deal to Thomson's own researches on the subject.

Thomson's work has practical as well as theoretical utility. An 1842 paper that had described the flow of heat through a solid wire was invaluable for solving the problems of transmitting current through a 3000-mile trans-Atlantic submarine cable. Thomson was eventually hired as chief consultant by the Atlantic Telegraph Company and at great risk to himself oversaw early attempts to lay such a cable. He invented several devices, including a telegraph receiver, that improved the efficiency of the cable so much that great amounts of time and money were saved. Thomson also joined two engineering firms that laid submarine cables. These activities made him rich enough to buy an oceangoing yacht and an estate at Largs, North Ayrshire, Scotland.

Although he could afford to rest on his laurels, having been the recipient of hundreds of honorary degrees and awards, Thomson continued to patent a number of devices, including a maritime compass and instruments for measuring tides and sounding depths. Yet Thomson's energetic mind, until the very end, did not accept or even seriously consider the possibility that his scientific worldview might be flawed; he was thus unprepared for the developments toward the end of his life that led to the theories of relativity and quanta. However, like Newton before him and Einstein after him, Thomson unified much of contemporary physics; unknowingly, he sowed some of the seeds that eventually undermined his own orderly conception of the universe.

To understand how Clausius and Kelvin formulated the second law, we must study various examples of irreversible processes in detail to find some common property that expresses the essence of the law. Clausius based his formulation of the second law on the spontaneous irreversibility of the flow of heat from a hot body to a cold body and stated the second law as follows: A spontaneous process that has as its only final result the net transfer of heat from a body at a given temperature to one at a higher temperature is impossible. Lord Kelvin, however, expressed the law in terms of the transformation of heat into work: A spontaneous process that has as its only final result the transformation into work of a given amount of heat taken from a source (an environment) that is at the same temperature throughout is impossible. In short, no work can be done unless heat flows, which can occur only if temperature differences exist.

It is easy to show that Clausius's and Kelvin's statement of the second law are equivalent, but neither of them points immediately to some physical quantity that can be measured and that changes in only one direction (for example, a quantity that always increases) when irreversible processes occur. Clausius finally discovered such a quantity, which he called the entropy of a system and which is now universally accepted as the physical entity in terms of which the second law is best and most easily expressed. Clausius also showed how the entropy of (or rather its change during a process in) a system is to be measured. The second law is now most generally stated as follows: The entropy of an isolated system can never decrease; at most, it remains constant, which occurs when the system reaches thermodynamic equilibrium (when its temperature is the same throughout). The second law thus indicates that change can (and will) go on in an isolated system (a system that cannot gain energy from or lose energy to the outside). Two conditions are thus imposed on a changing system by the two laws of thermodynamics: Its total energy cannot change and its total entropy must increase. These two conditions are extremely important in chemical dynamics in determining whether a particular chemical reaction will occur or not under given conditions (pressure and temperature). Two kinds of chemical reactions occur in nature naturally: exothermic reactions, which liberate heat, and endothermic reactions, which absorb heat. It is difficult to see in general why endothermic reactions occur at all, but the law of increasing entropy explains why.

One other feature of the second law, that of increasing disorder, may, at first sight, and to the uninitiated, seem to contradict our experience, because we know that spontaneous organization goes on continuously in the universe: Gaseous matter and dust organize spontaneously into stars and planets, stars organize themselves into galaxies, light nuclei in the very hot interiors of stars are organized into very complex heavy nuclei, atoms in cold interstellar space and on the warm planets organize themselves into complex molecules, and finally, here on the earth, complex molecules combine to form the most highly organized structures in the universe—living cells. Creationists, misunderstanding the second law, argue that such highly organized structures are the result of an act of creation, incorrectly believing that the disorder property of the second law forbids the spontaneous formation (evolution) of higher forms of matter from simpler forms.

That this belief is wrong and indicates a misunderstanding of the second law is clear from the following simple example: Consider a closed container with two kinds of atoms, A and B, moving about freely at a given temperature. Initially, before these atoms have a chance to interact with each other, no other structures are present, but in time other structures, the molecules AB, are formed. Hence, some of the atoms have organized themselves into molecules that are a higher form of order than the free atoms. Indeed, the entropy of the matter has decreased. Does this mean that the second law is violated? No, because in determining or calculating the increase in the entropy of a changing system, all constituents of the system must be taken into account, and this was not done above. When molecules AB are formed from atoms A and B, energy must be released, and the entropy of the energy must be added to that of the matter. One then finds that the total entropy of the system after molecules are formed is always greater than that of the system when only atoms are present; the entropy of the released energy more than compensates for the reduced entropy of the molecules. The total order is thus decreased—disorganization increases.

We have seen that processes that go on naturally (spontaneously) are irreversible and accompanied by an increase in entropy and that such processes can continue as long as the entropy for them can still increase (has not reached a maximum), but we have not described how the change in the entropy of a system can be measured. Note that only the change in the system's entropy is significant and not its magnitude. Here, we can give only a brief and elementary discussion of this point, and we follow Clausius's analysis because he probably based his definition of the change of entropy on Carnot's study of the efficiency of heat engines. Carnot had observed that no matter how much friction is reduced in a thermal engine, the efficiency is always less than 100%, and therefore reasoned that something in the very nature of the engine (that is, in the way heat is changed into work) limits its efficiency. With this observation, he had almost stumbled on the second law of thermodynamics and probably would have done so if the first law had been known at the time of his work. Nevertheless, Carnot is considered by many historians of science to be the father of thermodynamics.

To see how Carnot's analysis of the heat engine led to the entropy concept, we return to our piston and cylinder of gas, which we again treat as a very simple heat engine and let it operate through a complete cycle from some initial state through a series of steps back to the initial state. In doing this, we point out that when a changing system returns to its initial state, all its characteristics (pressure, volume, temperature, internal energy, and, of course, its entropy) must return to their initial values. To get our engine to do work, we place weights on the pistons to keep the pressure in the gas constant and place the cylinder in contact with a hot plate (for example, the top of a stove) at a definite high temperature. As the heat from the plate flows into the gas in the cylinder, the gas expands, the piston rises, and the engine (that is, the gas) does work, while its temperature remains constant because it is in constant contact with the hot plate, the temperature of which does not change; the temperature of the gas is equal to that of the hot plate at all times. At this stage of our cycle, all the heat has been changed into work, and so it appears, without further analysis, that our simple engine is *100%* efficient—all the heat has been transformed into work, showing up as an increase of the potential energy of the weights, which can thus be lifted to any desired height. But such an engine is useless if we stop at this point in the cycle, since we can use it only once; if we want to lift other weights (use the engine again), we must bring the piston back to its initial position.

Let us then go on with our cycle and remove the cylinder from the hot plate; the gas, of course, being hot, will continue to expand, still doing work, but it will cool during this phase of its cycle. We now remove the weights from the piston and push it down very slowly to bring it back to its original position, so that we can get it to transform some heat into work again. We are doing work while compressing the gas, of course, so that we are giving back some of the work obtained from the heat, but not all of it. The work we do shows up as heat, which we allow to escape from the cylinder so that the temperature of the gas remains constant while we do this work. Since we cannot get the gas back to its original temperature and the piston back to its original position by doing this work, we continue compressing the gas, but now allowing no heat to escape until the piston is back in its original position and the temperature is at its original value; the engine has now completed one cycle and is ready to do work again.

In analyzing this cycle, Carnot saw that no matter how carefully he operated the engine and how much he reduced the friction between the piston and the walls of the cylinder, he could never operate the engine (piston and cylinder) at 100% efficiency because it always had to give up some of the heat it had taken from the hot source. Thus, the work done by an engine can never equal the heat it absorbs in the first step of its cycle; the work equals the difference between that and the heat it releases to the cold environment in the reverse stages of the cycle (the second half of the cycle in the return of the engine). Clausius, starting from this point, carried the analysis further; he saw that if no engine were present and the heat from the hot source flowed directly into the cold environment, the result would be quite different, since none of the heat would then be changed into work. In both cases, a certain amount of heat (more in the second case than in the first) flows from the heat source (heat reservoir) into the cold environment (heat sink), but heat does useful work in the first case and not in the second, so that the amount of disorder (and hence entropy increase) produced by the heat transfer from reservoir to environment is smaller in the first case than in the second. Now, the only difference between the heat flow in the first and second cases is that in the former the heat entered the engine at a high temperature and a different amount of heat left the engine at a low temperature. Clearly, then, the heat flow and the temperature at which it occurs must determine the change in entropy, and Clausius defined the change in entropy of a system as the amount of heat it gains or loses divided by its absolute temperature while gaining or losing the heat; a loss of heat leads to a decrease in entropy and a gain of heat to an increase in entropy. The reason that a gain or loss of heat is associated with entropy change is that heat, as we shall see in our discussion of kinetic theory, leads to random motion or disorder.

THE KINETIC THEORY

The kinetic theory (often called the kinetic theory of gases) was an inevitable outgrowth of the first law of thermodynamics, for it is obvious that if energy starting as heat is conserved and appears as internal energy when a gas is heated, this internal energy must be the energy of the constituents (molecules) of the gas. It was, however, extremely difficult for the kinetic theory advocates in those early years to get a fair hearing from their scientific communities, let alone an acceptance of their ideas; even as late as the first decade of the 20th century, the chemist Ostwald and the physicist Mach rejected the atomic and molecular theory of matter. But kinetic theory concepts were in the minds of some of the greatest scientists and mathematicians of the late 17th and early 18th centuries. Thus, in his *Opticks,* Newton states:

> It seems probable to me, that God in the Beginning form'd Matter in solid, massy, hard, impenetrable, moveable Particles, of such Sizes and Figures, and with such other Properties, and in such Proportion to Space, as most conduced to the End for which he form'd them; and that these primitive Particles being Solids, are incomparably harder than any porous Bodies compounded of them; even so very hard, as never to wear or break in pieces, no ordinary Power being able to divide what God himself made one in the first Creation. While the Particles continue entire, they may compose Bodies of one and the same Nature and Texture in all Ages: But should they wear away, or break into pieces, the Nature of Things depending on them, would be changed. Water and Earth, composed of old worn Particles and Fragments of Particles, would not be of the same Nature and Texture now, with Water and Earth composed of entire Particles in the Beginning.[8]

In this passage, Newton emphasizes the permanence and indestructibility of molecules (the "particles" of matter).

Later, in 1738, the great Swiss mathematician and physicist Daniel Bournoulli published a paper on hydrodynamics in which the essential features of kinetic theory are clearly presented, as indicated in the following brief excerpt from that paper:

> 1. In the consideration of elastic fluids we may assign to them such a constitution as will be consistent with all their known properties, so we may approach the study of their other properties, which have not yet been sufficiently investigated. The particular properties of elastic fluids are as follows: 1. They are heavy; 2. they expand in all directions unless they are restrained; and 3. they are continually more and more compressed when the force of compression increases. Air is a body of this sort, to which especially the present investigation pertains.
>
> 2. Consider a cylindrical vessel set vertically, and a movable piston in it, on which is placed a weight P: let the cavity contain very minute corpuscles, which are driven hither and thither with a very rapid motion; so that these corpuscles, when they strike against the piston and sustain it by their repeated impacts, form

an elastic fluid which will expand of itself if the weight P is removed or diminished, which will be condensed if the weight is increased, and which gravitates towards the horizontal bottom just as if it were endowed with no elastic powers: for whether the corpuscles are at rest or are agitated they do not lose their weight, so that the bottom sustains not only the weight but the elasticity of the fluid. Such therefore is the fluid which we shall substitute for air. Its properties agree with those we already assumed for elastic fluids, and by them we shall explain other properties which have been found for air and shall point out others which have not yet been sufficiently considered.

3. We consider the corpuscles which are contained in the cylindrical cavity as practically infinite in number, and when they occupy the space we assume that they constitute ordinary air, to which as a standard all our measurements are to be referred: and so the weight P holding the piston in the position does not differ from the pressure [weight] of the superincumbent atmosphere, which therefore we shall designate by P in what follows. . . .[9]

It is remarkable that Bernoulli's ideas are so close to ours, but still more remarkable is the work of the British physicist John James Waterston, who submitted to the Royal Society in 1845, a paper in which he deduced many of the well-known properties of gases and the basic relationships between the temperature and pressure of a gas and the motions of its molecules. He was the first to show that the temperature of a gas is determined by the square of the average speed of its molecules and that the gas pressure is proportional to the product of the number of its molecules in a cubic centimeter (molecular density) and this squared average molecular speed. Unfortunately, Waterston's paper was not published, but was hidden away in the society's archives, owing to the negative report of the two referees to whom the paper was referred.[10] One of them stated that "the paper is nonsense, unfit even for reading before the society" and the other, more open-minded, noted that the paper "exhibits much skill and many remarkable accordances with the general facts . . . . but the original principle is . . . by no means a satisfactory basis for a mathematical theory." These statements indicate in what low esteem the molecular concept (the kinetic theory) was generally held at that time. Waterston's paper was finally published in 1892 at the urging of the great British physicist Lord Rayleigh, who wrote an introduction to it.

The big breakthrough in kinetic theory came in 1860 with the publication in *Philosophical Magazine* of James Clerk Maxwell's paper, "Illustration of the Dynamical Theory of Gases," with the subtitle, "On the Motion and Collisions of Perfectly Elastic Spheres." Maxwell states the essence of the kinetic theory of gases in the first paragraph as follows:

So many of the properties of matter, especially when in gaseous form, can be deduced from the hypothesis that their minute parts are in rapid motion, the velocity increasing with the temperature, that the precise nature of this motion becomes a subject of rational curiosity. Daniel Bernoulli, Herapath, Joule, Kröni, Clausius, &c. have shown that the relations between pressure, temperature, and density in a perfect gas can be explained by supposing the particles to move with uniform velocity in straight lines, striking against the sides of the containing vessel and thus producing pressure. It is not necessary to suppose each part to travel to any great distance in the same straight line; for the effect in producing pressure will be the same if the particles strike against each other; so that the straight line described may be very short. M. Clausius has determined the mean length of path in terms of the average distance between the centres of two particles when collision takes place. We have at present no means of ascertaining either of these distances; but certain phenomena, such as the internal friction of gases, the conduction of heat through a gas, and the diffusion of one gas through another, seem to indicate the possibility of determining accurately the mean length of path which a particle describes between two successive collisions. In order to lay the foundation of such investigations on strict mathematical principles, I shall demonstrate the laws of motion of an indefinite number of small, hard, and perfectly elastic spheres acting on one another only during impact.

If the properties of such a system of bodies are found to correspond to those of gases, an important physical analogy will be established, which may lead to more accurate knowledge of the properties of matter. If experiments on gases are inconsistent with the hypothesis of these propositions, then our theory, though consistent with itself, is proved to be incapable of explaining the phenomena of gases. In either case it is necessary to follow out the consequences of the hypothesis.

Instead of saying that the particles are hard, spherical, and elastic, we may if we please say that the particles are centres of force, of which the action is insensible except at a certain small distance, when it suddenly appears as a repulsive force of very great intensity. It is evident that either assumption will lead to the same results. For the sake of avoiding the repetition of a long phrase about these repulsive forces, I shall proceed upon the assumption of perfectly elastic spherical bodies. If we suppose those aggregate molecules which move together to have a bounding surface which is not spherical, then the rotary motion of the system will store up a certain proportion of the whole *vis viva*, as has been shown by Clausius, and in this way we may account for the value of the specific heat being greater than on the more simple hypothesis. . . .

The whole idea behind the kinetic theory is to show that the gross (macroscopic) properties of a gas such as its internal energy, its pressure, and its temperature can be traced back to its microscopic properties, that is, to the random motions of its constituent particles (molecules), which

Maxwell assumed to be tiny, hard, elastic spheres moving randomly about and colliding with each other. By "elastic" is meant the property that if two such spheres collide, they bounce off each other without any loss of energy; that is, no heat is generated during the collisions, so that there is no decrease in the total kinetic energy of the two molecules. It is clear that, in general, this must be so (at least if the temperature is not too high), for if molecules lost some of their kinetic energy during each collision, all the molecules in the gas would quickly come to rest and drop to the ground, which, of course, is ridiculous.

The kinetic theory easily accounts for the pressure of a gas in terms of the rate at which the gas molecules strike the walls of the gas container. Two things enter into this analysis: the number of molecular collisions against a unit area of a wall that occur per second and the effectiveness or average strength of a collision; the greater the number of collisions on a given area of wall per second and the stronger each collision, the greater the pressure. The frequency of molecular collisions clearly depends on the density of molecules and the average molecular velocity, for the greater the number of molecules and the faster on the average a molecule is moving, the more numerous will be the collisions with the walls of the container. The strength of violence of a single collision depends on the product of the speed and the mass (called the momentum) of the colliding molecule. Thus, the kinetic theory tells us that the pressure of a gas against the walls of its container depends on the product of the molecular density (number of molecules per cubic centimeter), the molecular mass, and the square of the average speed of a molecule.

This result, which stems from the microscopic description of matter (in this case gases), ties in to the macroscopic (thermodynamic) picture in two ways. First, we note that thermodynamics leads to a relationship between the pressure and temperature of a gas—the pressure is proportional to the temperature—and kinetic theory gives us a relationship between pressure and the average kinetic energy of a molecule of the gas (kinetic energy equals one half the product of the mass of a molecule and the square of its velocity). We thus deduce a relationship between the temperature of a gas and the average kinetic energy of a molecule of gas: The temperature is proportional to the average molecular kinetic energy.

The kinetic theory leads to another important conclusion: When energy in any form (heat or mechanical) is transmitted to a gas, it is distributed equally (on the average) among all the molecules. This is called the "theorem, or law, of the equipartition of energy." This does not mean that every molecule in the gas has the same kinetic energy; the kinetic energy varies from molecule to molecule, but only by small amounts, and the kinetic energies on the whole lie close to an average value that was first calculated by Maxwell in the paper quoted above. This paper, which gives a formula for the distribution of molecular velocities in a gas, greatly strengthened the belief in molecules and the kinetic theory of gases, while demonstrating the power of the kinetic theory as an analytical tool in theoretical physics.

In reducing the pressure, the temperature, the internal energy, the entropy, and other macroscopic properties of a gas to the dynamics of molecular motions, Maxwell and other proponents of kinetic theory showed that the gas laws are essentially consequences of Newton's laws of motion. It follows, then, that if discrepancies are found between certain observed properties of gases, such as their specific heats, and those calculated from the gas laws, the validity of Newton's laws of motion must be questioned. As we shall see in Chapter 12, Maxwell pointed to such a discrepancy, which was explained only with the advent of the quantum theory.

STATISTICAL MECHANICS

Four great names are associated with the development of statistical mechanics as an important branch of physics: James Clerk Maxwell, Josiah Willard Gibbs, Ludwig Boltzmann, and Albert Einstein. As the name implies, statistical mechanics is the application of statistical methods to the analysis of the properties, structure, and behavior of physical systems. It is based on the same concepts as kinetic theory and is thus closely related to kinetic theory; indeed, one may consider Maxwell's paper on the calculation of the average speed of the molecules in a gas and the distribution of molecular velocities as an important contribution to kinetic theory or as the first step in the development of statistical mechanics. Maxwell's formula for the distribution of molecular velocities, which gives the number of molecules with velocities in a given range, is in fact called the "Maxwell–Boltzmann distribution"—the basic formula in classical statistical mechanics as derived by Boltzmann from the purely statistical point of view some 30 years after Maxwell's work.

Ludwig Boltzmann (1844–1906) was an Austrian physicist who utilized statistical mechanics to show how the second law of thermodynamics can be explained by applying the laws of mechanics and the theory of probability to the motions of atoms. This statistical interpretation of the second law shows that a condition of thermodynamic equilibrium is the most probable state of a system and led Boltzmann to derive the theorem of equipartition of energy, which shows that on average the energy of an atom moving in any direction is the same.

Boltzmann was the son of a Vienna civil servant. He studied at Linz and Vienna before receiving his doctorate from the University of Vienna in 1866. As a student, he had worked for the physicist Josef Stefan, who showed that the total radiation given off by a hot body is proportional to the 4th power of its absolute temperature (so that if the absolute temperature is doubled, the radiation rate increases 16 times). Boltzmann derived the same principle from thermodynamics to complete what is now known as the Stefan–Boltzmann law.[11] Although this law figured prominently in the development of the quantum theory, it was also used by the British astonomer Sir Arthur Eddington in the 1920s to calculate the equilibrium of stellar atmospheres.

Boltzmann received successive appointments as professor of physics and mathematics at Vienna, Graz, and Munich. He was also one of the earliest continental scientists to recognize the importance of Maxwell's theory of the electromagnetic field. He also developed Maxwell's distribution function and eliminated the need to assume that all atoms move around at about the same velocity and are equally spaced. Instead of trying to give the coordinates of any one atom, the distribution function gives the probability of finding an atom in a given range of velocity and position. His analysis of Maxwell's function shows that a fixed total amount of energy in a system will be distributed equally among the molecules that constitute the system.[12]

Boltzmann's researches in probability and statistical mechanics led him to formulate his H-theorem, which highlighted the apparent contradiction between the reversibility of collisions by individual atoms and the irreversibility predicted by the theorem for a system of many molecules.[13] Boltzmann argued that while there might be random increases in the order (or a reversal in the entropy) of a few collisions, the majority of possible collisions would tend toward greater disorder. The recurrence of a system's initial state was statistically guaranteed only if one waited for an unimaginably long period of time.[14] For example, if the molecules in a perfume bottle were allowed to fill a room, an estimated 10^{60} years would pass before enough collisions occurred to return all the molecules back to the bottle. On the cosmic scale, Boltzmann suggested that there could be local decreases in entropy even while the universe itself moved inexorably toward an entropic maximum.[14]

Despite Boltzmann's invaluable contribution to atomic theory, his work was attacked by "energeticists" such as Ernst Mach, who accepted atoms only as convenient mathematical fictions but not as physically extant objects, and Georg Helm, who denied both their existence and their utility. The energeticists assumed that all matter is actually made of energy and viewed the law of the conservation of energy as the supreme tenet of nature. The dispute was intensified by the rise in Vienna of the logical positivism school, which sought to eliminate from science all phenomena that cannot be directly detected by the senses. Unfortunately, Boltzmann's brilliant career was cut short when he committed suicide during a period of severe mental depression just before his ideas received general acceptance by physicists.

Like all statistical methods, statistical mechanics derives its validity from the large number of individual components in any system it deals with—the larger the number, the smaller the percentage error inherent in such methods. In fact, this percentage error diminishes as the square root of the number of individuals in a system. Thus, in a system consisting of one million individuals (for example, the population in a medium-size city), the error in any statistical analysis is of the order of one tenth of one percent. Since the number of molecules even in a very good vacuum is many thousands of trillions per cubic centimeter, the results obtained from statistical mechanics are very reliable.

The philosophy and motivation behind statistical mechanics is that the gross or macroscopic properties of a system consisting of many individual microscopic members should be deducible from the distribution of the individual members among the various possible energy states of the system. The great contribution of Gibbs (who applied it to the chemical dynamics and equilibrium of interacting atoms and molecules) to this discipline and type of analysis was to define the state of a system precisely in terms of the distribution of the system (molecules and atoms) among the various possible states of energy.

Josiah Willard Gibbs (1839–1903) was perhaps the greatest American scientist of the 19th century. He was born in New Haven, Connecticut, the only son of a professor of literature at Yale. He received his early education at a local grammar school, then entered Yale in 1854. After receiving many awards during his college years, he pursued postgraduate research in engineering. His dissertation was eminently practical, dealing with the design of gearing. Gibbs received his doctorate in engineering in 1863, the first in that discipline to be awarded in the United States.

He began his academic career as a tutor and then used a modest

Josiah Willard Gibbs (1839–1903)

inheritance to finance three years of study of mathematics and physics in Europe. While overseas, he became interested in the principles governing the Watt steam engine, which led him to thermodynamics and the search for a method to calculate the thermal equilibria of chemical processes. He was appointed professor of mathematical physics at Yale in 1871, but received no salary for nine years until a lucrative offer by Johns Hopkins prompted Yale to begin paying Gibbs a stipend. Aside from his summer holidays in New England and occasional scientific conferences, Gibbs rarely left New Haven. He spent his entire life in the house he was born in, sharing it and the household chores with two of his sisters.

Gibbs's scientific papers were quite mathematical and terse. The sparseness of his style made it difficult for many of his colleagues to understand his papers. Yet his writing style was both methodical and self-assured. In his very first published paper, he assumed that entropy is as necessary a concept as temperature, pressure, volume, or energy for dealing with thermodynamic systems, even though there was still great confusion among many European physicists as to its physical significance and the meaning of the word entropy.[15]

Gibbs's great contribution was to apply the thermodynamic principles developed by Carnot, Joule, Clausius, and Kelvin in a rigorously mathematical manner to chemical reactions. His major work, "On the Equilibrium of Heterogeneous Substances," extended the principles of thermodynamics to chemical, elastic, electromagnetic, and electrochemical systems. Although his brilliance was recognized by both Maxwell and Boltzmann, his work went largely unnoticed, in part because it was all published in an obscure Connecticut science journal. Only after Wilhelm Ostwald had translated some of Gibbs's papers into German did Gibbs become widely known in Europe and were his contributions uniformly recognized.

To discuss Gibbs's work in its simplest form, we consider a gas consisting of individual noninteracting atoms (a so-called "monatomic molecular gas"—that is, each molecule of the gas contains only one atom as against polyatomic molecules like CO_2) and introduce the concept of the degree of freedom of a single molecule. We also assume here that the molecules (atoms) have no internal structure or, more accurately, that the internal structure plays no role in the dynamics or gross properties of the gas. This is equivalent to the assumption that a molecule is a rigid, perfectly elastic sphere.

With this understood, we now define the concept of the degrees of freedom of a single molecule, which is associated with the number of independent motions to which energy can be assigned. In this simple case, we see that each molecule has just three degrees of freedom because it can move along any one of the three independent (mutually perpendicular) dimensions of space with a certain amount of kinetic energy in each case. Notice that this is not so for a molecule composed of two or more atoms, for such molecules can have, in addition to kinetic energy of translation, rotational and vibrational energies and hence more than three degrees of freedom. Briefly, then, the number of degrees of freedom of a molecule is the number of its energy modes. Consider a system such as a gas consisting of a large number, N, of such molecules. We may say that such an ensemble, as the system is called, has $3N$ degrees of freedom.

If the system is in equilibrium, its temperature and its total internal energy remain constant regardless of what happens. Statistical mechanics tries to answer the following question: What is the probability for any particular state of the system to occur, which, of course, depends on the number of ways that a particular state can be realized by interchanging the molecules in every possible way.

To be precise, suppose that we divide all the molecules into different energy groups, placing all the molecules that have very nearly the same energy in the same group. We may picture this by representing each energy group by a box that is numbered according to the energy group it represents. Thus, we may number the boxes by the integers 1, 2, 3 . . ., etc., to indicate that box 1 has all the molecules with very nearly 1 unit of energy (whatever the unit may be), box 2 all those with very nearly 2 units of energy, and so on. Here, we must be careful to note that the boxes are not to be pictured as physical containers with definite walls, each occupying its own physical space (volume); this cannot be, because different molecules with the same energy, and therefore assigned to the same energy box, will in general be in different spatial parts of the actual volume occupied by the gas. Thus, each energy box must be divided into the same number of tiny subboxes, each of which is at a different distance from some reference point such as the center of the cylinder containing the gas. The total number of subboxes we must use in this procedure equals the product of the number of energy boxes and the number of spatial subboxes in each energy box. Notice that the spatial volume encompassed by each energy box exactly equals the total volume occupied by the gas.

This procedure of dividing an ensemble of particles into subgroups consisting of particles that lie in a given small region of actual space and have very nearly the same energy is the basis of all statistical mechanics; it has been applied with incredible success, for example, to molecular and atomic ensembles (for example, gases, crystals, liquids), photon ensembles (radiation), electron ensembles (atoms, white dwarf stars), and stellar systems (globular clusters, galaxies). Mathematical physicists have refined the preceding description of assigning the individual members of an ensemble to particular subboxes of momentum and space by combining the space concept with the momentum concept and thus going from three to six dimensions. This is based on the following idea: Since space is three-dimensional, three numbers are required to locate (specify the position of) any molecule in a gas at a given moment. But such a molecule can be moving at a given speed along any one of the three independent directions of space, and so three other numbers are needed to specify its motion (or its momentum). Thus, six numbers are needed to specify the state of a molecule completely, and that is why this extended space, called by Gibbs the "phase space" of the gas, is said to be six-dimensional. Introducing all the tiny subboxes described above is thus equivalent to dividing the six-dimensional phase space into tiny cells, each of which represents a certain amount of momentum and a small region of space. Statistical mechanics is based on the idea (now fully confirmed experimentally) that every macroscopic state of a system (for example, our gas) is the result of one or more microscopic states, that is, of some particular distribution of the particles (molecules) in the system among the available cells (subboxes) of the phase space. The macroscopic state that then actually exists for a given set of macroscopic parameters (pressure, volume, temperature) is the one that corresponds to the largest number of possible microscopic states (distributions of particles among the cells).

The actual procedure for determining the maximum number of microscopic states that define a particular macroscopic state is quite simple (consider our subboxes again and picture distributing all the molecules of our gas among them in any arbitrary way we may wish, keeping in mind one very important restriction: The total energy (the sum of the energies of all the molecules of any distribution must be the same as that of any other). Now the only thing that counts in the correlation of microscopic states (partitions of all the molecules among the phase space cells) and macroscopic states is the number of molecules in each cell, and not which particular molecule lies in any particular cell. It is clear, then, that we can get many different microscopic states to correspond to the same macroscopic state by rearranging the molecules among the various cells *without* altering the number in any cell. The problem is a combinatorial one—namely, of all possible partitions of the molecules among the various cells (subboxes), which one can be achieved most often by shuffling the molecules among the cells without altering the numbers of molecules in the cells? The calculus can be used to find this maximum or optimum distribution, with the built-in constraints that the sum of all the individual molecule numbers in the separate cells must always equal the total number of molecules in the gas and the sum of all the energies of these molecules must always equal the total energy of the gas.

The formula that gives this distribution (also called the "partition function" in the language of statistical mechanics) is one of the most important theoretical discoveries ever made in physics. It was first deduced in the manner outlined above by Boltzmann and is equivalent in a limited sense to Maxwell's formula for the distribution of molecular velocities in a gas. Gibbs introduced this distribution function in his treatment of chemical dynamics before Boltzmann's work. The importance of the distribution (or partition) function for physics is that it enables one to calculate the macroscopic observable properties such as the internal energy, the pressure, the entropy, and others, of a system consisting of many particles.

CHAPTER 12

Origin of the Quantum Theory

Earthly minds, like mud walls, resist the strongest batteries; and though, perhaps, sometimes the force of a clear argument may make some impression, yet they nevertheless stand firm, keep out the enemy, truth, that would captivate or disturb them.

—JOHN LOCKE

The story of the quantum theory begins with Max Planck, whose pioneering studies of the nature of radiation marked the end of what James Jeans called the ''mechanical age of science'' and ushered in a new era of discoveries that shook the very foundations of classical physics. According to the Danish physicist Niels Bohr, Planck's quantum theory ''brought about a radical revolution in the scientific interpretation of natural phenomena [because] the picture of the universe formed on the lines of quantum physics must be looked upon as a generalization that is independent of classical physics, with which it compares favorably for its beauty of conception and the inner harmony of its logic.''[1] Werner Heisenberg, the discoverer of the uncertainty principle, was equally lavish in his praise of Planck's work: ''At the time he [Planck] could scarcely have foreseen that within a span of less then thirty years this theory, which flatly contradicted the principles of physics hitherto known, would have developed into a doctrine of atomic structure which, for its scientific comprehensiveness and mathematical simplicity, is not a whit inferior to the classical scheme of theoretical physics.''[2] Finally, Albert Einstein wrote that Planck's work ''has given one of the most powerful of all impulses to the progress of science,'' an impulse that Einstein believed would be effective ''as long as physical science lasts.''[3]

Max Karl Ernst Ludwig Planck was born in Kiel, Germany, in 1858. His father was a professor of constitutional law at the university there who

Max Karl Ernst Ludwig Planck (1858–1947)

was best known for his co-authorship of the Prussian Civil Code. Young Planck was said to have inherited some of his father's legal talents, including the abilities to sift through large amounts of evidence and distinguish relevant and irrelevant facts.[4] Doubtless Max Planck's belief that physical science was an integral part of human knowledge that would ultimately shape the destiny of mankind was prompted by his family associations.

Planck attended the Maximilan Gymnasium in Munich, where his interest in science was first kindled by his mathematics teacher, Hermann Müller, who was, according to Planck, a "master at the art of making his pupils visualize and understand the meaning of the laws of physics."[5] Müller used the image of a bricklayer lifting with great effort a heavy block of stone to show that the energy used to lift the stone is not lost but stored up until the stone is dislodged and falls back to earth. This principle of conservation of energy astounded Planck because it was the first physical law he learned that possessed "absolute, universal validity."[5] From that day, Planck considered the search for absolute or fundamental laws of nature to be the noblest task any scientist could undertake.

After graduating from the Gymnasium, Planck attended the universities at Munich and Berlin, where he studied experimental physics and mathematics. Only after Planck went to Berlin was he able to attend the classes of world-renowned physicists such as Hermann von Helmholtz and Gustav Kirchhoff, who, Planck said, was responsible for his initial interest in thermodynamics.

If Planck was awed by the reputations of his renowned professors, he soon overcame his wonderment. Planck later confessed in his *Scientific Autobiography* that "the lectures of these men netted me no perceptible gain."[6] Helmholtz, in Planck's opinion, never prepared his lectures properly and gave the impression that "the class bored him as much as he did [the class]."[6] Kirchhoff, by contrast, was remembered by Planck as one who delivered carefully prepared lectures that "sounded like a memorized text, dry and monotonous."[7]

Owing to the less than ideal academic environment at Berlin, Planck studied the subjects he was interested in on his own. His scientific career was to be forever charted the day he stumbled across the thermodynamic treatises of Rudolf Clausius, "whose lucid style and enlightening clarity of reasoning made an important impression"[7] on Planck and helped to regenerate what had been his flagging enthusiasm for science. The study of the second law of thermodynamics was the subject of Planck's doctoral dissertation at Munich, which he presented in 1879.

Although Planck's work had little effect on his professors (he suspected Helmholtz did not even bother to read it), he continued his studies of entropy, driven by the belief that it was exceeded in importance only by energy as a property of physical systems. Planck became a university lecturer at Munich, where he formulated several useful theorems in studies on gas mixtures. Unfortunately for Planck, he found that these very same theorems had previously been announced by the American physicist Josiah Gibbs, so that he received no recognition for his efforts.[8]

Planck accepted an appointment as associate professor in theoretical physics at the University of Kiel in 1885. Despite the scarcity of academic positions in theoretical physics at that time, Planck felt especially fortunate because the appointment enabled him to move out of his parents' house.[9] After moving to Kiel, Planck finished a paper entitled "The Nature of Energy" that was ultimately awarded second prize in 1887 by the Philosophical Faculty at Göttingen. Planck then wrote a series of monographs dealing with principles of entropy, especially as they relate to the laws of chemical reactions. He went to the University of Berlin in 1889 to take the place of his old professor Kirchhoff, who had recently passed away. Planck not only developed enduring friendships with the physicists there, including Helmholtz, but also began extensive correspondences with physicists such as Wilhelm Ostwald about Ostwald's controversial (and, in Planck's opinion, untenable) division of energy into three types corresponding to the three spatial dimensions. He also became embroiled in the controversy over the accuracy of the analogy most physicists at that time drew between the passage of heat from higher to lower temperatures and the falling of a stone from a higher to a lower height. Planck criticized the so-called "energetics" who supported this view:

> This theory [does] not take into consideration the essential fact that a weight can rise as well as fall, and that a pendulum has reached its greatest velocity at the moment when it has attained its lowest position and therefore, by virtue of its inertia, passes the position of equilibrium and moves to the other side. A transference of heat from a warmer body to a colder body, on the contrary, diminishes with the diminution of the difference in temperature, while, of course, there is no such thing as any passing beyond the state of temperature equilibrium by reason of some kind of inertia.[10]

Planck was greatly frustrated by the unfavorable response of his colleagues to his criticisms of the analogy, but he admitted that "it was simply impossible to be heard against the authority of men like Ostwald, Helm and Mach."[11] Although his suspicion of the incompatability of the analogy would eventually be proved to be right, he did not have the satisfaction of seeing himself vindicated, because his view was ultimately accepted only because it was incorporated into the atomic theory later proposed by Ludwig Boltzmann. Planck sided with Boltzmann in his ongoing battle with Ostwald, but Boltzmann evidently did not appreciate Planck's support, "for Boltzmann knew very well that my viewpoint was basically different from his."[12] Planck did not think much of the atomic theory that had formed the basis for Boltzmann's research, because at the time Planck regarded "the principle of the increase in entropy as no less immutably valid than the principle of conservation of energy itself, whereas Boltzmann treated the former merely as a law of probabilities—in other words, as a principle that could admit of exceptions."[12] Boltzmann's eventual triumph and the general acceptance of the differences between the conduction of heat and a purely mechanical process caused Planck to note wryly that "[a] new scientific truth does not triumph by convincing its opponents and making them see the light, but rather because its opponents eventually die, and a new generation grows up that is familiar with it.''[13]

Planck's interest in radiation was prompted by the realization that classical physics was increasingly unable to explain the results yielded by experiments. However, Planck's work was to be carried on not in optics, but in thermodynamics, and arose from measurements of radiation in the emission spectrum of black bodies.

Despite the many honors that followed Planck's introduction of his quantum theory, including the 1919 Nobel Prize for Physics, his secure position in the scientific world did not protect him from the tragedies of his personal life and the ravages of two world wars. Both of his daughters died soon after getting married. His eldest son, Karl, perished in the battle of Verdun in 1916, while his second son, Erwin, was murdered in January 1945 during the rule of terror. His house was destroyed in an air raid, and Planck himself was once buried in the rubble of an air raid shelter for several hours. In May 1945, the advancing American forces took him from his estate on Rogätz on the Elbe, which was then a war zone, to Göttingen.[14] His own death came at the age of 89 in 1947.

Planck maintained a quiet dignity throughout his life despite the tragedy of his children. The painful memories seemed ''to invoke a certain wistful quality which is profound in his nature and gives it the warmer glow that one is inclined to call mystic.''[15] But Planck dealt with his pain by his wholehearted devotion to his work, though he remained ''a perfectly practical man of the world and an up-to-date gentleman in manner and dress and also a sportsman, who climbed the Jungfrau to celebrate his seventy-second birthday.''[16] Like Einstein, he was an accomplished musician who preferred Beethoven. Certainly his scientific efforts reflected the imagination of an artist, for Planck's search for the hidden harmonies in nature enabled him singlehandedly to bridge the chasm between the classical world of Newtonian mechanics and the modern world of quantum physics and relativity theory. We now turn to the story of Planck's great discovery.

On October 19, 1900, Max Planck, then a 42-year-old German theoretical physicist, submitted to the Berliner Physikalische Gesellschaft a new radiation formula that was to revolutionize science. Planck's proposal was to signal both the ending of what is now called classical physics and the beginning of a new epoch in man's thinking and view of the laws of nature. Though some years were to elapse before scientists accepted Planck's quantum concept as anything more than a ''working hypothesis'' that would ultimately be eliminated in a proper theory, Planck was well aware of the importance of his discovery: On a walk with his son Erwin, that same day, he stated: "Today I have made a discovery as important as that of Newton." Two months later, at a meeting of the German Physical Society, Planck expanded on this discovery by stating that it means the "quantization of action" and that, therefore, all action is an integral multiple of the "elementary quantum of action," which, denoted as h by Planck (the famous Planck constant), takes its place with the gravitational constant G, the speed of light c, the Boltzmann constant k, and the unit electric charge q in nature's great hall of constants.

Since the quantization of action is a basic law of nature, Planck's constant, like the speed of light c, must enter all the laws of nature; any "law" that does not take account of h is therefore incomplete. As a result, Newton's second law of motion, which relates the rate of change in the state of motion of a body to the forces acting on the body, is not correct because it does not contain h. As we shall see, the completion of Newton's laws of motion to conform to the quantization of action led to what is now called the quantum mechanics.

Since the quantum theory represents an enormous conceptual departure from classical theory, one must wonder how such a revolutionary concept as quantization, which leads to a basically discontinuous description of phenomena, emerged from continuous classical theories. The wonderment must be all the greater if one contemplates the state of physics when Planck was driven, much against his will and better judgment, to propose the existence of the quantum of action as an explanation of the strange properties of "black-body" radiation. Classical physics appeared to be in an impregnable position at the peak of its achievements. On one hand, the Newtonian laws of motion and law of gravity gave a wonderfully precise, and what appeared to be complete, description of planetary motion and of moving bodies on the surface of the earth; even the rise and fall of the tides were explained. On the other hand, Maxwell's brilliant electromagnetic theory of light, as delineated in his famous equations of the electromagnetic field, did for radiation, in general, what Newton had done for the motions of material bodies. This dualism was extended to a remarkable trinity with the discovery of the laws of thermodynamics: the conservation of energy and the law of entropy.

It seemed in the last decade of the 19th century, when Planck was being led, step by step, to the quantum theory, that physics was a complete discipline that had no gap to be filled by a new theory—particularly a theory of discontinuity. But a close examination of the facts by critically minded physicists at that time revealed the vulnerability of the beautiful theoretical edifice that Newton, Maxwell, and their followers had constructed: Two sets of phenomena could not be explained. The first set consisted of the negative results of the Michelson–Morley experiment (that is, their inability to detect the motion of the earth by observing light beams moving in different directions) and the non-Newtonian advance of the perihelion of Mercury. Both puzzles were ultimately to be explained by the special and general theories of relativity, which we discuss in Chapter 15. The second set consisted of the discrepancy between the observed and calculated values of the specific heats of gases, the properties of the radiation emitted from a small aperture in the wall of a furnace (black-body or thermal radiation), and the discrete bright lines in the spectra of atoms. These were to be explained by the quantum theory.

In 1858, Maxwell had already shown that his kinetic theory of gases leads to a serious discrepancy between the calculated values of the specific heats of gases and the observed (measured) values. The kinetic theory, as introduced by Maxwell, treats a gas as an ensemble of randomly moving, submicroscopic particles (molecules), which are governed by Newton's laws of motion. Using this kinetic model, Maxwell showed that the absolute temperature of a gas is determined by (actually is proportional to) the average kinetic energy of a molecule in the gas. The total energy content of the gas is then given by this average kinetic energy multiplied by the number of molecules in the gas. If, then, certain of the observed properties of a gas do not agree with those calculated from the kinetic theory, Newton's laws of motion are suspect. This is precisely the situation for the specific heats of gases.

The specific heat of a substance is defined as the amount of heat required to increase the temperature of 1 gram of the substance by 1°C (exactly 1 calorie for water). From the kinetic theory, we see that to increase the temperature, the average kinetic energy of a molecule must be increased. The amount of heat required to increase the temperature by a given amount should be the same for all temperatures according to the kinetic theory. But this conclusion is contrary to the observations, which show that the specific heat of a gas decreases as the gas gets colder. Maxwell pointed to this result as a serious flaw in the theory, stating, "I consider this to be the greatest difficulty yet encountered by the molecular theory." The validity of Newton's law of motion was thus questioned more than a century ago, since the kinetic theory, from which this contradiction stems, is a direct consequence of the laws of motion.

Another difficulty confronting classical physics near the end of the 19th century was revealed in the investigation of the properties of radiation emitted by hot bodies. The spectroscope, which separates radiation into its constituent wavelengths (frequencies or colors), had already been used extensively to study the radiation flowing from hot solid bodies and from stars. It was known that the spectrum (the spread of colors or wavelengths produced by the spectroscope) of the light from a glowing gas (fluorescent light) consists of a discrete set of sharp, bright-colored lines (a different set for each kind of gas). The spectrum of the light from an incandescent solid (for example, the hot filament of an electric bulb) is a continuum of colors from the red to the violet. A number of questions arose concerning these two types of spectra. Physicists tried to deduce the answers to these questions from the basic physical laws known at that time.

Since the nature of the radiation from hot solid bodies or glowing gases seemed to depend not only on the nature of the body (gas or solid) but also on the temperature of the body, and since radiation, according to Maxwell's electromagnetic theory, is an electromagnetic phenomenon, physicists were convinced that a proper application of the laws of electricity and magnetism and the laws of thermodynamics to hot bodies and glowing gases would yield the answers to the questions presented by the observations. Two kinds of questions arose: (1) Since gross matter is electrically neutral, how can it generate, emit, and absorb electromagnetic radiation? (2) Why is the radiation emitted by a glowing gas so vastly different from that emitted by a hot body, and what is the relationship between the properties of hot-body radiation and the temperature of the body? During this historical period, there was little hope of answering questions of the first kind, since very little, beyond what Faraday had discovered in his studies of electrochemistry and electromagnetism, was known about the electrical constitutents of matter. The great Dutch theoretical physicist Hendrik Antoon Lorentz had begun to develop an "electron theory of matter," but the structure of the atom was still a deep mystery, so that painting a detailed picture of the emission and absorption of radiation by atoms was out of the question. But finding general relationships between the thermal properties of a body and the radiation it emits or absorbs seemed an achievable goal, and Planck took that direction.

Before Planck began his pioneering research that led to the quantum theory, Gustav Robert Kirchhoff, his teacher, in 1860 deduced from thermodynamics an important radiation law that relates the rate at which 1 square centimeter of the surface of a body at a given temperature emits radiation to the rate at which it absorbs it. In his investigations, Kirchhoff drew a sharp distinction between surfaces that reflect radiation readily and those surfaces that absorb radiation. It is clear that these two properties are mutually exclusive; if a surface absorbs a large fraction of the radiation striking it, it can reflect only the small fraction it does not absorb, and vice versa. The two extreme cases are a perfect reflector that reflects all wavelengths that strike it and therefore absorbs none and a perfect absorber that absorbs all wavelengths. The perfect reflector is called a "white body" and the perfect absorber a "black body," the latter of which played a very important role in the history of the quantum theory. Of course, there are no perfect reflectors or perfect absorbers, since all surfaces both absorb and reflect radiation to some extent.

To explain fully the nature of Kirchhoff's law, we must also define the concept of the emission of radiation (the emissivity of a surface). Just as all bodies reflect and absorb radiation, they also emit radiation at a certain rate. In short, they radiate. Here, we must be careful not to confuse emission with reflection, which is a process in which the radiation bounces off the surface without being changed in any way (red light remains red and blue light remains blue). Thus, a perfect reflector cannot emit any radiation. An absorbing surface, however, is also an emitter of radiation, but the radiation it emits is quite different from the radiation it absorbs.

To make the distinction between reflection and emission quite clear, we consider a particular wavelength of light (for example, a red beam) striking a perfectly reflecting surface and the same kind of ray striking a perfect absorber. In the first case, the light is reflected and is exactly the same after leaving the surface as it was before incidence, so that none of the energy is transferred to the surface; the temperature of the surface remains unaltered, so that the rate at which it emits energy is not affected by the incident radiation. Note that the rate at which the perfect reflector reflects radiation exactly equals the rate at which it receives the radiation. The history of the same radiation striking a perfectly absorbing surface (that is, a "black body") is quite different, however, as is that of the surface itself. First, the radiation itself completely disappears, becoming, in a sense, part of the surface, the temperature of which then increases (conservation of energy or the first law of thermodynamics). This temperature increase then causes the surface to emit, or radiate, energy faster than it did before the same beam of red light struck it. In fact, it must emit, in a unit time, exactly the same amount of energy (radiation) that it absorbed in that time. But this emitted radiation is quite different from the incident radiation: It is a mixture of a continuum of wavelengths with a total energy equal to that of the incident red beam. Clearly, the wavelength of only a small fraction of the total emitted radiation equals that of the incident red light. In other words, an absorbing surface spreads incident radiation into a continuous spectrum.

The following question immediately arose in connection with the emission and absorption of radiation by any surface: What is the relationship between the quantity (intensity or brightness) of radiation at a given wavelength (of a given color) that 1 square centimeter of a surface at a given temperature emits per second and the quantity of radiation of the same wavelength the square centimeter of surface absorbs per second? In 1860, the German physicist Kirchhoff deduced the answer from the laws of thermodynamics: The rate of emission of radiation of a given wavelength from a unit area of any surface at a definite temperature divided by the rate of absorption of the same wavelength by the same unit area at the same temperature is a universal quantity (the same for all surfaces) that depends only on the wavelength of the radiation and the temperature of the surface (Kirchhoff's law). It does not depend on the nature of the surface at all. Since the absorption rate of all wavelengths for a black body is unity, this universal quantity is just the emission rate of radiation of the given wavelength for a black body! The whole question as to the rate of emission and absorption of radiation of a given wavelength by 1 square centimeter of any kind of surface was thus reduced to the question as to the nature of the radiation emitted by a black body at a given temperature. The task that Planck undertook was to deduce from the basic laws of physics known at the time (the laws of thermodynamics and electromagnetism) the mathematical formula that correctly gives the intensity of each color (wavelength) in the total radiation emitted each second by 1 square centimeter of a black body at a given temperature.

Although careful experiments had made it possible to obtain experimental data, something analogous to a perfect black body was necessary. Since no perfectly black object was known, an artifact had to be found that can approximate a perfectly black body to any desired degree. The solution to this problem was simple: A tiny hole in an opaque wall surrounding a cavity behaves exactly like a black body. Any radiation that enters the hole has little chance of getting out; it is reflected back and forth within the cavity until it is completely absorbed. This phenomenon is illustrated by the appearance of the human eye. The pupil of the eye is essentially a tiny opening in the head that appears pitch black. If a tiny hole of the sort described above then absorbs all the radiation that strikes it (that is, it absorbs like a black body), it must radiate like a black body.

Imagine a container with opaque walls (such as a furnace) that are at a definite temperature so that they radiate energy into the container. If the temperature of the walls is kept constant, the total radiant energy in the container remains constant and in equlibrium with the walls; that is, the rate at which the walls reabsorb the radiation in the container exactly equals the rate at which they radiate energy back into the container. We may then say that the temperature of the radiation in the container equals the temperature of the walls. It is customary to speak of this cavity radiation as "thermal radiation." If we now punch a tiny hole in a wall of the container, the radiation streaming out of it is "black-body" radiation, the temperature of which is that of the inner surfaces of the walls of the container.

A number of questions concerning this radiation immediately confronted both experimentalists and theoreticians: (1) How much radiation of all kinds (of all wavelengths combined) is emitted per second from a 1-cm^2 hole in the wall of a furnace at a given temperature? (2) How is the wavelength of the most intense color (the wavelength in which more of the radiation is concentrated than in any other wavelength) related to the temperature of the furnace? (3) How does the intensity of the radiation of any wavelength depend on the wavelength and on the temperature of the furnace (that is, are the intensity of the radiation for a given wavelength, the given wavelength, and the temperature tied together by a simple algebraic formula)?

The experimentalists answered these questions easily, for they simply collected on a radiometer all the radiation (all the wavelengths combined) emitted per second from the 1-cm^2 hole to answer the first question empirically. They used a spectroscope to spread the radiation into its individual wavelengths (a continuous spectrum) and measured the individual intensities to answer the other two questions. These experimental results were a guide to the theoreticians, giving them three distinct tests against which their theoretical models or deductions were to be measured. At that time, of course, the theoretical physicists had only the classical laws of motion (Newtonian mechanics), the two laws of thermodynamics, and Maxwell's electromagnetic theory of radiation to guide them in their search for an overall law of black-body radiation that contains the answer to all these questions. This search, however, was fruitless, although the correct formula that gives the total energy (radiation) emitted in all colors at a given temperature of the furnace was obtained from thermodynamics. But this is as much as the classical physics could yield. A whole new concept and a new basic principle were needed. Planck supplied them with his quantum theory. We shall see in the following chapter just how he did that, but we first consider some important and essential preliminary ideas.

We noted previously that Planck introduced the concept of the quantum of action, which is basic to all the developments in quantum theory that followed. It is important to understand what action means in physics and how it was brought into physics as well as its origin. To answer this question, we must go back to the 18th-century French astronomer and mathematician Pierre Louis Moreau de Maupertuis, who first introduced the principle of least action as a ''rule of economy'' to which all actions of nature must conform. To understand this principle, we must define the Maupertuis concept of action. We have already treated this concept in some detail in Chapter 8 in relation to Hamilton's contributions to physics, but we recall it here to establish continuity with Planck's work. Consider a particle of a given mass moving along a tiny stretch of its path at a given speed; the product of the particle's mass and speed and the length of the tiny piece of its path is called the ''Maupertuis action'' of the particle in that stretch. The Maupertuis principle of least action states that if all these little pieces of action along the entire path of the particle are added, the sum must be smaller along the particle's true path than along any other conceivable curve (path) connecting the starting point and the end point of the particle's motion.

Having defined the concept of action, we must now show how and why it played such an important role in Planck's pioneering work and why he quantized it to obtain the correct black-body radiation formula. Since we discuss these points in detail in the following chapter, we limit our discussion here to a brief survey to show Planck's motivation. He applied the laws of thermodynamics to the analysis of the radiation problem, for two reasons: (1) Like most German physicists, he was much more comfortable with thermodynamics than with Maxwell's electrodynamics, which up to that time had acquired few supporters outside Great Britain. (2) The laws of thermodynamics, as we saw in the previous chapter, are extremely general and require, for their application, no special assumptions about the internal atomic structure or the electrical properties of matter, about which very little was known at the time. In any case, Planck was an expert thermodynamicist and felt that if he derived the correct black-body radiation law from such general principles as the first and second laws of thermodynamics, nobody could object to his derivation.

Planck proceeded very cautiously, making only the most general assumptions that introduced very little about the structure of matter. In line with the allowed procedures in thermodynamics, he assumed that the radiation in a closed furnace at any given temperature T is in equilibrium with the walls of the furnace, which means that the temperature and the total intensity of the radiation, consisting of a continuum of wavelengths (colors) of different intensities, remain constant. Moreover, the intensity of each color or wavelength also remains constant. This equilibrium is not static, but is dynamic, in the sense that the walls of the furnace absorb and radiate energy (radiation) at the same time, and the rate at which each square centimeter of the furnace emits energy of a given wavelength exactly equals its rate of absorption of such energy. This is all very good thermodynamics, as we saw in the previous chapter, but Planck soon discovered, in his analysis, that he had to make some assumptions about the way the radiation in the furnace interacts with the material on the walls of the furnace. He knew, of course, that the radiation in a furnace consists of standing waves (that is, waves like those along a plucked violin string). Since the end points of such a string are fixed, the standing wave along the string consists of a row of equally spaced stationary points (nodes) with transverse oscillations of equal amplitude between the points (antinodes). Now, such periodic oscillations are identical to the oscillations or vibrations of a particle attached to the end of a perfect spring that is allowed to vibrate freely with the spring. This kind of oscillation is called "simple harmonic motion" and is thoroughly described by Newtonian mechanics. Planck therefore made the very safe assumption that the walls of the furnace consist of little springs (harmonic oscillators) that are coupled to the radiation in the furnace, and here is where Planck had to consider the action concept, for the energy (amplitude) of a harmonic oscillator is related in a simple way to its action. Following Maupertuis's prescription for calculating the action of an oscillating harmonic oscillator, one finds that this action, for a single oscillation (a single back-and-forth movement), is just the energy of the oscillator (the sum of its kinetic and potential energies) divided by its frequency (the number of oscillations it performs per second). The energy of a given frequency (or wavelength) emitted by a single oscillator is thus its action multiplied by its frequency. Planck knew from thermodynamics that if an ensemble (aggregate) of objects (for example, oscillators) that could both absorb and emit energy of different frequencies were immersed in a bath of radiation of all frequencies at a given temperature (like a collection of corks of different sizes floating in turbulent water), each object (oscillator) would have, on the average, the same energy that would be determined entirely by the temperature of the radiation bath. He reversed this idea or thermodynamic argument and reasoned that if one can deduce the average energy of an oscillator in a radiation bath, one should be able to deduce the type of radiation emitted by oscillators (of all frequencies) raised to a given temperature.

To Planck, this seemed a fairly simple and straightforward problem: Given an oscillator capable of oscillating at only one frequency, the energy it emits at a given temperature is its action multiplied by its frequency, and only radiation of this frequency can be emitted by this oscillator. If the temperature is raised, the only way the oscillator can emit more energy is by increasing its own action, which is measured by its amplitude (the size of its oscillations) and which must vary continuously with temperature variation if an oscillator can have any amount of action, as is assumed classically. Using this idea and noting that each oscillator regardless of its frequency must on the average have the same amount of energy, Planck tried to deduce the correct radiation formula for an aggregate of oscillators at a given temperature. He failed in this attempt and noted that no matter what mathematical trick he tried or classical principle he applied, he would always fail. After much travail, anguish, and doubt, Planck had to give up the classical concept that the action of an oscillator (essentially its amplitude) can have any value and replace it with the concept of quantized action. In the next chapter, we see how this concept led Planck to the correct radiation formula and led Einstein to the very fruitful concept of the photon.

CHAPTER 13

Planck's Black-Body Radiation Formula and Einstein's Photon

How often have I said to you that when you have eliminated the impossible, whatever remains, however improbable, must be the truth.

—SIR ARTHUR CONAN DOYLE

In the previous chapter, we saw that the quantum theory of radiation arose because classical physics, as contained in Newton's laws of motion, in thermodynamics, and in Maxwell's electromagnetic theory of radiation, cannot correctly describe "black-body" radition, that is, the continuous spectrum of the energy emitted per second from a 1-cm^2 hole in the wall of a furnace at a given absolute temperature. This spectrum had been studied experimentally for some years before Planck obtained its correct algebraic formula, so that the theoretical physicists, like Planck, who were trying to deduce the formula from first principles knew their goal: to find a formula from which the correct intensity of the emitted radiation at a given frequency for a given furnace temperature can be calculated. This formula for the intensity of the radiation must contain algebraically only the absolute temperature of the furnace and the frequency of the emitted radiation in such a combination that the intensity for a given frequency, as calculated from the formula, agrees with the observed intensity for that frequency.

To understand the problem, we briefly discuss the state of experimental and theoretical "black-body" research when Planck began the theoretical investigation that led to his revolutionary discovery of the quantum of action. The experimental work was quite straightforward and involved nothing more than sending the "black-body" radiation through a prism to spread it out into its various wavelengths ranging from the infrared (long wavelengths) to the ultraviolet (very short wavelengths) and then using a very sensitive photometer to measure the intensity (amount of energy) in each wavelength (actually in each wavelength band). These data are best represented by plotting the intensities along a vertical axis against the various wavelengths along a horizontal axis. Connecting all such plotted points on this graph by a continuous line gives a curve (the famous Planck radiation curve) that shows at a glance the basic observable features of black-body radiation, which fit into our superficial knowledge of the kind of radiation one expects a furnace, at a given temperature, to emit. At the very-long-wavelength end (the infrared and red colors), the curve lies close to the wavelength axis (low intensity) and rises slowly, reaching a maximum value (a hill) for a given color; it then turns around and drops very sharply to zero in the ultraviolet region.

Three features of this graph are important, since they give us some insight into the properties of the radiation that should be deducible from basic theory. First, the total energy emitted per second, in all colors, is given by the area under the plotted curve; as the temperature of the furnace rises, the plotted curve is shifted upward, so that the area under it increases, indicating that the rate at which the furnace radiates increases with absolute temperature. By comparing areas under curves for different temperatures, one finds that the rate at which the furnace radiates energy increases as the 4th power of the temperature. This feature of black-body radiation was first discovered empirically from measurements of the areas by the German experimental physicist Stefan. Later, the great Austrian theorist Ludwig Boltzmann deduced the Stefan formula from classical thermodynamics, so that it is known as the Stefan–Boltzmann law.

Second, the hill on the curve, that is, the wavelength (or the color band) for which the intensity of black-body radiation is a maximum, is shifted to the left (to shorter wavelengths) as the temperature increases, and one can see from the curves for different temperatures that the product of the wavelength of maximum intensity for a given curve multiplied by the temperature for that curve is always the same number. The German physicist Wilhelm Wien deduced this relationship (now known as the Wien displacement law) from Maxwell's radiation laws, but that is as far as theoretical physicists could push the classical theories in their attempt to obtain an algebraic formula that fits the entire black-body radiation curve. This brings us to the third feature of the graph, its exact shape.

The first two features of black-body radiation agree with our superficial observations, which tell us that the hotter a furnace is, the more rapidly it radiates in all colors and the more its radiation changes from infrared (which we can feel but cannot see) to cherry red, and finally to blue white, as the temperature rises from a few hundred degrees absolute to thousands of degrees. The question that is naturally prompted by these observations is why the furnace temperature controls the color (frequency) of the emitted radiation and why, in particular, high temperatures are required for blue colors. Two attempts, one by the late-19th-century physicist Lord Rayleigh and the other by the German physicist Wien, were made to answer these questions using the framework of classical physics, and these attempts were only partially successful.

Lord Rayleigh (1842–1919) was the last of the great classical physicists in Britain. He was born John William Strutt in Witham, Essex, England. His father, who had the same name, was a member of the aristocracy; his mother was the daughter of a decorated soldier. Although Strutt demonstrated his intellectual abilities at an early age, he was a frail child whose education was often interrupted by illness. He attended Eton briefly, but spent most of his time there in the infirmary.[1] Strutt managed to complete three years at a private school at Wimbledon before entering Harrow, but was again forced to withdraw owing to poor health. He completed his preparatory education in four years with the aid of a private tutor. In 1861, he enrolled at Trinity College, Cambridge. Although not as gifted in mathematics as some of his fellow students, Strutt was diligent in his studies and became so proficient in his subjects that he graduated as senior wrangler, in the mathematics tripos, in 1865. Strutt was awarded a fellowship to continue his studies at Cambridge, which he completed in 1871. During that time, he studied mathematics and physics and conducted a number of experiments in electricity and magnetism.

Strutt's scientific work was interrupted by an attack of rheumatic fever that occurred shortly after his marriage in 1871. For a time, it seemed that Strutt would not survive through the year, but he gradually regained his health while on an extended holiday in Greece and Egypt, which included a houseboat journey up the Nile.[2] The change of scenery evidently encouraged him to begin working on his most important book, *The Theory of Sound,* which appeared in two volumes in 1877 and 1878.[2] Strutt's book examined the phenomena of vibration, resonance, and acoustics and "has remained the foremost monument of acoustical literature."[2]

Soon after Strutt returned to England, his father passed away, so Strutt became the new Baron Rayleigh. With the title came the responsibility of managing the 7000-acre family estate, so Rayleigh took up residence there. He built a laboratory near the house so he could continue his scientific work in electromagnetism and acoustics: "Perhaps his most

John William Strutt Lord Rayleigh (1842–1919)

significant early work was his theory explaining the blue color of the sky as the result of scattering of sunlight by small particles in the atmosphere."[2] Rayleigh also combined his scientific knowledge with the most advanced agricultural techniques known at that time to increase the efficiency of the estate. Although he was quite successful as a gentleman farmer, he turned over the management of the estate to his younger brother in 1876 to devote himself fully to science.[3]

In 1879, Rayleigh succeeded Maxwell at Cambridge and took over the administration of the Cavendish laboratory. Rayleigh published a number of important papers while at Cambridge and oversaw an ongoing research program to standardize electrical measurements. He also implemented programs to improve the quality of instruction given to students at Cambridge. However, he found the administrative burdens of his office to be more onerous than rewarding and resigned his position in 1884 to return to his country estate. Soon after his departure, he was elected secretary of the Royal Society. In 1887, he succeeded Tyndall as professor of natural philosophy at the Royal Institution of Great Britain and retained that position until 1905.[3]

Rayleigh's scientific interests were varied. Unlike many of his contemporaries, Rayleigh was free to investigate the subjects that interested him most, since he spent much of his scientific career outside the university environment: "[His] later work ranged over almost the whole field of physics, covering sound, wave theory, colour vision, electrodynamics, electromagnetism, light scattering, flow of liquids, hydrodynamics, density of gases, viscosity, capillarity, elasticity, and photography."[4] Rayleigh preferred to work in several areas of physics at the same time, finding it more invigorating to shift from subject to subject as he chose. Fortunately, his eclectic tastes did not dilute the quality of his research or impair the brilliant clarity of his papers.

Rayleigh is probably most famous for his discovery of argon, which he finally succeeded in isolating in 1895 after years of laborious experiments. Rayleigh shared the credit for the discovery of argon with the chemist William Ramsay, who had discovered it independently, though Ramsay had begun his work only after Rayleigh had published his earlier experimental results. In any event, Rayleigh's discovery resulted in his being awarded the 1904 Nobel Prize in Physics; his counterpart Ramsay received the same award in Chemistry for his role in the discovery.

Rayleigh served as an advisor to the British government in several capacities and also sat in the House of Lords, though he spoke only on rare occasions. He served as Chancellor of Cambridge University and was the recipient of numerous honorary degrees. He also received the Order of Merit and every important honor that the Royal Society could bestow. He published nearly 500 papers during his career and retained his mental faculties to the very end of his life.

Rayleigh, who, with Lord Kelvin, dominated British physics for the last quarter of the 19th century, used a very clever method to deduce the black-body radiation formula by comparing the vibrations of the waves of the radiation in a cavity, at a given temperature, to the vibrations of harmonic oscillators and then assigning to each such vibration the energy of an appropriate harmonic oscillator. To follow Rayleigh's reasoning, we must consider black-body radiation in equilibrium with the walls of a furnace at a fixed temperature, so that the amount (intensity) of such radiation of a given frequency (or wavelength) in each unit volume of the furnace is constant even though the walls are constantly absorbing and emitting radiation of all frequencies. This radiation is present within the cavity in the form of transverse waves, but not advancing waves, which are not bounded in any way, but in the form of standing waves owing to the confining walls of the cavity. A standing wave is produced between two confining surfaces if a wave advancing from one to the other is reflected back and forth between the surfaces so that interference between an advancing and a reflected wave occurs. Such a wave is produced by plucking a tautly stretched string along which the standing wave appears as a series of transverse vibrations (all in phase) separated by equally spaced stationary points; the vibrations are called "antinodes" and the stationary points "nodes." Since the two end points of the string are fixed, the number of equally spaced antinodes for any given frequency is just the length of the string divided by half the wavelength of the wave.

Rayleigh applied this standing-wave construction to the black-body radiation in the cavity by picturing the radiation of a given frequency as consisting of standing waves along a line between any two points on opposite walls of the cavity, if we picture the cavity as a cube. The number of vibrations associated with any standing wave of the radiation is then the length of the side of a cube divided by half the wavelength of the wave. In this way, Rayleigh calculated the number of distinct radiation vibrations a cube of a given volume contains. This favors the high-frequency (short-wavelength) vibrations more than the low-frequency vibrations, because more small half-wavelengths can be arranged along a line of given length than long ones. Arguing that each such standing vibration is a harmonic oscillator, he assigned to each of these oscillators an amount of energy dictated by the theorem of the equipartition of energy, namely, Boltzmann's constant multiplied by the absolute temperature in the cavity. This classical formula for black-body radiation is called the "Rayleigh–Jeans formula," although James Jeans, who proposed a slight correction, had little to do with the original concept. Following Rayleigh's reasoning and his clever derivation of this black-body radiation formula, we see that the intensity of black-body radiation emitted by a furnace at any temperature should, according to the formula, increase with the frequency of the emitted radiation, since the intensity, as given by the formula, is proportional to the square of the frequency. This is clearly false, since it states that all furnaces should blow up with a vast outflow of ultraviolet radiation.

Although the Rayleigh–Jeans formula gives nonsensical results for the high-frequency end of the black-body radiation spectrum, it agrees very well with the low-frequency end, which means that the classical physics of Newton and Maxwell can be used to describe long wavelengths of radiation fairly accurately, but is hopelessly inadequate for high-frequency radiation. The question that immediately arises concerning this breakdown of classical physics is where the flaw is and how it is to be corrected. We saw in our discussion of the gas law that the Newtonian laws of motion also break down; they lead to certain discrepancies between the theoretical and measured specific heats of gases. One may surmise that the same flaw in classical physics causes both discrepancies; we shall see later that this is indeed so. The question then becomes: What is the flaw and how is it to be corrected? We see that it does not lie in the way Rayleigh counted the number of vibratory degrees of freedom associated with the standing waves he assigned to the black-body radiation; the only conclusion left is that it stems from the classical equipartition theorem, which states that in an ensemble of dynamical systems (particles in a gas or standing waves in black-body radiation) at a given absolute temperature, each degree of freedom must have on the average an amount of energy equal to Boltzmann's universal constant multiplied by the temperature. Wien, who proposed his own classical formula for the intensity of black-body radiation, probably understood that something was wrong with the classical equipartition theorem, but he did not produce his formula from that standpoint, since he did not know how to do so. Knowing that the correct formula must suppress high-frequency radiation in the black-body radiation spectrum, he used Boltzmann's statistical physics as a guide to his own formula. Statistical physics (statistical mechanics) was first applied successfully by Maxwell and Boltzmann (independently) to the determination of the average speed of molecules in a gas at a given temperature and to the number of molecules in the gas with a given energy. The analysis shows that this number falls off exponentially as the energy increases, but the rate of this falloff decreases with increasing temperature. This makes sense because it tells us that the higher the temperature of an ensemble, the greater the number of high-energy particles in the ensemble. However, the sum of the energies of all the particles cannot exceed the total energy of the ensemble as dictated by the first law of thermodynamics.

Wien took this idea and applied it to the various frequencies rather than to energies in the black-body radiation; he applied an exponential frequency barrier to the number of vibrational nodes with high frequencies with the temperature. This suppresses the high frequencies, so that his formula correctly represents the black-body radiation at the high-frequency end of the spectrum, but it gives wrong results for the low-frequency end. But physicists were puzzled by the formula, for they could not understand why frequency and energy should behave in the same way in nature and why a statistical formula, which was developed for material particles, is applicable to waves.

This was the state of the theory of black-body radiation when Max Planck began his historic investigations of the subject with the purpose of deducing the correct algebraic formula for the entire spectral range from classical thermodynamics. To accomplish this task, he concentrated on the relationship between the entropy of the radiation and its energy instead of the temperature of the radiation, because he pictured the radiation as being in equilibrium with real harmonic oscillators of varying frequencies in the walls of the furnace. His idea was that these oscillators absorb radiation of any given frequency and then re-emit it over a complete range of frequencies, so that the black-body character of the radiation is maintained. To do this, he had to have some idea of the average energy of an oscillator, and so he assumed that he could calculate this average value by the usual procedure of obtaining averages: The procedure involved adding the energies of all the oscillators, with each given the proper weight (the probability of the oscillator's having that energy), and dividing by the total number of oscillators. The entropy enters into this procedure because, as Boltzmann had shown, the probability of an oscillator's having a given energy depends on the entropy in an exponential way. Since entropy and energy are related through the temperature, Planck was convinced that he could obtain the correct radiation formula through thermodynamics. But to do this, he had to have a formula for the energy of a harmonic oscillator of a given frequency, which is obtained in a straightforward way from Newtonian dynamics; it is the frequency of the oscillator multiplied by its action associated with a single oscillation. Planck called the action h and wrote for the energy $h\nu$, where ν is the frequency of the oscillator. With this equation and the Boltzmann probability formula, he then calculated the average energy of an oscillator, which is not the classical expression obtained from the equipartition theorem.

The classical equipartition theorem states that the average energy of an oscillator in an ensemble at temperature T does not depend on the frequency of the oscillator, but varies directly as the temperature; this leads to the incorrect Rayleigh–Jeans law. Planck's average value for the energy of a harmonic oscillator depends in a complicated way on the frequency and the temperature as long as h, the action, is a finite nonzero value. Planck was unhappy with this condition because he felt it to be a flaw in his basic concepts, even though the black-body radiation formula he obtained with his expression for the average energy of an oscillator is correct. Allowing h to go to zero always led him to the Rayleigh–Jeans law, so he left h finite, which means that the black-body radiation spectrum can be explained only if no action less than h can occur in nature. This means that action is quantized and h, which has a numerical value in basic units of 6.625 divided by one thousand trillion trillion (6.625×10^{-27}) erg seconds, is the unit or quantum of action.

Planck's unhappiness with this result stemmed from his perception that the quantization of action conflicts with Maxwell's electromagnetic theory of radiation, since it means that radiation is corpuscular and not wavelike. To overcome this repugnant idea, Planck did not reject his black-body radiation formula, which agrees in every aspect with his observations, but suggested that black-body radiation is emitted in discrete lumps (quanta) that immediately change into waves. Planck was certain about the correctness of his formula because he had obtained exactly the same formula, some months previously, by simply combining the Rayleigh–Jeans formula and Wien's formula into a single algebraic expression. Planck perceived this to be only a mathematical trick with no physical basis, so when he obtained the same formula using thermodynamics with the additional quantization of action condition, he was convinced of its correctness, even though he did not grasp the full significance of his discovery.

EINSTEIN'S CONTRIBUTION TO THE QUANTUM THEORY

Albert Einstein published a series of remarkable papers on radiation, the first of which appeared in 1905 as one of three revolutionary articles in the same issue of *Annalen der Physik.* The modest title of this paper, "Concerning a Heuristic Point of View About the Creation and Transformation of Light," gives no indication of the great impact it has had on physics and, in particular, on our understanding of the quantum theory and the nature of radiation. Einstein began this paper by pointing out the profound differences between the laws that govern material bodies (Newton's laws of motion) and those that govern radiation (Maxwell's electromagnetic wave theory); he noted that this difference is clearly delineated by the possibility of localizing particles (position and momentum) and the impossibility of localizing waves (radiation). He then indicated that this difference may not be as definitive and sharp as it appears and that certain radiation phenomena can be explained only if the wave theory gives a partial picture of the nature of radiation and that one must consider the possibility that "radiation is distributed discontinuously in space," rather than as an unbroken continuous wave.

Starting from this basic idea, Einstein considered radiation enclosed in a container with perfectly reflecting walls and showed that if the spectral distribution is governed by Planck's radiation formula (that is, if the intensity as a function of the frequency is given by Planck's formula), then the radiation behaves, in every respect (for example, pressure, entropy), like a gas consisting of pellets (quanta), each having an energy $h\nu$ (Planck's constant times the frequency). This clearly established the existence of light quanta, later called "photons," a concept that Planck rejected. Einstein went further, in his analysis of radiation, to show that the fluctuation of the energy of radiation in a small volume of the container (the energy in a small volume changes erratically from moment to moment) is a sum of two terms, one of which can be deduced from Maxwell's wave theory of light (it is the standard classical derivation), but the second of which is pure quantum theory (nonclassical) and stems from the presence of quanta in the radiation. This shows that neither the wave nor the particle (photon) aspect of radiation may be neglected without introducing errors in one's analysis. This is the first example of the wave–particle duality that now permeates all of physics. We say more about this duality in our discussion of quantum mechanics and the wave theory of the electron.

Einstein completed his 1905 paper by showing, in a very simple way, that the quantum theory of light explains the famous photoelectric effect completely. As we mentioned in our discussion of Hertz's experiments that proved the correctness of Maxwell's electromagnetic theory of light, Hertz discovered that when a charged metal sphere is irradiated with ultraviolet radiation, it quickly loses its charge, which does not happen when red light strikes the sphere. With J. J. Thomson's discovery of the electron, it was clear that the ultraviolet radiation tears the electrons from the surface of the sphere and thus discharges it, but it was not clear why red light cannot do the same thing in accordance with the classical electromagnetic wave theory. Since the electrons are swept off the surface of the metal sphere because they absorb enough energy from the radiation to break the bonds that keep them attached to the sphere, they should, according to the classical wave theory, be removed by red light as well as by ultraviolet radiation; classical theory states that the energy transported to the sphere per second depends only on the intensity of the wave (its amplitude and not on its wavelength (or frequency). But the observations deny this theory; no matter how intense the red light is, it removes no electrons, but even a very weak beam of ultraviolet light removes electrons.

Einstein explained this result using the concept of the light quantum (photon): Each electron is removed by absorbing a single photon; hence, this photon must have enough energy to do the work required to remove the electron and to give it its observed kinetic energy. But the frequency of a red photon is too low for it to have enough energy, according to Planck's formula, to do the required work. Hence, no matter how many red photons strike the metal sphere, they remove no electrons. On the other hand, the ultraviolet frequency is large enough for each ultraviolet photon to tear an electron out of its metal base. Einstein pointed out one other feature of the photoelectric process: The electrons emerge from discrete points of the metal surface, indicating that each such point receives a pellet (quantum) of radiation. A wave would spread over the entire surface, so that all points on the surface could emit electrons. Einstein set down a simple equation that gives the frequency that a photon must have to release an electron from the metal surface and give it its observed kinetic energy. He received the Nobel Prize in Physics in 1921 for this equation (completely verified by Robert Millikan's experiments), which he presented in about one page of his 1905 paper.

Most physicists at the time considered Planck's quantum theory as an artificial mathematical technique for explaining the continuous black-body radiation spectrum, without any real physical content; the concept of the photon was particularly repugnant in view of the great successes achieved by Maxwell's electromagnetic wave theory. Einstein, however, accepted the quantum theory as a profound scientific truth, to be applied to all phases of physics. Acting on this belief, he investigated the specific heats of solids and applied the quantum theory to them. He thus explained why the measured specific heats, at very low temperatures, disagree drastically with the values calculated from classical theory. According to the classical theory, one treats a solid as a collection of harmonic oscillators (the molecules are pictured as vibrating elastically owing to the forces of attraction among them which keep them attached to each other) and assigns to each mode of vibration the classical equipartition average energy (Boltzmann's constant multiplied by the absolute temperature). This gives the incorrect value for the specific heat at low temperatures. Einstein obtained the correct value by assigning the average energy value to an oscillator as calculated from the quantum theory.

In his last great paper on radiation in 1917, Einstein pointed out two remarkable features of the photon that have had a profound influence on physics and on optical technology. In this paper, he deduced the Planck radiation formula without referring to black-body radiation, but simply by considering the emission and absorption of radiation by an electron in an atom. He accepted Bohr's theory of the atom, with its electrons emitting and absorbing radiation by jumping from one to another of a series of discrete, permissible orbits, and assumed that in each such jump, a photon of a definite frequency is either emitted or absorbed by the jumping electron. To complete his derivation of the Planck formula by this method, he had to assume not only that an electron jumps spontaneously from a higher to a lower orbit with the emission of a single well-defined photon, but also that if such a photon passes the electron before it has a chance to jump down, this photon stimulates the electron (it gives the electron a push, so to speak) to jump, forcing it to emit an identical photon moving in exactly the same direction as the passing photon so that the two photons move off together, intimately bound to each other. Einstein called this the stimulated emission of radiation, which is the basis of the laser, an acronym for "light amplification by stimulated emission of radiation."

In discussing the concept of the stimulated emission of radiation, Einstein pointed out that one must assign momentum as well as energy to a photon to give it all the properties of a particle. However, the momentum of the photon cannot be its mass times its velocity, since it has no mass in the usual sense of the term. This presented no difficulty to Einstein, who defined the momentum of the photon as its energy (Planck's constant times its frequency) divided by its speed (the speed of light), or Planck's constant divided by the wavelength of the photon. This formula was later completely verified by Compton's famous experiment on the scattering of light by electrons, which we discuss later.

All of quantum theory and the quantum mechanics stems from the formula for the energy of a photon—Planck's constant times the photon's frequency—which was presented to physicists as a sort of gift without any attempt to deduce it from basic principles. But one of the authors (Motz) has shown, in a previously published paper, that if one assumes the existence of photons, one can derive the Planck formula from classical thermodynamics and the Doppler effect. Picture compressing radiation in a container with perfectly reflecting walls by slowly pushing in a piston and thus doing work. This work appears as internal energy of the radiation that is equally distributed among all the photons, so that each photon's energy is increased by the same percentage (the same small fraction) of its initial energy if the piston is pushed in a very small distance. At the same time, the frequency of each photon increases by the Doppler effect as the photon is reflected from the moving piston, and the percentage increase of the frequency exactly equals its percentage energy increase. Hence, the ratio of the energy of any photon to its frequency must be a universal constant known as Planck's constant.

CHAPTER 14

Experimental Physics at the Close of the Nineteenth Century

The true worth of an experimenter consists in his pursuing not only what he seeks in his experiment, but also what he did not seek.

—CLAUDE BERNARD

A curious dichotomy divided physicists at the end of the 19th century into those who believed that physics had reached a kind of ultimate state, with very little left to be done from a fundamental point of view, and those who were troubled by certain unexplained discrepancies between classical theory and experiment. The experimentalists were generally in the first group, because they saw in Newtonian physics and Maxwell's electromagnetic theory of light the completion of theoretical physics; as far as they were concerned, physics was essentially only a matter of collecting increasingly accurate data about phenomena that could be fully explained by the known laws. The troubled physicists, those in the second group, were primarily theoreticians like Planck and Einstein, who saw flaws in classical physics that could be explained only by revolutionary ideas.

The contented physicists in the first group appeared to have the best of the controversy before the year 1900, for the successes achieved by experimental physics were intoxicating because they all seemed to fit into either Newtonian or Maxwellian theory. Such difficulties as the failure of the Michelson–Morley experiment to detect the earth's motion around the sun using beams of light, the discrepancy between the measured and calculated specific heats of gases, the failure of classical physics to account for the black-body radiation spectrum, and the failure of Newtonian gravity to give a correct orbit for Mercury were considered minor matters that presented no threat or challenge to classical physics. A few years later, with Max Planck's discovery of the quantum of action, the magnificent intellectual edifice of classical physics began to crumble; its destruction was completed some five years later with Albert Einstein's promulgation of his special theory of relativity, which, together with the quantum theory, revolutionized the way in which physicists probe the universe. Before we discuss the theory of relativity and how it has changed physics and our concept of the universe, we consider the important experimental discoveries in the last quarter of the 19th century and the first decade of the 20th century that ushered in the atomic theory of matter.

A number of physicists and chemists in the 19th century developed atomic models to account for the structure and properties of matter, before conclusive experimental evidence for the atom, such as the existence of the electron and proton, was available. In 1802–1803, the British chemist John Dalton published a series of papers propounding his atomic hypothesis, which introduced the concept of interatomic forces: All matter consists of different kinds of atoms (elementary particles with different masses) that attract each other to combine in appropriate numbers and kinds and thus form all the known compounds. Since Dalton did not indicate the nature of the interatomic forces, his theory did not lead to any new developments, and so it remained fallow. Nevertheless, chemists used it to explain simple chemical reactions and to strengthen the molecular concept, which was extremely useful in the study of gases. In fact, Joseph-Louis Gay-Lussac, who, with Jacques Alexandre César Charles, had discovered the law of gases, showed that Dalton's atomic theory explains the quantitative aspects of the chemical reactions of gases: In any chemical reaction involving two different gases, the ratio of the quantities of the two gases that combine to form a definite quantity of a third gas is always the same, no matter how much of the third gas is formed. To Gay-Lussac, this meant that each molecule of any one of the three gases consists of a fixed number of Dalton's atoms.

A big step in the development of the idea of the existence of a primordial substance, of which all matter is constituted, was taken by William Prout, who suggested in two papers in 1815 and 1816 that all atoms consist of combinations of different numbers of hydrogen atoms (Prout's hypothesis). This concept, which gives hydrogen the primordial role, was based on the observations, made by many chemists skillful in the technique of measuring the atomic weights (masses) of the known chemical elements, that these weights are all very nearly integer multiples of the atomic weight of hydrogen. To Prout, this underlying regularity in the atomic weights meant that all atoms have a common base, hydrogen, and differ only in the number of hydrogen atoms they contain. Although this idea was very attractive, it was not readily accepted and, in fact, remained on the outskirts of physics until the work of Ernest Rutherford and Niels Bohr placed atomic theory on a sound basis.

The first big break in the barrier to atomic theory came with Michael Faraday's important electrochemical experiments, which were discussed in Chapter 10. These experiments showed clearly that matter consists of heavy positive particles (the nuclei) of atoms and negative particles (the light electrons), but it was not clear just how these two kinds of particles combine to form an atom; indeed, the very idea of the electron as a particle was not known and did not come until Sir Joseph John Thomson isolated a negative elementary charge that was later dubbed the ''electron.''

On the chemical side of the atomic theory, the most important accomplishment was Dimitri Mendeleev's discovery of the periodic table of the chemical elements, which exhibits the periodicity of the chemical properties of the elements. To explain Mendeleev's discovery, we define the atomic number and the atomic weight (mass), which increase as we go from the light to the heavy elements. The atomic number is the position assigned to an element in the table, starting with hydrogen (in the first position with atomic number 1) and proceeding one step at a time to the heavier elements. Thus, helium, in the second position, has atomic number 2; lithium, in the third position, has atomic number 3; and so on. The atomic mass (or weight) is the number of times the mass of an atom of any element is greater than that of an atom of hydrogen, which is assigned the atomic mass 1. Roughly speaking, the atomic mass (at least for the light elements) is about twice the atomic number. Mendeleev discovered that the chemical properties of elements with atomic numbers differing by 8, 18, 32, and so on have very similar chemical properties and can therefore be grouped in chemical families (as shown in the table of chemical elements) and arranged in such a way that elements differing in atomic number by 8, 18, 32, and so on are arranged in vertical columns, such as lithium, sodium, potassium, and so on (the alkali metals); fluorine, chlorine, bromine, and so on (the halogens); and helium, neon, argon, and so on (the noble gases). This numerical regularity was very useful to the late-18th-century and early-19th-century chemists and very provocative to the physicists, who saw in it an indication of some kind of atomic structure, but had no way of deciphering the riddle of the periodicity contained in Mendeleev's table without some knowledge of the basic constituents of the atom, which began with Thomson's discovery of the electron in 1897.

In his early experiments, as far back as the 1830s, Faraday had studied cathode rays (particles emitted from the cathode—the negative electrode) in a vacuum tube across which (via two electrodes in the tube—the anode and the cathode) he had placed a high voltage. The cathode rays, as the emanations from the cathode were called (although it was not known then whether these emanations are electromagnetic waves or particles), were studied with ever-increasing care and technical sophistication as it became possible to obtain ever-better vacuums in the discharge tubes.

Using the best vacuums that were then possible and the most advanced tools, Thomson showed that cathode rays are indeed material particles carrying a negative charge. He accomplished this task by placing the cathode-ray tube between two condenser plates and noting that the cathode rays are deflected from the negative to the positive plate. He then measured the magnitude of the deflection of a beam that traveled the length of the tube, which depends on how fast the particles in the beam travel and on the ratio of the charge e on a particle to its mass m—that is, on the quantity e/m. By placing a magnetic field across the tube perpendicular to the velocity of the deflected cathode-ray particle, he brought the deflected beam back to its original straight-line path by making the magnetic field (which he could easily control) strong enough. Since the electrostatic force exerted on the particles by the condenser plates must then equal the force produced by the magnetic field, he equated the two forces and thus eliminated the particles' velocity. This left in his equation only the quantity e/m (unknown) and the length of the tube, the strength of the electric field, and that of the magnetic field (known). He thus obtained a value for e/m that is larger than Faraday's e/m for the hydrogen ion by a factor of about 1836. Thomson called these particles "electrons," and in 1905, Robert Millikan, using his famous oil drop experiment, found their charge to be 4.77 divided by ten billion (4.77×10^{-10}) electrostatic units. The work of Thomson and Millikan was the first step in constructing a correct model of an atom.

From the measured value of the charge e on the electron and Thomson's measured value of its e/m, its mass can be calculated as the first of these measured quantities divided by the second; its value in grams is 9 divided by ten thousand trillion trillion (9×10^{-28}). Simply expressed, this means that a clump of about one thousand trillion trillion electrons has a mass of 1 gram. These numbers tell us that the electrons contribute very little to the total mass of an atom or a collection of atoms. Nevertheless, electrons are absolutely essential for the structure of neutral (uncharged) atoms, for they contribute the negative electric charge to compensate the positive charge in an atom and thus to make the atom electrically neutral.

As we shall see, the electrons in an atom account for the atom's chemical properties; all chemical reactions are generated and controlled by the electrons.

Shortly after electrons were discovered as the negatively charged constituents of cathode rays, Thomson and others discovered "anode rays," which are positively charged particles moving at much lower speeds than the electrons from the anode to the cathode in the cathode ray tube. The e/m's for these particles were thousands of times smaller than that of the electrons and depended on the nature of the anode. The largest e/m was found for the positively charged particles coming from an anode coated with a hydrogen compound; these were the protons, with a mass about 1840 times that of the electron and a positive charge equal in magnitude to the electron's charge. The protons and the later-discovered neutrons account for the atoms' masses.

With the discovery of the electron, theoretical physicists began to construct atomic models to account for the observed properties of matter, but severe and what appeared to be insurmountable difficulties plagued all such attempts until Niels Bohr constructed the famous "Bohr model" of the atom. The great Dutch theoretical physicist Lorentz did the most notable, and partially successful, theoretical work in incorporating the electron, as a basic constituent of matter, into the Newtonian–Maxwellian physics of the time. He began by changing (extending) Maxwell's equations of the electromagnetic field to include the interaction of Maxwellian electromagnetic waves (as well as changing electric and magnetic fields) with fundamental, charged particles, which he called electrons. This mathematical feat occurred just before Thomson discovered the electron, so that Lorentz's theoretical "electron" was not identified with Thomson's experimental electron until about 1899. But that did not matter as far as Lorentz's theoretical work was concerned; its validity did not depend on Thomson's discovery.

Hendrik Antoon Lorentz (1853–1928) was born in Arnhem, Holland, into a family that had shown no particular talent for the sciences. His father was a nursery owner who tried to provide a stable household when Hendrik's mother died in 1857. By the time his father remarried five years later, Hendrik had already begun to demonstrate his aptitude for the sciences in grade school. He was a gifted student in the Arnhem high school, where he concentrated on the classics and mathematics. Hendrik entered the University of Leyden in 1870 and obtained his baccalaureate degree in mathematics and physics the following year. He taught classes at night to support himself until he obtained his doctorate in physics at the

Hendrik Antoon Lorentz (1853–1928)

age of 22; his thesis dealt with the nature of light and marked the beginning of his work in optics and electricity, which led him to the concept of the electron.

In 1878, Lorentz accepted the chair of theoretical physics at Leyden, where he remained throughout his career. That same year, he published "an essay on the relation between the velocity of light in a medium and the density and composition" of the medium which is now known as the Lorenz–Lorentz formula.[1] He studied Maxwell's equations of the electromagnetic field and searched for ways that this magnificent mathematical edifice could be extended to other areas of physics. At that time, he was not alone in his belief that the twin tools of Newtonian mechanics and Maxwell's equations for the electromagnetic field were the only tools needed to probe the universe, since the only two known factors in nature at that time were gravity and electromagnetism.

Lorentz made substantial contributions to the electrodynamics of moving bodies, but did not discard the hypothesis that there is an all-pervasive ether acting as a medium for all electromagnetic waves in space: "From Lorentz stems the conception of the electron; his view that this minute, electrically charged particle plays a role during electromagnetic phenomena in ponderable matter made it possible to apply the molecular theory to the theory of electricity, and to explain the behavior of light waves passing through moving, transparent bodies."[1] Lorentz also found that the electromagnetic forces between charges become altered when the charges are moving, causing electrons to become slightly compressed; this so-called "Lorentz transformation" proved to be a consequence of Einstein's special theory of relativity.[1]

Apart from his theoretical work, Lorentz also presided over a committee formed to study the movements of the seawater along the shores of newly reclaimed areas of the country.[2] In the course of his administrative career, Lorentz made a number of calculations that have continued to be of substantial value to the field of hydraulic engineering. He chaired the Solvay Conferences, which served as forums for the great physicists of the world to discuss and debate the newest ideas in physics, and in 1923 he was one of seven world-renowned scholars elected to the International Committee of Intellectual Cooperation of the League of Nations.[2] Lorentz also used his considerable influence to bring about the creation of an organization to direct applied scientific research in Holland so as to coordinate existing research programs.

Lorentz extended Maxwell's equations by including a term that describes the current arising in a conductor owing to magnetic induction as the flow of electrons produced by the force of the changing magnetic field on the free electrons in the conductor: This term, now called the "Lorentz force," takes into account not only the force of the electric field on the electron (which is just the Coulomb force) but also the interaction of the magnetic field with the electron as mentioned above. Lorentz assumed the magnitude of this magnetic interaction to be the product of three terms: the electric charge on the electron, its speed divided by the speed of light, and the strength (intensity) of the magnetic field at the electron's position. The direction of this force is at right angles to the velocity of the electron and the direction of the magnetic field at the position of the electron. This Lorentz force has stood the test of time very well and is universally accepted as correctly describing the interaction of any electric charge with an electromagnetic field.

Lorentz extended his theory of the electron to explain a series of phenomena that cannot be understood if the concept of a basic electric charge as a constituent of all matter is not accepted. Lorentz's first success was the explanation of the so-called "normal Zeeman pattern" of the spectral lines of atoms placed in a magnetic field; this pattern consists of the usual spectral pattern (no magnetic field) accompanied by additional spectral lines close to the original lines. Lorentz explained these additional lines by showing that the motions of the electrons that produce the original pattern are altered by the magnetic field so that new lines are produced. This dramatic success of Lorentz's theory gave the concept of the electron enormous appeal and respectability.

Lorentz also explained such phenomena as electrical conductivity (electrons move about freely in a conductor, but not in an insulator), heat conductivity, refraction, the scattering of light, and other properties of light in a medium with his electron theory, but despite all these successes, a correct model of the atom, incorporating protons and electrons, eluded theoreticians. But other exciting experimental discoveries came rapidly as the 19th century was ending.

One set of experiments dealt with the penetrating electromagnetic rays discovered accidentally in 1895 by Wilhelm Roentgen (x rays, also called Roentgen rays), and the other set dealt with Antoine-Henri Becquerel's discovery in 1896 of very energetic rays (electromagnetic and particle) emitted spontaneously by heavy atoms (radioactivity). Roentgen, along with many other experimental physicists, was working with various types of discharge tubes (essentially cathode-ray tubes or Crookes tubes, such as Thomson had used in his discovery of the electron) and noticed that a piece of black cardboard with which he had covered the tube and which was opaque to ordinary light was transparent to some other kind of radiation coming from the tube. He detected this penetrating radiation when he observed that a piece of paper, coated with barium platino-cyanide, acquired a black streak across it from some kind of emanation from the tube when either the coated or the uncoated surface of the paper faced the tube. Moreover, the coating was affected by the radiation even if the paper was placed at a distance of up to 6 meters from the tube. In a darkened room, Roentgen observed that the barium platino-cyanide fluoresced every time the discharge tube (Crookes tube) was activated. Later, Roentgen showed that these rays can penetrate many different substances and that their penetrating power diminishes rapidly with increasing density of the material they strike. He discovered also that they ionize the air through which they pass, ripping electrons out of the atoms they encounter.

Becquerel's discovery of the phenomenon of radioactivity was in some sense the culmination of three generations of family research on phosphorescence and fluorescence. His grandfather was the physicist Antoine-César, who had made several important discoveries in electrochemistry and been awarded the Copley Medal by the Royal Society. His father, Alexander-Edmond, was a renowned and prolific physicist who published useful papers on his field of study for nearly half a century.[3] Both men were members of the French Academy of Sciences; each held the chair of professor of physics at the Musée d'Histoire Naturelle during his academic career, a tradition that was continued by Antoine-Henri and, later, his son, Jean Becquerel.

Because of the positions held by both his father and grandfather, it was inevitable that Henri would become familiar at an early age with the laboratory in which his father worked and the daily routine of the teaching scientist. Henri was born in 1852 in Paris and received his early formal education at the Lycée Louis le Grand. He entered the Ecole Polytechnique in 1872 before his 20th birthday and, two years later, began his studies at the Ecole des Ponts et Chaussées, where he spent four years receiving a thorough grounding in mathematics and civil engineering. Henri joined the staff of the Administration of Bridges and Highways as an engineer in 1877, one year after becoming *démonstrateur* at the Ecole Polytechnique and one year before he became *aide-naturaliste* at the Musée d'Histoire Naturelle. Because both positions involved working closely with his father, it is fortunate that Henri was a motivated young man who was highly conscious of his distinguished lineage and determined to perform whatever arduous tasks were deemed necessary by his father. Indeed, Henri appears to have considered his own discovery of radioactivity to be the logical conclusion of the work of his father and, like Newton before him, believed that his own achievements were due in large part to his having stood on the shoulders of others—namely, his father and grandfather.

Becquerel's early research dealt with the effects of magnetic fields on polarized light as well as the absorption of light by crystals.[3] His optical researches enabled him to obtain his doctorate in 1888 from the Faculty of Sciences of the University of Paris, and he was elected to the Academy of Sciences the following year. After the death of his father in 1891, Henri succeeded to his father's chair at the Musée and began giving lectures in physics at the École Polytechnique. At the same time, he continued to work as an engineer at the Ponts et Chaussées and was named chief engineer at that institution in 1894.

Becquerel's discovery of radioactivity did not occur in a vacuum, but was given its initial impetus by Roentgen's discovery of x rays, which was

Antoine-Henri Becquerel (1852–1908)

announced at the beginning of 1896. That x rays are not charged particles was deduced from their passage through magnetic and electric fields in straight lines without deviation; the paths of charged particles deviate from straight lines in such fields. By carefully observing the source of the x rays in the discharge tube, Roentgen discovered that they are emitted from a small bright spot on the metal surface struck by the cathode rays (by the electrons coming from the cathode), and this explained the mystery, in full accord with Maxwell's electromagnetic theory and Lorentz's electron theory, according to which a charged particle radiates electromagnetic waves when it is decelerated. This deceleration occurs when the electron strikes the metal surface; it radiates away its energy in the form of electromagnetic radiation.

Taking the quantum theory into account, one can easily calculate the energy (or the frequency) of the x ray emitted. If the electron, on striking the metal surface in its path, transforms all its kinetic energy into a single quantum of energy (an x-ray photon), one obtains this photon's frequency by dividing the electron's kinetic energy by Planck's constant h (in a sense, this is just the reverse of the photoelectric effect—electrons produce photons on striking a metal surface instead of photons producing electrons).

This simple quantum relationship between the frequency (energy, which is just frequency times Planck's constant) of the x-ray photon and the kinetic energy of the electron producing it permits one to alter the energy of x rays produced by altering the kinetic energy of the cathode rays producing them, and this can be done by altering the voltage across the discharge tube.

Roentgen's discovery of x rays prompted Becquerel to perform a series of experiments with fluorescent molecules, which either fluoresce spontaneously or are stimulated to fluoresce by absorbing sunlight. Such phenomena were well known, but the fluorescent mechanism was a complete mystery. Becquerel thought that this mechanism was somehow related to x-ray production. In describing his x-ray observations, Roentgen had spoken of the "fluorescence" of the barium plantino-cyanide coating, and so Becquerel assumed that x rays are somehow related to fluorescence. This assumption was wrong, but the idea initiated the series of observations that led to his discovery of radioactivity.

Becquerel had worked with phosphorescence and fluorescence (the terms are used interchangeably) for some years previously and was convinced that phosphorescent substances, if properly treated, electrically or by exposure to sunlight, produce x rays. But testing several such substances over a period of weeks showed no x rays. Becquerel did not give up, however, and was rewarded for his persistence when he worked with the complex uranium crystal potassium uranyl disulfate, which he placed on a photographic plate, protected from sunlight by two sheets of very thick black paper. He then placed the whole thing in bright sunlight to "stimulate the phosphorescence" of the crystal and discovered that the part of the photographic plate on which the crystal lay had become blackened by rays that had passed through the thick black sheets of paper in which the photographic plate was wrapped. Becquerel accepted this result as strong support for his incorrect assumption that fluorescence or phosphorescence generated penetrating rays similar to Roentgen rays. A few days later, however, on examining a photographic plate near some of his potassium uranyl disulfate that had not been irradiated by sunlight but had been placed in a drawer waiting for a sunny day, he discovered that the photographic plate had been darkened by some radiation from the crystal, just as in his previous experiment. Becquerel understood immediately that phosphorescence or fluorescence had nothing to do with the rays that came from his crystals, and he later traced the radiation to the uranium atoms in the crystal. Becquerel's detective work culminated with his discovery of the radioactivity of heavy atoms. Since very little was known about the structure of the atom, nothing could be said about how the atom produces these Becquerel rays, which remained for Lord Ernest Rutherford to discover. Since his discovery did not raise much of a stir in scientific circles, Becquerel lost interest and went on to far less important research, not understanding the revolutionary nature of his discovery.

Because Becquerel's talent lay primarily in experimental physics and was perhaps best exemplified by his talent for carrying out tedious experiments, he resisted the temptation to construct a theoretical explanation for the phenomenon of radioactivity, preferring to leave that task to others. While it is undeniable that there is a mutual interdependency between theory and experiment, Becquerel seems to have had no delusions about his ability to formulate an intellectual edifice that would explain the radioactivity of particles. In any event, his discovery did eventually get the scientific recognition it deserved, even though it occurred only after Marie and Pierre Curie had shown the importance of radioactivity in nuclear physics. Becquerel was awarded the Rumford Medal of the Royal Society and was made a foreign member of the Royal Society in 1900. He received honorary doctorates from Oxford and Cambridge and shared the 1903 Nobel Prize in Physics with the Curies: "It was an appropriate division [because] Becquerel's pioneer investigations had opened the way to the Curies' discoveries, and their discoveries had validated and shown the importance of his."[4] Becquerel received what may have been his most personally rewarding honor when he became the president of the Academy of Sciences in 1908. Unfortunately, Becquerel enjoyed his office for only a few months; he died of a heart attack on August 25, 1908, while on holiday at Le Croisic, Brittany, France.

Although Becquerel's work on radioactivity made no great impression on established physicists initially, it stimulated two of the most brilliant members of the younger generation of physicists, Marie Curie

Marie Sklodowska Curie (1867–1934)

and Ernest Rutherford, to continue Becquerel's work. Marie Sklodowska Curie (1867–1934) was born in Warsaw, Poland. Her father taught mathematics and physics in a government secondary school, and her mother managed a girls' boarding school. Despite the long hours worked by both parents, there was barely enough money in the household to pay for necessities, because Marie's father was denied the opportunity to apply for the more lucrative teaching positions because of his political beliefs. Because of their strained finances, the Curie family did not have access to good medical care. Marie's eldest sister Sophia died of typhus in 1876 and her mother died two years later after a long bout with tuberculosis.[2]

Marie's father was forced to take in boarders to earn some extra money, so Marie often slept in the living room. Despite the less than ideal study conditions, Marie was a diligent student who won a gold medal in Russian while at high school, even though the Czarist government was responsible for the political repression in Poland at that time. She supplemented her formal education by reading many literary and political works by authors ranging from Dostoevsky to Marx.[3] In 1886, Marie became a governess for the children of a wealthy Polish administrator who lived near Pzasnysz to pay for her sister Bronia to study in France. Three years later, she returned to Warsaw to be closer to her family. By this time, she had developed an interest in chemistry, and her sister Bronia had become a medical doctor in France. Bronia insisted that Marie come to Paris, so Marie packed her few belongings and traveled by train to France in 1891. She received a scholarship and began studying mathematics and physics at the Sorbonne. Although she originally lived with her sister and her sister's husband, she found that the distance she needed to travel to the Sorbonne and the frequent interruptions of her studies at home necessitated that she rent a small room near the university. Because Marie had to live on a tight budget of 100 francs a month, she often had to go without food or coal for her stove even during the coldest months of winter. She fainted from hunger on more than one occasion, and she suffered various illnesses because of her malnourishment. Despite her personal hardships, however, Marie pursued her studies with a single-minded dedication and passed the grueling license examinations in physics and mathematics with honors in 1893 and 1894, respectively. During this time, she met Pierre Curie, who had made a name for himself investigating magnetic phenomena and discovering "piezoelectricity" and worked as the chief of the laboratory at the School of Physics and Chemistry in Paris. Within a year, they were married, and Marie joined Pierre in his laboratory. She heard of Becquerel's work and decided to devote her life to the study of radioactivity. In this effort, she was supported by her husband Pierre, and they began a systematic investigation of the radioactivity of the heavy elements. They not only confirmed Becquerel's discoveries but also made some important new discoveries, namely, that thorium oxide is even more radioactive, gram for gram, than metallic uranium, and that the intensity of the rays emitted increases with but is not proportional to the concentration of the uranium or thorium in a compound. This nonproportionality puzzled physicists for some time until they found the correct answer—that some of the rays emitted by a radioactive atom are reabsorbed by other atoms in the materials that have already emitted rays. The Curies also showed conclusively that radioactivity is a process of individual atoms and not a collective process; each atom emits a single ray and no more, so that radioactivity is an intrinsic property of the atom.

The Curies' greatest success came, however, with their discovery of the radioactivity of radium and their isolation of that element from pitchblende. Marie Curie had noted that the mineral pitchblende, a uranium oxide mineral, is far more radioactive than an equal amount of pure uranium. She therefore concluded that pitchblende contains a far more radioactive element than uranium. With Pierre's help, she verified this assumption by isolating radium from pitchblende in 1898, a momentous event that gave birth to the science of radiochemistry.

The Curies carried out their work under difficult conditions that can not be appreciated by a chemist working in a modern laboratory. Their researches on polonium and radium had necessitated that each of the elements be isolated and their atomic weights determined, a task that was accomplished only by processing tons of pitchblende ore. After the ore was processed, the Curies had to carry out a number of chemical investigations, but the absence of any space in the laboratory meant that they had to set up a makeshift laboratory in an abandoned wooden shed across a courtyard from the laboratory Marie had previously used: "It was in these quarters, stifling hot in summer and freezing cold in winter, that the fanatically dedicated couple endured forty-five months of almost unremitting labor to prepare a sample of pure radium chloride and to determine its atomic weight."[7] The herculean effort left them both exhausted, especially since they had taken on additional teaching duties during that time to supplement their meager income.[7] Marie did not realize it then, but her persistent anemia and physical exhaustion were due to her exposure to the debilitating effects of the radiation.

Although Marie had begun her work in radioactivity to complete her doctoral requirements at the Sorbonne, she had neglected to go through the formalities of applying for a degree. The obvious merit of her work convinced the administrators to relax the formal requirements; they gave her the doctoral examination in June 1903 and soon thereafter conferred the degree. What is especially remarkable is that within six months of receiving her degree, Marie, along with her husband Pierre, shared in the 1903 Nobel Prize in Physics with Henri Becquerel. However, both Marie and Pierre were too ill to go to Stockholm to the formal ceremony. Even if they had been healthy, it is not certain that the Curies would have made the trip to Stockholm, because they abhorred receptions and social functions, believing them to be an unjustifiable distraction from their work in the laboratory.

The next few years saw the Curies continue their investigations of the properties of radium. In 1905, a chair in physics was created for Pierre at the Sorbonne. Soon afterward, he was elected to the Academy of Sciences. Unfortunately, he was unable to enjoy his new status for very long, as he was killed on April 19, 1906, by a horse-drawn van while trying to cross a street in inclement weather. His death at the age of 46 brought even greater pressures on Marie, who had to raise the couple's two young daughters by herself.

Despite the conservative attitudes of several of the faculty members of the Sorbonne who questioned whether any woman was fit to be a physics professor, Marie was chosen by a unanimous vote of the Faculty Council to succeed to Pierre's chair at the Sorbonne, where she continued her work in the laboratory.[5] Her appointment was something of a milestone, for it was the first time a woman had been elected to teach at the Sorbonne. Even though she did not enjoy her classroom lectures very much, believing herself better suited for conducting experiments in the laboratory, her use of demonstration experiments coupled with her worldwide fame helped to ensure the popularity of her lectures.

Marie continued her selfless dedication to science, declining the Legion of Honor in 1910, as she had an earlier pension offer by the Ministry of Education.[6] She was persuaded to stand as a candidate for the Academy of Sciences, but a scandalous newspaper article alleging that she had had an affair with an assistant caused the academy to reject her application for admission in favor of another, relatively undistinguished scientist.[6] Undaunted, Marie turned her attention to helping to create a radium institute and to establishing official standards for radium. The 1910 Radiology Congress in Belgium charged her with the task of processing 20 milligrams of radium metal, which was to be deposited with the International Bureau of Weights and Measures in Paris.[6] That same congress also honored the memory of her husband by defining the curie as a unit measure of radioactive emanation from radium.[6]

In 1911, Marie received the Nobel Prize in Chemistry, becoming the first person to receive two Nobel Prizes. Most of the money was used to fund research projects. In 1914, the Conseil de l'Institut du Radium was formed, and its building was completed later that year.[6] Marie served as a member of the Conseil, but she found the administrators to be unsympathetic to many of the research projects she proposed. Although she was soft-spoken and almost shy around nonscientists, the outbreak of World War I and the heavy medical casualties suffered by the French armies bearing the brunt of the German invasion in northern France convinced Marie to make public pleas for funds to equip ambulances in the field with radiology equipment. The funding effort was successful, and Marie was elected by the Red Cross to be the official head of its Radiological Service.[6] With the help of her daughter Irène, Marie devised advanced courses in radiology and taught doctors new techniques for locating foreign objects in the human body.[6]

Following the Armistice in 1918, the Radium Institute began to operate, and Marie joined its staff. In 1921, a private fund-raising campaign in the United States solicited enough contributions to purchase 1 gram of radium, which was officially presented to Marie by President Warren G. Harding during her visit to the United States that same year.[7] Although Marie never cared for the publicity that had followed her since the receipt of her first Nobel Prize, she was deeply touched by the presentation, as it had been instigated by a journalist who was impressed with Marie's sense of dedication and national service.[7]

Marie spent the remaining years of her scientific career supervising the expanding staff at the Radium Institute, which by 1933 included 17 different nationalities.[7] However, she was forced to cut back on her official duties owing to poor health. She underwent four cataract operations and often suffered from lesions on the fingers. Owing to her inability to care for herself adequately, Marie entered a nursing home in Paris.[7] Her health continued to decline, and she died peacefully on July 4, 1939. Although she was given many honors and tributes after her death, perhaps the most moving was one offered by Albert Einstein, who wrote: "Her strength, her purity of will, her austerity toward herself, her objectivity, her incorruptible judgment—all these were of a kind seldom found joined in a single individual. . . . The greatest scientific deed of her life—proving the existence of radioactive elements and isolating them—owes its accomplishment not merely to bold intuition but to a devotion and tenacity in execution under the most extreme hardships imaginable, such as the history of science has not often witnessed."[8]

Ernest Rutherford (1871–1937), who brought order out of the chaos of the apparently unrelated radioactive phenomena that the research of the Curies had discovered, became one of the greatest experimental physicists of all time. He was born at Brightwater, near Nelson, on the north coast of South Island, New Zealand. Ernest was the fourth of 12 children; both his parents were first-generation New Zealanders, having been brought to that

Sir Ernest Rutherford (1871–1937)

country as children from Scotland by their parents. When Ernest was four, the family moved about a dozen miles to the town of Foxhill, where Ernest received his early primary education. The family stayed in Foxhill until 1882, when they moved to the town of Havelock, where Ernest won an academic scholarship that made it possible for him to enroll at Nelson College.[9] Although Rutherford was particularly interested in science and mathematics, he excelled in all subjects. His fine academic performance at Nelson earned him a scholarship to Canterbury College at Christchurch, New Zealand, where he earned both his baccalaureate and master's degrees, the latter with honors in mathematics and physical science.[10] During this time, he constructed a homemade apparatus in his basement and investigated the propagation of electromagnetic waves. Rutherford also won a two-year scholarship to study at Cambridge as a research assistant; he was digging potatoes in a field when his mother told him that he had won the fellowship. On hearing the news, he reportedly tossed the shovel aside and said, "That is the last potato I will dig."[10]

Rutherford began his work on x rays in 1896 as a research assistant to

J. J. Thomson at the Cavendish Laboratory. He was particularly interested in the ionization of atoms by x rays and the recombination of such atoms with free electrons. It was therefore quite natural for him to go on to study the ionizing properties of Becquerel rays; his most important radioactivity discoveries stemmed from this work. He showed, first, that the Becquerel rays (the radiation from radioactive atoms) are not homogeneous but consist of at least two different components, which he called "alpha rays" and "beta rays," the alpha particles being the components in the radiation from uranium. He found the alpha rays to be far less penetrating than the beta rays and both rays to consist of charged particles. Later, he showed that the beta rays are electrons and the alpha rays are positively charged (doubly ionized) helium atoms (the helium nuclei). The difference in the speed with which the alpha and beta rays are emitted stems from their mass difference; the alpha particle mass is about 8000 times that of the beta particle mass. From his analysis of the behavior of alpha and beta particles in electric and magnetic fields, Rutherford deduced their electric charges, which are very close to today's accepted values.

Rutherford spent several years working with Thomson at Cambridge. In 1898, he was offered a professorship at McGill University in Montreal, Canada, and, despite some misgivings at leaving what was then perhaps the finest laboratory in the world and his mentor Thomson, who was then among the foremost investigators of atomic phenomena, he accepted the job. Soon after he began working at McGill, Rutherford discovered that the radioactivity of thorium decreases exponentially over time.[11] At this point, he met Frederick Soddy, a talented chemist who had recently joined the chemistry department as a demonstrator. With the help of Soddy, Rutherford soon determined that atoms are not stable, as had long been supposed by chemists, and that radioactive elements undergo a series of transformations, as indicated by radioactive emanations.[12] By a series of experiments with thorium nitrate, Rutherford and Soddy "reached the general conclusion that radioactivity is a manifestation of subatomic change—a conclusion that seems prophetic in view of the nuclear atom, which Rutherford was to propose nine years later."[13] The experimental work of Rutherford and Soddy revolutionized chemistry because it altered the basic view of matter as immutable by showing that all radioactive elements undergo spontaneous transformation into new elements.

Despite the importance of Rutherford's work at McGill University, he felt that he was somewhat isolated from the important work then being done in atomic physics. When Arthur Schuster, the head of the physics department of Manchester University, retired in 1907, Rutherford was offered the opportunity to take his place.[14] He accepted the invitation immediately and, after arriving in England later that year, began his researches in atomic physics, which culminated in his discovery of the nuclear atom. One of his most remarkable achievements at Manchester was to prove that alpha particles are actually ionized helium atoms, a hypothesis that had been raised some years earlier but that had evaded the efforts of many scientists to provide direct experimental proof of until Rutherford's own work at Manchester.[14] His fame as an experimental physicist had grown, so that he was now in great demand as a lecturer for both scientific and lay audiences. Although Rutherford found the demands on his schedule to be onerous at times, he welcomed the opportunities to inform both his colleagues and the wider audience about the revolutionary discoveries then being made about the atom. One of his most famous lectures occurred in 1904, when he visited the Royal Society in London and demonstrated, using measures of radioactive decay, that the earth's age is not the few million years suggested by Lord Kelvin, but instead is at least several hundred times that figure.[15] Kelvin's estimate had assumed that the source of the sun's power is the heat generated by its gravitational contraction, but Rutherford showed that this could not be, because such a form of power generation would not be sufficient for the sun to continue burning for a period equal to Rutherford's estimate of the earth's age based on the rate of decay of certain radioactive elements.

Rutherford's arrival in Manchester began the most productive stage of his scientific career. He had already published some 50 important papers and was recognized as an eminent physicist, but he was not content to rest on past achievements.[16] Invigorated by the almost daily discussions he had with other noteworthy physicists and chemists, Rutherford plunged into his research and teaching duties with renewed enthusiasm. Within a year of his arrival at Manchester, he was awarded the Bressa Prize by the Turin Academy of Sciences for his demonstration of the mutability of matter and an honorary degree from Trinity College, Dublin.[17] His greatest honor and irony came later in 1908, when he was rewarded the Nobel Prize in Chemistry—an award that Rutherford found both annoying and amusing, since he considered himself to be a physicist.

Rutherford's investigations of the thorium family of radioactive elements convinced him that the the alpha particles emitted during radioactive decay can be used as a probe to study the structure of the atom.[17] His own experiments had shown that the atom consists almost entirely of empty space, and he found Hans Geiger's work on the scattering of alpha particles to be crucial to his formulation of an atomic model consisting of a positively charged nucleus.[18] Rutherford's own experiments with scattering alpha particles hinted at the vast energies bound up inside individual atoms, and his measurements showed him that the atomic nucleus is very compact, 3×10^{-10} centimeters in diameter; these results represented the beginning of what is now known as the nuclear atom.[18]

In 1914, Rutherford's scientific achievements were officially recognized by the Crown, and he was knighted. He was still a relatively young man who was somewhat ill at ease with being addressed as "Sir Ernest Rutherford." However, he took the award in stride, as he had done with the Nobel Prize six years earlier. Rutherford continued his own research while carrying out extensive correspondence with other scientists, such as Otto Hahn and Lise Meitner, who were studying the nature of gamma rays.[19] He also supervised the growing variety of experiments being carried out by his research assistants, although the outbreak of World War I in 1914 ended the collegial atmosphere that had existed up to that time as Rutherford's assistants were assigned to different research projects to help assist the British government's war against Germany. Rutherford himself did his part for the war effort by trying to find ways to detect submarines as well as spending a year as a member of the British mission to the United States.[19]

After the war, Rutherford completed his famous experiments on the artificial disintegration of nitrogen by alpha-particle bombardment and, shortly thereafter, left Manchester to become the Director of the Cavendish Laboratory at Cambridge and a fellow of Trinity College.[20] In 1920, Rutherford gave another lecture to the Royal Society and presented his theory of radioactivity, "suggesting the existence of a neutral particle of unit mass"—which was later called the "neutron" by its discoverer, James Chadwick.[20] Later that year, Rutherford proposed that the nucleus of the hydrogen atom—owing to its positive charge—be referred to as the "proton," a term that remains in use to this day.[20]

Rutherford's administrative duties were interspersed with frequent lectures and awards ceremonies. He was given many honorary degrees and the Order of Merit. He became president of the Royal Society in 1925 and was made a baron by the Crown in 1931, so that he was now formally known as Baron Rutherford of Nelson. Although he enjoyed his many honors and the attention lavished on him by the popular press, Rutherford continued to devote himself to supervising the work being done at Cavendish as well as keeping up with the newest discoveries of nuclear physics, including Chadwick's discovery of the neutron, Cockroft and Wilson's use of proton bombardment to induce artificial radioactive decay, and

Anderson's discovery of the positron.[21] His death at the age of 66 from a strangulated hernia occurred before the first successful nuclear chain reaction was completed, but it is likely that his own experimental work provided him with some suggestion of the destructive power that could be unleashed by the atom.[21]

A third component of the radioactivity rays—the gamma rays—had been discovered in 1900 by Paul Villard, who showed that these rays are very penetrating and are not deflected by magnetic fields. Villard correctly concluded that these rays are electromagnetic waves and that, qualitatively, they are the same as x rays, differing only in wavelength. Using the angstrom (one hundred millionth of a centimeter: 10^{-8} cm) as the unit of length, the wavelength of x rays is of the order of a few angstroms and that of gamma rays is 100–1000 times smaller.

Rutherford introduced a number of important concepts in the study of radioactivity, the most important of which is probably the half-life of a radioactive element. He defined the half-life as the time required for half a given quantity of a radioactive element to decay, that is, to change into another element by emitting a beta or an alpha particle. He measured the half-life of thorium (the first half-life to be measured) by carefully monitoring the rate at which the intensity of the emitted radiation diminished with time and deduced from his measurement that this intensity falls off (decays) exponentially. The half-lives of radioactive elements are important in geological dating. Thus, knowledge of the half-life of uranium together with the fraction of uranium in a gram of uranium mixed with lead gives us the age of the earth, if the lead in the gram is the end product of a series of radioactive decays that began with uranium. This leads us to another important discovery contained in Rutherford's and Frederick Soddy's famous transformation theory of radioactive elements.

In those early days, when no one knew the source of the alpha, beta, and gamma rays, radioactivity was the most mysterious and puzzling of all phenomena confronting physicists, for it clearly appeared to violate the principle of conservation of energy. A radioactive element like uranium seems to emit energy continuously, without any apparent chemical change and uninfluenced by external agencies; no wonder the physicists of the early 1900s considered it a kind of miracle that defies basic physical laws. Marie Curie described it as "a contradiction, or so it seems, with the principle of Carnot" (the first law of thermodynamics), and Lord Kelvin stated that for the first time in the history of the principle of the conservation of energy, Becquerel's discovery had placed the first question mark against it.

The difficulty presented by radioactivity arose becuse these early physicists thought of radioactivity as arising from a continuum of matter rather than from individual atoms in the matter. Indeed, the atomic concept was only vaguely understood and was vigorously rejected by some outstanding chemists and physicists, but even if it had been universally accepted at that time, radioactivity would still have presented serious problems, because the atomic concept means indivisibility, contrary to radioactivity, which implies that the radioactive atom is not indivisible because it emits a particle, which implies transformation of elements and a subatomic world.

All these difficulties were swept away by the Rutherford and Soddy transformation theory, which starts from the idea (revolutionary at that time) that some atoms are unstable and undergo spontaneous transmutations, emitting rays and changing into other kinds of atoms. This did not eliminate the energy difficulty and debate entirely, because the amount of energy emitted in the form of alpha, beta, and gamma rays in radioactivity far exceeds that produced in any chemical reaction. Pierre Curie and a colleague found in 1903 that 1 gram of radium can raise the temperature of 1.3 grams of water from the melting point to the boiling point in an hour and can go on doing this almost endlessly. These measurements raised the important question as to the source of this energy. This question became critical with the discovery that one of the daughter products of uranium, the radioactive element radon, emits, per unit time, a million times as much energy as is released in the explosive chemical combination of the same volume of hydrogen and oxygen to form water. Most physicists at the time tried to explain this vast radioactive energy as energy from an external source that the atom stored up within itself until it became radioactive and re-emitted it. This external-source hypothesis persisted for some time and was discarded entirely only when nuclear structure was understood.

The Rutherford–Soddy transformation theory is quite straightforward and gives a complete picture of the transmutation of heavy nuclei to light ones as a step-by-step process produced by individual atoms in the various elements involved. The end product of a radioactive process depends on whether an alpha, a beta, or a gamma ray is emitted in the process. If an alpha ray (a helium nucleus) is emitted, as in the decay of a uranium atom, the atomic weight of the daughter atom is 4 units less than that of the original atom, and its atomic number is reduced by 2. Thus if A' is the new atomic weight and N' is the new atomic number, then $A' = A - 4$ and $N' = N - 2$, where A and N are the atomic weight and number of the original atom. If a beta particle is emitted, the atomic weight is unchanged, and the atomic number is increased by 1; that is $A' = A$ and $N' = N + 1$. If a gamma ray (a very high-frequency photon) is emitted, the atomic weight and atomic number are unchanged. Clearly, in the emission of gamma rays, the nature of the atom does not change; the atom changes from a given energy state to a lower energy state, the difference in energy being the energy of the emitted photon.

It should be clearly understood that the alpha-, beta-, and gamma-ray processes are in no way related to each other and that a given kind of radioactive atom emits only one kind of particle; that is, a radioactive atom is either an alpha-particle emitter, a beta-particle emitter, or a gamma-ray emitter and never changes its character. In an aggregate of radioactive atoms of a given kind, one can never predict which atom will decay in any time interval; one can only predict the total number of atoms that will do so, and the rate at which this happens determines the element's half-life. The longer the half-life of a radioactive element, the weaker, in general, is its radioactivity.

Although the work of Rutherford and Soddy brought enormous order to the physics of radioactivity, the basic questions of why and how atoms become radioactive and what determines radioactive lifetimes could not be answered until the nucleus was discovered and nuclear theory was developed. The theory of the emission of alpha particles required, in addition, the development of the quantum mechanics. Before we come to that discussion, however, we must consider Albert Einstein's contributions to the development of modern physics, which influenced every phase of it profoundly.

CHAPTER 15

Albert Einstein and the Theory of Relativity

If a man will begin with certainties, he shall end with doubts; but if he will be content to begin with doubts, he shall end in certainties.

—FRANCIS BACON

Albert Einstein (1879–1955) was born in Ulm, Germany, the son of Hermann Einstein and the former Pauline Koch. Like many German Jews who were descendants of peasants, the Einsteins were not overly religious and seldom attended the local synagogue. Both parents had pleasant dispositions; Alfred's father, who ran an electrical workshop in Ulm that had been financed by some of Pauline's relatives, was a jovial man who preferred to take his family out into the country for a picnic rather than to attend to the more mundane activities of running a business. The failure of Hermann's business before Albert turned one year old caused Hermann to pack up the family and move to Munich, where he formed a more successful partnership with his brother Jakob running a small electrochemical factory. A year later, Albert's sister Maja was born; she became his closest confidant, and her death, when Einstein was 70 years old, distressed him more visibly than the deaths of either of his parents or his two wives.[1]

Although Munich was an overwhelmingly Catholic city, the Einsteins encountered few instances of the anti-Semitism that became so notorious in Bavaria with the rise of Nazis 40 years later. The family's life-style can be summed up as a gentle defiance of convention; they rarely observed the Sabbath and did not follow the dietary guidelines of their faith. Hermann Einstein considered most of the customs of his faith to be mere superstitions. His attitude toward religious authority was passed down to his son, who displayed an almost stubborn disregard for the

Albert Einstein (1879–1955)

conventions of society in his dress and his views about religion, politics, and, indeed, physics.

Like Isaac Newton, Albert was not a precocious child. He was slow to learn to speak and did not become fluent in German until the age of ten. His parents feared that he was retarded, but it seems more likely that he was a daydreamer who was withdrawn from the world.[2] Because Albert showed little interest in his lessons at the Catholic school he attended from the age of five to ten, his teachers did not think very highly of his abilities or his prospects; one instructor reportedly told Hermann that it did not matter what field Albert chose because he would never succeed in it.[2] Albert transferred in 1889 to the Luitpold Gymnasium, a typical German school where the teachers were concerned as much with maintaining the obedience of the students as they were with teaching them their subjects. This rigid and coercive environment drove home to Albert the virtue of distrusting authority, especially educational authority. Although the Luitpold Gymnasium was probably no better or worse than the other secondary schools in Germany, the willingness of its teachers—many of whom, Albert later asserted, were unqualified for their jobs—to use force or coercion to bring unruly or talkative students into line showed Albert that one should always be skeptical of convention. Doubtless, this questioning of authority was to figure greatly in the way he looked at the intellectual edifice of classical physics at the end of the 19th century, when many physicists were urging their students to go into other fields because little work remained to be done. However, one incident did occur while he was at the Gymnasium that made a profound impression on Albert and pointed him toward the subject that he would one day dominate: At the age of 12, he picked up a mathematics textbook and taught himself geometry before it was presented to his class: "For Einstein, the orderliness and the logic of the theorems made an impression that was never lost."[3]

The failure of Hermann's business in 1894 preceded a move by Albert's parents and sister to Milan, Italy. As Albert had not yet completed the requirements for his diploma (which was necessary for admission to a university), he was left behind in the care of his relatives. Unhappy at school and at home, Albert paid less and less attention to his studies; his indifference to his work became so great that one of the teachers eventually asked him to leave the Gymnasium, advice that Albert happily accepted. He journeyed to Milan to rejoin his family and, owing to the precarious nature of the family finances, began to consider seriously what sort of career he should pursue. His lack of a diploma prevented him from matriculating at any of the universities in Italy, but he soon learned of the Swiss Polytechnique Institute in Zurich, where a diploma was not required for admission; a prospective candidate had only to pass the entrance examinations.

Einstein traveled to Switzerland and took the examination, but did not pass. His failure was due more to inadequate preparation than to lack of knowledge about the sciences and basic mathematics, so he enrolled in the gymnasium at Aarau and boned up on his weaker subjects, such as biology and the languages. Einstein found his year at Aarau to be a pleasant contrast to the Luitpold Gymnasium; most of the teachers seemed more interested in teaching the students to think for themselves than in bullying them. In any event, he retook the entrance examination in 1896 and was duly admitted to a four-year program of study that, if completed, would qualify him as a teacher. Einstein's admission to the Swiss Polytechnique Institute in the fall of 1896 occurred six months after he had officially given up his German citizenship, a decision he made owing to his association of military authority with Prussian authority and his unfavorable opinion of Germans in general. Because Einstein had not yet applied for Swiss citizenship, he became a stateless person.

His years as a student at the institute were not extraordinary. He carried on lengthy discussions with his friends about subjects ranging from politics and religion to science and mathematics. He dressed casually and played the violin frequently in evening recitals. Einstein took long walks in the countryside and learned to sail; after the gloominess of his academic career in Germany, he found life at the institute to be uplifting and gentle. However, his lackadaisical attitude toward classes did not change; he rarely attended the lectures. He read his books in his room and borrowed notes from his classmates to pass the examinations. Even so, Einstein found the academic requirements demanding, because his desire to become a mathematical physicist required that he receive a thorough grounding in mathematics and the sciences. The demands of the curriculum did convince Einstein of the virtue of self-discipline in learning; he forced himself to concentrate on mastering the basic principles of each subject even though he continued to believe that class attendance was an impediment to his education.

When Einstein graduated in 1900, he failed to secure a position as an assistant at the institute. Not surprisingly, his lack of interest in his classes did not flatter any of the professors, who felt that Einstein would probably show a similar lack of interest in his job. Disappointed, Einstein secured a job in Zurich with Alfred Wolfer, the director of the Swiss Federal Observatory; his employment there made it possible for Einstein to meet the requirements for obtaining Swiss citizenship.

In December 1900, Einstein's first published paper appeared in *Annalen der Physik,* a work that had been inspired by the pioneering work of the chemist Wilhelm Ostwald on the principles of catalysis.[4] Although the paper did not gain him a research job, Einstein eked out a living from substitute teaching and tutoring. During this time of uncertain employment, he completed his thesis on the kinetic theory of gases and sent it to the University of Zurich to satisfy his doctoral requirements. In 1902, Einstein finally found steady employment as a junior patent examiner in the Swiss Patent Office. Although he had to work six days a week, he found the solitude of his post to be ideal for thinking about space and time and the nature of the physical world.

For the next three years, Einstein developed his revolutionary ideas about the relationship of space and time in the back room of his tiny apartment in Berne. His marriage in 1903 to a former classmate of his, Mileva Maric, enabled him to avoid the time-consuming daily chores of cooking and cleaning, though it is doubtful that Einstein cared whether his trousers were pressed or his dinner served warm. In any event, he spent most of his free time thinking about Newtonian physics and gradually developed a theoretical framework that convinced him that Newton's concept of absolute space and absolute time was erroneous. What is especially remarkable about this three-year period is that Einstein did not discuss his ideas with any professional physicists, but developed them entirely on his own. Because Einstein had not yet received his doctorate when he submitted the three papers he had written to *Annalen der Physik,* he faced the additional hurdle of having his papers considered seriously, since he did not possess outstanding academic qualifications. Fortunately, the revolutionary character of his papers was recognized, and Wilhelm Wien, the editor of *Annalen der Physik,* saw that they were the work of a young man—then 26 years old—who possessed an extraordinarily superb intuitive "feel" for physics. The papers were "comparatively short, and all contained the foundations for new theories even though they did not elaborate on them—'blazing rockets which in the dark of the night suddenly cast a brief but powerful illumination over an immense unknown region,' as they have been described by Louis de Broglie."[5] Even so, Einstein's ideas were not universally accepted from the start, because it is clear that a number of conservative scholars resisted his more sweeping conclusions until definitive proof of his theories was gathered by experimental physicists: "The first of [his papers] revived the corpuscular theory of light by introducing the revolutionary idea of the free photon—the atom of light—as an explanation for radiation phenomena (other than black body radiation) and especially as an explanation for the photoelectric effect. His ideas in this latter field were verified by [Robert] Millikan's experiments between 1912 and 1915. The second paper was a mathematical theory of Brownian motion that provided further proof of the reality of gas molecules on the basis that particles suspended in a fluid should behave as large gas molecules. This prediction was verified by the beautiful experiments of [Jean] Perrin in 1909. Finally, the third paper, his first publication on the theory of relativity, dealt with that branch of the subject now called the special theory which has been most useful in atomic physics."[6]

Although Einstein's work was favorably received by some of the most distinguished European scientists, such as Max Planck, the one man whom Einstein idolized, it did not catapult him to overnight fame. One reason it did not is that even though his papers were clearly written and lacked the ponderous footnotes that usually pervade most papers appearing in scientific journals, they required scientists who had built their careers propping up aspects of Newtonian mechanics and Maxwell's theory of electromagnetism, hitherto considered distinct and essentially unrelated branches of physics, to consider that they might somehow be related by the speed of light—the so-called "cosmic speed limit." Since Newtonian mechanics assumes that an object can travel at any speed so long as the necessary force is used to accelerate that object, Einstein's conclusion that nothing can travel faster than the speed of light upset a cornerstone of classical physics. In direct opposition to Newton, Einstein showed that an infinite amount of energy must be transferred to an object for it to travel at the speed of light—an impossible task, since the amount of available energy in the universe is finite.

As news of Einstein's work slowly spread through the universities of the world, Einstein continued to work at the Swiss Patent Office until 1909. He was now one of its most technically proficient examiners, and his value was recognized by his employers in the form of successive wage increases. As he had always regarded his job in the patent office as a temporary one even though it provided him with the time he needed to formulate his basic theories, he had no second thoughts about leaving his post when he accepted the position of associate professor of physics at the University of Zurich in 1909.

Einstein's formal entry into the academic world did not greatly alter his life-style, since his salary at Zurich was the same as that at the Swiss Patent Office. Although he earned some extra money by giving lectures, the higher costs of living in Zurich absorbed most of this gain, so that far from being content with his situation, Einstein was receptive to the informal offers that were made to him by representatives of several other institutions, most notably the German University in Prague. The promise of a full professorship, more money, and fewer expenses was an important factor in his decision to move to Prague in 1910 for two years, but he was especially impressed with the facilities, especially the library at Prague.

When Einstein arrived in Prague, he began to formulate his ideas on the curvature of time and space by mass and energy, which became the basis of the general theory of relativity. He soon found that his official university duties, especially the routine experimental work and student lectures, took up much more of his time than he was willing to give.

Within a few months, his initial enthusiasm about moving to Prague had evaporated. In 1912, he left the German University and returned to Zurich, but he was in Switzerland for only one year before he accepted the post of director of the Kaiser Wilhelm Institute in Berlin.

Although Einstein had never cared for Germany or its people, as shown by his forfeiture of his German citizenship as a teenager, he had been swayed by the personal pleas of Planck and Walter Nernst, two of the most distinguished physicists at that time, to join them in Berlin. Not only was Einstein promised the directorate of the institute, but also he was offered a nominal professorship at the University of Berlin that would free him from the mundane duties of lectures and laboratory sessions and permit him to devote himself fully to his research. As the move to Berlin also meant a doubling of the salary Einstein had been receiving at Zurich, the offer was irresistible. The Einsteins moved to Berlin in April 1914, amid the threatening clouds of war, but Mileva could not stand living in Berlin, so she left Einstein to return to Switzerland, taking the couple's two sons with her.

Einstein's marriage had been in trouble for some time, and, since he had always put his work before his family, the end of his marriage did not cause him a great deal of personal grief, particularly since he was immersed in a massive effort to correct the mathematical errors in his general theory of relativity. The outbreak of the Great War substantially altered the university atmosphere as many of Einstein's colleagues began devoting themselves to government research that promoted the war effort. Einstein avoided being drawn into such work and saw the German invasions of Russia and France as a folly that would only bring great suffering to all the European states, with little to show for the bloodshed. The German sweep through neutral Belgium also intensified Einstein's dislike for what he saw as the particularly militaristic character of Germany, even though much of his salary was paid from endowments originally set up by German industrialists. Einstein believed that a European government was the only alternative to the conflagration that was destroying an entire generation of young men, but his views were largely ignored. Einstein considered himself a pacifist, and his views about wars and self-defense were somewhat naive; his belief that no war could be justified did not change until he reluctantly concluded in the 1930s that a war was necessary to rid the world of Adolf Hitler. In any event, Einstein's Swiss nationality protected him from much of an official backlash for his "unpatriotic" behavior and enabled him to travel to Switzerland on several occasions during the war.

In 1916, Einstein's paper outlining his general theory of relativity appeared in *Annalen der Physik.* In less than 60 pages, Einstein showed that space is not merely a backdrop against which the events of the universe unfold, but that space itself has a fundamental structure that is affected by the energy and masses of the bodies it contains. Of Einstein's work, Max Born remarked that the "theory appears to me to be the greatest feat of human thinking about nature, the most amazing combination of philosophical penetration, physical intuition, and mathematical skill. But its connections with experience were slender. It appealed to me like a great work of art, to be enjoyed and admired from a distance."[7] Einstein's paper incorporated the positively curved geometry of Georg Riemann, casting aside the flat planes and straight lines of Euclid, so that the curvature of space by energy and matter and the bending of light rays by gravity could be mathematically described.

Einstein's general theory required an intensity of concentration involving months of agonizingly complex calculations as well as a total neglect of his own physical health. He suffered a nervous breakdown in 1917 after publishing two other significant papers dealing with the stimulated emission of light and the structure of the universe, the former of which provided the theoretical basis for the laser and the latter of which would found the subject of modern cosmology. Einstein slowly regained his health with the help of his second cousin Elsa, whom he married in 1919.

Even though Einstein was now recognized as one of the greatest physicists of the 20th century, if not the greatest, his fame did not extend far beyond scientific circles. This situation changed drastically in 1919, when a British expedition led by Sir Arthur Eddington journeyed to Principe Island in the Gulf of Guinea, where they photographed a solar eclipse; their analysis of the plates, six months later, showed that the path of the light of a distant star passing near the rim of the sun during the eclipse was in fact bent by the gravitational field of the sun, thus confirming Einstein's theory.

The restructuring of space and time that was completed with the confirmation of Einstein's general theory led to a torrent of articles and books about relativity and its creator. The predictions of his theory of relativity, which often seemed at odds with common sense, intrigued the public, and his charismatic personality and "common man" appearance were attractive subjects. With this worldwide attention came thousands of invitations to lecture and write, most of which Einstein turned down, since they would take too much time from his work. However, he did become actively involved in the Zionist movement to establish a Jewish nation in Palestine and lent his name to their fund-raising efforts. In 1921, he was awarded the Nobel Prize in Physics for "his contributions to mathematical physics and especially for his discovery of the law of the photoelectric effect." No mention was made of relativity because of Alfred Nobel's stipulation in his will that the awards be given for discoveries that benefited mankind; it was difficult for the Nobel committee to agree how the theory of relativity had improved the condition of mankind.

Einstein met Niels Bohr for the first time in 1920, and while the two men admired each other greatly from the start, they were inflexible opponents who debated the implications of the quantum theory for the next three decades, each convinced that the other was wrong: "For it was Einstein who, fifteen years earlier, had first brought an air of unexpected respectability to the idea that light might conceivably consist both of wave and of particle and to the notion that Planck's quantum theory might be applied not only to radiation but to matter itself. It was Bohr who was to bring scientific plausibility to the first of these ideas with his principle of complementarity and substance for the second with his explanation of Rutherford's nuclear atom. Yet these very ideas were to create not a unity between the two men but a chasm."[8] Although Bohr was eventually acknowledged the victor by most physicists and his argument that causality is not necessary in quantum physics accepted, Einstein's belief in a deterministic universe, as evidenced by his oft-quoted remark that "God does not play dice," led him to continue arguing against the probabilistic statistics favored by Bohr until the end of his life. Einstein's philosophical views about the quantum physics that he had done so much to create and nurture caused him to be left behind by the physics community in the later years of his life and hastened his withdrawal from the mainstream of modern physics.

Until Adolf Hitler came to power in 1933, Einstein remained at Berlin. Although greatly admired by portions of the population, who saw him as a symbol of the "new German" and German scientific prowess, he was vilified by other groups such as the Fascists, who criticized his pacifism and his cultural heritage. He was the subject of death threats and numerous staged meetings in which "distinguished scientists" met to discuss the fundamental flaws of the theory of relativity. Einstein found the anti-Semitic displays to be more pathetic than threatening, but he complained that the tumult only increased the fears felt by the Jewish community in Germany. On more than one occasion, however, he seriously considered leaving Germany for good, only to be persuaded by

Planck to stay because his departure would be a tremendous loss to the University of Berlin and the German nation itself, which was still trying to dig itself out of the disaster of the Great War. Despite his dislike of Germany, Einstein repeatedly passed up tempting offers from other European universities because he felt that it was his duty to stay behind while the shaky Weimar government established itself. More important, he knew that he was already at the most distinguished scientific center in the world and that his departure would complicate the efforts he had begun in 1920 to formulate a mathematical framework that would unite both electromagnetism and gravitation—the so-called "unified field theory."

The collapse of the world economy in 1929 and the subsequent rise to power of the Nazis spelled the end of Einstein's career at Berlin. It also marked an evolution of his political views from that of absolute pacifism to qualified support for defensive wars. Einstein saw Hitler as the threat to peace in Europe and realized that if his anti-Semitic speeches were any indication of his true intentions, Einstein's head would be among the first to be served on a platter. The venom of the Fascists intensified his support for the Zionist movement—even though he distrusted some of its more radical leaders—and encouraged him to begin thinking about where he would go if and when Hitler came to power.

Einstein did not have long to wait for an answer. In the early 1930s, he was a visiting professor at the California Institute of Technology in Pasadena for two winters. His third visit to Pasadena in 1933 preceded Adolf Hitler's appointment as chancellor of Germany and brought forth the statement by Einstein that he would never return to Germany. After a well-publicized tour of America that was not without controversy owing to Einstein's left-wing political sympathies, he traveled to Belgium, where he gave his German passport to a German embassy representative in Brussels and settled at Oersted while he considered offers made by several universities, including California, Oxford, and the Institute for Advanced Study at Princeton, the latter of which he joined at the end of 1933. Aside from occasional trips around the United States, Einstein remained at Princeton until his death in 1955.

Although the final years of Einstein's career were spent in a fruitless search for the unified-field equations, he did not regret the time he spent on this work, since he felt that he had established his own scientific reputation and had nothing to lose even if the entire effort came to naught. Given the skepticism of his colleagues that a single set of equations could be found that would explain all physical processes in the universe, Einstein believed that he was the only one who could perform such a towering intellectual feat. He pursued this goal with a single-minded determination for nearly 30 years despite the death of his second wife Elsa in 1936 and his growing suspicion that his intuitive mastery of physics was eroding.

Perhaps Einstein's most significant act while at Princeton was to send his famous letter to President Franklin Roosevelt in the spring of 1940 warning that a fission bomb might be constructed by the Nazis and urging a coordinated research effort by the United States government. Although fission research was proceeding at a number of American universities before Einstein became involved, his letter helped to spur government interest in nuclear research. Einstein also worked for the United States Navy's Bureau of Ordnance as a consultant from 1943 to 1946.

After the war, Einstein divided his time between his work and his efforts to speak out against the threat of nuclear war. He correctly saw that America's participation in an international effort to control the development and proliferation of nuclear weapons during the postwar era was a prerequisite for maintaining some sort of global peace. Einstein also continued to speak out in favor of the Zionist movement and was offered the presidency of the state of Israel, a largely ceremonial post, in 1952 following the death of Chaim Weizmann. Although Einstein was deeply honored, he declined the offer, saying that he was too old to move to Israel. His health, which had steadily worsened, limited his activity, and while he retained his confidence in himself to the very end, he found his stomach cramps and nausea to be a persistent warning of his own mortality. His last important act, a few days before his death, was to sign a declaration, originally sponsored by Bertrand Russell, outlining the dangers of nuclear war and urging all nations to resolve their disputes peacefully.

THE REVOLUTIONARY NATURE OF THE THEORY OF RELATIVITY

No single event in the history of science has had so profound an effect on man's thinking as the promulgation of the theory of relativity, which occurred in two vast intellectual steps, as the special theory in 1905 and the general theory in 1915. To many people, this statement may appear presumptuous and somewhat drastic, for they may point to Darwin's theory of evolution or to Planck's quantum theory as having influenced our thinking more than the theory of relativity. That evolution and the quantum theory have played very important roles in our understanding of the universe is true, but on balance, the theory of relativity is the most important product of science, for it influences every phase of philosophical and scientific thinking.

The merging of three-dimensional space and one-dimensional time into a four-dimensional space–time manifold as demanded by relativity theory has altered philosophy enormously, so that owing to relativity theory, modern philosophy is as different from Kantian philosophy as modern physics is from classical physics. Insofar as physics itself is concerned, the theory of relativity stands above all other theories, in that it is a kind of master theory against which all other theories must be measured. An acceptable physical theory must be "relativistically invariant," in that it must conform to certain constraints imposed by the theory. This requirement applies to the quantum theory as it does to all other theories, so that in a sense, the quantum theory is "subservient" to the theory of relativity. But the relationship of the quantum theory to the relativity theory goes deeper than that, for the theoretical basis of the quantum theory indicates that it may be derivable from relativity theory.

Putting these considerations aside for the time, we note that to most people, the essence of modern physics is contained in the theory of relativity, and at the same time, the theory is also the great scientific mystique that they must accept without question. The very word "relativity" conveys something of that mystery, for nothing in it gives the slightest indication of its subject matter or any clue as to its relationship to a theory of nature. Indeed, most people think of relativity from the general point of view that the appearances and apparent sizes of objects vary as the observer's position with respect to these objects changes. This variation is evident, but if this were all there is to relativity, classical physics would not have been affected by it and there would have been no need for Einstein. But Einstein's theory of relativity is a theory of the relativity of motion, and therein lies its great impact on physics and on our thinking in general.

Although Einstein's first paper on relativity, together with two other revolutionary papers (one of which we have already discussed in Chapter 12), appeared in *Annalen der Physik* in 1905, it is clear from Einstein's later writings that the ideas for the special theory of relativity began to germinate in about 1900 when certain features of the behavior of light puzzled him. Physics, as described in the previous chapter, was growing rapidly and revealing many new features of matter and energy, but one experiment brought consternation and confusion—the famous Michelson–Morley experiment. Since this experiment had a direct bearing on the acceptance of the theory of relativity, although historical evidence indicates that Einstein did not know about it when he wrote his relativity paper, it is useful to examine this experiment.

In 1886, when Alfred Michelson was in the physics department of the University of Chicago, he began an experiment with his collaborator, Edward Morley. Michelson's great interest in the propagation of light and measurements of its speed in a vacuum indicated to him that he could, from such measurements, determine the speed of the earth (in its solar orbit) with respect to the "all-pervasive ether," which at that time was one of the accepted features of the universe. Maxwell's electromagnetic theory of light had shown that light is propagated as a wave through the vacuum, and since wave propagation was thought to require a medium, the "luminiferous ether" was proposed as such a medium, even though no experimental evidence for it had ever been obtained. Michelson proposed to detect the ether by comparing the speed of light moving in the direction of the earth's motion with the speed of a beam moving at right angles (perpendicular) to the earth's motion. The difference between these two speeds should then not only demonstrate the earth's motion but also actually give the speed of the earth in its orbit.

The theoretical basis of the experiment is that if an ether exists, the earth's motion creates a stream of ether, opposite to the earth's velocity, just as a moving vehicle creates a stream of air that flows past the vehicle. The speed of light as measured on the earth should either be or not be affected by the ether stream, depending on whether the light is moving parallel to the stream (with it or against it) or perpendicular to the stream. The analysis here is exactly the same as that applied to two equally fast swimmers in a river; one swims a given distance downstream and back, and the other, starting from the same point at the same time, swims the same distance across the river and back. The two swimmers cannot return to their starting point at the same time; the cross-stream swimmer always returns first, as one can show from the simple law of the arithmetic addition of speeds. If light is propagated through an all-pervasive fixed ether in space, the ether stream created by the earth's motion should delay a beam of light that, moving in the direction of the earth's motion, strikes a mirror at a definite distance from the light source and returns, compared to an identical beam moving to an equidistant mirror and back at right angles to the earth's motion. The Michelson and Morley apparatus was extremely sensitive and had been designed to detect a time-of-return difference between the two beams even if the earth's speed around the sun had been 1 mile per second instead of its true speed of $18\frac{1}{2}$ miles per second; they detected no difference whatsoever, to Michelson's bitter disappointment, for he considered the experiment a failure. He concluded that the experiment, as designed, could not detect the earth's motion.

Although Michelson rejected his experiment and dismissed it as of no significance, other physicists at that time saw in its null-result a very important statement about nature, although they were not aware of its importance. The physicist Hendrik Antoon Lorentz made the boldest attempt within the framework of classical physics to explain the Michelson–Morley null-result, using his electron theory of matter. The complex details of his very brilliant analysis are not important, but his result was startling: His analysis showed that a moving spherical electron is flattened somewhat in the direction of its motion owing to its electrical properties, and the faster it moves, the more it is flattened. Lorentz therefore reasoned that matter, consisting of electrons, is flattened somewhat along the line of its motion if it is moving. He used this analysis to explain the negative result of the Michelson–Morley experiment and stated that the path to the mirror and back of the light moving parallel to the earth's motion shrinks and thus allows this beam to go and return at exactly the same time as the perpendicular beam. The remarkable part of this analysis is that it states that the shrinkage along the line of motion is exactly the right amount to cancel the delay produced by the ether stream in the time of flight of the beam of light moving parallel to the ether stream. This effect is known as the Fitzgerald–Lorentz contraction hypothesis, since the British theoretical physicist Fitzgerald had proposed a similar contraction hypothesis at about the same time as Lorentz.

The contraction hypothesis was not taken seriously, since it seemed like too much of a conspiracy for the electrostatic interactions among the charged particles constituting matter to reduce the length of one of the arms (the one parallel to the earth's motion) of the Michelson–Morley apparatus by just the right amount to give the null-result that the experimenters found. This null-result remained a thorn in the side of theoretical physicists until it was brilliantly explained in Einstein's first paper, which spelled out the special theory of relativity. This paper brought about a revolution of the first magnitude in our concepts of space and time and the laws of nature. Its impact on science was immeasurable.

Einstein did not develop the special theory of relativity to explain the Michelson–Morley null-result, for he did not know of that experiment when he produced his theory; indeed, he was out of touch with the mainstream of physics at the time and knew no physicists at all. But he did know Maxwell's electromagnetic theory and was deeply immersed in trying to understand the nature of light—particularly its motion. He was also prompted in his work by a sense of the unity of the laws of nature and his comprehension that this unity means that the laws of moving bodies (Newtonian mechanics) and the laws of optics (propagation of light) must be on the same footing in nature; that is, the two sets of laws must be governed by the same set of overall principles. In particular, he was convinced that if the laws of mechanics appear the same for all observers regardless of their motions with respect to each other, so should the laws of optics. This is the essence of Einstein's famous principle of the invariance of the laws of nature, which we explain more fully later.

To see the significance of this concept, we consider mechanical experiments (watching objects in flight, projecting objects, or even moving about ourselves) and try, by observing the behavior of these objects, to determine our motion through space. No matter how carefully we observe such objects, we discover nothing in their behavior to tell us whether or not we are on a moving planet. This is also true if we are on a train or in an airplane moving with constant speed in the same straight line—we cannot detect our uniform motion (motion at constant speed in a straight line) by any observation within our vehicle. The reason is that Newton's laws of motion are independent of the uniform motion of the observer, so that these laws cannot change when the observer goes from one uniformly moving frame of reference to another (transformation of coordinates). This is often taken for granted owing to constant exposure to such phenomena in our experience, particularly in a smoothly moving airplane; as far as anything we observe in the plane goes, we might as well be at rest.

Einstein carried over these ideas to optical phenomena and convinced himself that the laws of optics can no more reveal our uniform motion than can the laws of mechanics. This means, as outlined by Einstein, that Maxwell's equations that describe the propagation of electromagnetic waves (light) cannot depend on the motion of the observer. But these equations contain the speed of the electromagnetic wave, that is, the speed of light, and this speed cannot depend on the motion of the observer, for if it did, optical phenomena via Maxwell's equations could be used to determine one's absolute motion through space in the manner of the Michelson–Morley experiment or some other optical experiment. Einstein saw that the speed of light in a vacuum must be independent of the motion of the source of the light and of the motion of the observer. This means that the speed of light, as measured by any observer, cannot depend on the relative motion of the source and the observer (that is, the motion of one with respect to the other). The invariance of the speed of light had a tremendous impact on physics, especially the concepts of space and time.

Einstein now had two strong heuristic concepts (the constancy of the speed of light and the invariance of nature's laws) to lead him to his special (or restricted) theory of relativity, which has replaced Newtonian mechanics with relativistic mechanics. The difference between the two, from a formalistic point of view, is that Newton's laws do not contain or refer to the speed of light as a universal constant, whereas all relativistic laws do; indeed, the signature of special relativity is the presence of the speed of light in all its formulas. Just as the presence of Planck's constant of action, h, heralds the quantum theory, the presence of the speed of light, c, heralds the relativity theory. Since the laws of nature are in accord with both quantum theory and relativity theory, all expressions of these laws must contain both constants.

Einstein's assumption of the constancy of the speed of light led him to a thorough analysis of the accepted concepts of absolute space and absolute time, which convinced him that these concepts are untenable if the speed of light is constant. To prove this point to his own satisfaction, he performed one of his famous thought experiments to demonstrate that the concept of the simultaneity of two events, separated in space, has no absolute meaning, but depends on the observer's motion. Since the concepts in this experiment are somewhat subtle, we describe another thought experiment that shows that neither space nor time is absolute; distances and time intervals as measured by two observers moving with respect to each other are different.

Our thought experiment involves two observers moving with respect to each other who are to measure the speed of light. Each observer has an identical clock and a ruler 186,000 miles long (a fanciful idea, but it simplifies the analysis). One observer is at rest on a railroad siding, with his ruler mounted from left to right parallel to the tracks (which are pictured as extending in a straight line for at least 186,000 miles and therefore far out into space, which makes our thought experiment quite imaginary). The moving observer is to be in an open car moving from left to right as seen by the fixed observer along the tracks at 185,000 miles per second with his ruler also parallel to the tracks. Each observer is to measure the speed of light with his ruler and his clock and record his result. He is to do this by timing a beam of light as it moves from one end of his ruler to the other.

To make things simple, the measurement of each observer is to begin when the two left ends of the rulers as seen by the fixed observer coincide; at that moment, a laser beam from some distant source and moving from left to right strikes the coincident left ends of the two rulers and triggers the two clocks. What speed does each observer find for the laser beam? The fixed observer follows the beam and notes that it reaches the right end of his ruler (186,000 miles away) just as his clock ticks off 1 second. For him, the speed of light is 186,000 miles per second. What does the moving observer find? He too follows the beam along his ruler and notes that just as it reaches the distant end, his clock ticks off 1 second, and so he records the speed, as he measured it, as 186,000 miles per second. This is exactly in accord with the fact of nature that each observer must find the speed of light to be 186,000 miles per second, but if we were fixed observers along the track, watching the experiment, and knew nothing but Newtonian concepts, the moving observer's statement would appear nonsensical to us and contrary to elementary logic and "common sense."

To explain why we, as Newtonian observers, would react this way, we point out (without any observation and according to our Newtonian reckoning) that the right end of the moving rod has advanced 185,000 miles in one second of our time (the speed of the car) and that therefore the beam of light, which has reached the right end of the fixed ruler (186,000 miles down the track), still has, according to our Newtonian perception, another 185,000 miles to go to reach the right end of the moving ruler. Without observing the events in the moving car, we are therefore inclined to reject the moving observer's statement as grossly mistaken, since no other explanation seems reasonable in the Newtonian space–time framework.

To eliminate all doubts, the experiment is to be repeated, but now the fixed observer is to watch every aspect of the moving observer's actions and to keep the moving clock and rod in view constantly. Again, the laser beam strikes both left ends of the two rods simultaneously (the zero moment on each clock) and reaches the right end of the fixed rod when the fixed clock registers the passage of one second, but the right end of the moving rod is not 185,000 miles further to the right but only about one tenth that distance, and the moving clock registers the passage of only about one tenth of a second. Thus, distances and time in the space–time frame of the moving observer as seen by the fixed observer are not the same as they are in the frame of the fixed observer. Distances and time in every frame of reference must so adjust themselves as to give the same measured value for the speed of light. Moving rods shrink and moving clocks slow down, and this effect is completely reciprocal so far as the two observers are concerned, since only relative motion matters; either observer may be considered as being at rest and the other one as moving, and each observer notes that lengths and times in the other observer's frame shrink and slow down.

All these phenomena stem from the constancy of the speed of light in a vacuum for all observers moving with contant speed in the same straight line with respect to each other. To Einstein, this constancy meant that neither space nor time is an absolute entity. The distance and the time interval between two events depend on the state of motion of an observer with respect to the two events. To one observer, the two events are simultaneous; to another observer, one of the events, event A, for example, precedes the other, event B; and to a third observer, event B precedes A. Moreover, the distance between the events is different for each of the three observers. The constancy of the speed of light is thus one of the basic blocks in Einstein's foundation of the special theory of relativity. The importance of this constant as far as the laws of physics are concerned is that the speed of light c is one of the important constants, along with Planck's constant of action h and Newton's gravitational constant G, that nature has used to construct the universe and, as such, must be present in all of nature's laws. If the constant c (the speed of light) is absent from a law such as Newton's law of motion $F = ma$, the law is incomplete because it does not conform to the requirements of relativity and must be properly enlarged (or completed) to do so. We say that the law is then relativistically invariant and therefore enunciates a higher truth than it does without its proper relativistic completion, and this leads us to the second block in the foundation of Einstein's theory—the principle of invariance.

Einstein was not the first to introduce the concept of invariance, which goes back to Newton, but the way Einstein introduced it and used it to construct his special and general theory of relativity was quite new and extremely useful. To see this, we first define "invariance" as the term is used by physicists. To this end, we again consider the nature of a law in physics. We begin by defining an event as the coincidence of a particle (for example, an electron, a photon) with a point of space at a particular time. To specify an event, we must know where and when it occurs, which means we must have a frame of reference (a coordinate system) to locate it and a clock to time it. Since a frame of reference consists of three distinct sets of parallel intersecting lines (curved or straight) in space (a three-dimensional grid or mesh), we locate the event in space by giving its position in this grid relative to any point of the grid where three of the lines (one from each set) intersect; we call this point the "origin" of our coordinate system. Thus, to specify the position of an event, we must give three numbers (the coordinates of the event in our grid), just as we have to give three numbers to locate a person in a tall building—his room number in the building and the address of the building (which requires two other numbers—those of the street and the building); to specify the time t of the event, we need a clock. An event is thus specified by four numbers: its three spatial coordinates and the time of its occurrence. The motion of a particle is thus described as a collection of events (sets of four numbers), and the orbit of the particle is a curve that connects these events. A law, then, is a general statement, involving space and time, that enables us to correlate events and thereby deduce orbits of particles. Since a law does not deal with specific events, but with the intrinsic properties of nature, it must be the same for all observers regardless of their frames of reference. This is the essence of the principle of invariance.

To illustrate the principle of invariance as simply as possible, we picture two observers moving with constant velocity with respect to each other (that is, with constant speed along the same straight line). The limitation of the relative motion of the observers to constant velocity leads to the special theory of relativity (the restricted theory); the two observers are to study the same set of events to deduce the basic laws of motion, as Galileo and Newton did. Each observer describes the events in his own frame of reference and therefore in terms of his own sets of four numbers (that is, his own three spatial coordinates and time of occurrence of each event). These two descriptions of any event differ in general, but any law (for example, the law of motion) that each observer deduces must be identical in content and mathematical form if it is a true law. The principle of invariance implies that any statement about events in nature that remains unaltered when we express it in different frames of reference is an intrinsic truth about nature and therefore a law. Clearly, such a principle is a powerful intellectual tool for separating the intrinsic features of the universe (the basic truths or laws) from the purely apparent or superficial ones.

To delineate further the invariance concept, we consider a "law" as formulated by one of the observers (observer 1) and then see whether it remains unaltered or changes when we translate it into the "language" (frame of reference) of another observer (observer 2); only if it remains unaltered is it a law. To explain this further, we note that the first observer states this "law" in terms of his own frame of reference, that is, in terms of his set of four numbers (three of space and one of time). To express this law in the frame of the second observer, we must have some mathematical scheme for translating the space and time coordinates (observations) of observer 1 into those of observer 2; this transformation of coordinates is one of the most important concepts in physics.

The essential feature of such a transformation is a set of algebraic equations that relate the three space coordinates x, y, z and the time coordinate t of an event, as observed in one frame of reference, to the three space coordinates x', y', z' and the time t' as observed in another frame moving with velocity v with respect to the first frame. The nature of these transformation equations depends on the geometry of space and time; by adopting one or another concept of space, we obtain one set of transformations or another. In Newtonian physics, the space–time concept is Euclidean (flat geometry), and space and time are absolute. This is consistent with the belief (fully accepted in Newtonian physics) that the speed of light, as measured by any observer, depends on the motion of the observer relative to some fixed frame, which was assumed to be the ether (introduced as the medium that propagates light). With these assumptions, the transformation equations (called Galilean transformations) are very simple and contain only the relative speed v of the two observers; time is the same for both observers, and space coordinates of events are changed since, owing to the relative motions of the observers, the same events appear in different positions. Space, however, is absolute in the sense that distances between events are the same for both observers.

In Einsteinian or relativistic physics, which is based on the observed fact that the speed of light is the same for all observers, the transformation equations (the Einstein–Lorentz transformations) are more complex than the Euclidean transformations; in addition to the speed v, they also contain the speed of light c. In fact, these equations are marked by the presence of the factor $\sqrt{1 - v^2/c^2}$, which is the signature of the theory of relativity and the most famous and profound expression in the history of physics. These transformation equations apply equally to both space and time, so that in relativity theory, space and time are treated on the same footing and become intermixed in such a way that neither space by itself nor time by itself is absolute.

But the nonabsoluteness of space and time separately does not mean that relativity is not a theory of absolutes, because its absoluteness is on a higher level than in Newtonian physics, for it combines space and time into a space–time manifold of absolutes. To explain this, we note first that according to Newtonian theory, both the distance and the time interval separating two events are the same for all observers (separate absolutes), but in relativity, different observers (different in that they are moving with respect to each other) find different distances and different times. Relativity, however, teaches us that a certain combination of the space and time intervals between two events is the same for all observers. The square of this absolute space–time interval between any two events is obtained by squaring the distance r between the events and subtracting from that the square of the product ct, where c is the speed of light and t is the time interval; this entity, $r^2 - c^2t^2$, is absolute in the sense that all observers moving with constant velocity with respect to each other find the same value for it.

This simple statement, $r^2 - c^2t^2 =$ constant for all observers, contains all of special relativity and leads to the four-dimensional space–time physics of Einstein, which has replaced the three-dimensional physics of Newton. One passes from Newtonian physics over to relativity physics by replacing the Newtonian laws that involve relationships among three-dimensional vectors with laws that involve four-dimensional vectors, each of which consists of three dimensions of a space component and one dimension of a time component. The space–time interval is the basic example of this kind of vector.

All the remarkable consequences that stem from the special theory of relativity, such as the shrinkage of moving rods in the directions of the motion, the retardation of moving clocks, the increase of mass of a moving body, and the equivalence of mass and energy as expressed in Einstein's famous equation $E = mc^2$ (energy equals mass times the square of the speed of light), can be deduced from the constancy of the square of the space–time interval described above. Just how this theory affects the laws of physics is easily demonstrated by considering the conservation of energy and of momentum. In Newtonian physics, the momentum of a particle is conserved by itself and its energy is conserved separately, as is its mass. According to relativity, this cannot be, because momentum is three-dimensional and energy and mass are one-dimensional, which contradicts the relativity requirement that laws (which include conservation principles) must be statements about four-dimensional vectors. Momentum and energy combined give such a vector, so that in relativity physics, the three separate conservation principles are combined into a single energy–momentum–mass conservation principle. This immediately gives Einstein's mass energy equation, which has had such a tremendous effect on science and technology.

Since the special theory of relativity introduces four-dimensional space–time, we must consider the nature of space–time geometry, which is determined by the space–time interval between two neighboring events.

In Newtonian physics, the geometry is three-dimensional Euclidean (flat) geometry and determined entirely by spatial relationships—time plays no role in it. The square of the distance between two events, that is, r^2 expressed in terms of the coordinates of the events in the observer's coordinate system, is just given by the sum of the squares of the coordinates $x^2 + y^2 + z^2$ (the theorem of Pythagoras). This simple expression for the square of the distance is the signature of Euclidean (flat) geometry. We pass over to special relativity by adding the term $-c^2t^2$ to the aforestated sum to obtain the space–time interval $x^2 + y^2 + z^2 - c^2t^2$, which is the signature of Euclidean (flat) four-dimensional geometry, so that special relativity, like Newtonian physics, operates with Euclidean geometry. One slight difference between these two Euclidean geometries should be noted, however; the three space terms x^2, y^2, and z^2 in the distance formula in both geometries appear with positive signs, but the time term c^2t^2 appears with a negative sign; this negative sign is crucial for all the consequences of the special theory relativity. In this geometry, the path of a particle through space–time is called a "world-line."

THE GENERAL THEORY OF RELATIVITY

In 1916, some ten years after Einstein had published his paper on special relativity, his paper on general relativity appeared in the journal of the Berlin Academy of Sciences. It is a comparatively short paper, but it represents ten years of the most intense and penetrating thought of the greatest mind of the 20th century. As such, it stands at the very peak of the intellectual creations of humanity. To have achieved this result, Einstein must have been driven by a passion to achieve a great scientific synthesis. The reason for this drive was that the special (or restricted) theory, as indicated by its very name, left the story of space and time unfinished.

As we have noted, Einstein was seeking a unification of the laws of physics and saw a way of doing that in his principle of invariance; all laws must have the property of being the same for all observers, regardless of their frames of reference (their states of motion). To put it differently, the laws must not permit an observer, by using these laws, to determine his state of motion. But Einstein had not achieved this with his special theory, for the special theory applies only to so-called "inertial observers" (that is, observers moving with uniform velocity with respect to each other). The special theory thus singles out inertial frames of reference (frames moving with uniform velocity) as those preferred by nature for expressing natural laws. This restriction on the types of coordinate systems in which to formulate laws to inertial systems is, as Einstein saw, a deficiency in his theory; he was deeply convinced that all coordinate systems, regardless of how they are moving, are equivalent in the eyes of nature, whether they are in a state of uniform motion (constant velocity) or accelerated in any way whatsoever. This means that one should not be able, by applying any law to (that is, by making any observation of) events within or outside one's frame of reference, to determine one's state of motion.

Our first reaction to this proposition is that it cannot be right and that Einstein would fail in his attempts to generalize the principle of relativity by extending it to accelerated motion. This reaction is based on our experience, which tells us that while we cannot detect our unaccelerated motion because all things in our frame of reference move together (all things behave as though we were not moving), we can immediately detect our accelerated motion because objects all around us do not move in accordance with Newton's first law of motion. Things behave quite differently from the way they do when our system is at rest or moving with constant velocity. Accelerated motion seems to be absolute. If the room we are in begins to rotate, we are all thrown to the walls and, independently of any other observation, we conclude at once that our frame of reference is spinning. Otherwise, we would have to conclude from Newton's second law that our room is not spinning and that all of us in it are dragged to the walls by an invisible force of some sort, which, of course, we dismiss out of hand and draw the more "reasonable" conclusion that we are in an accelerated frame of reference. But Einstein did not accept this "commonsense" notion as a deterrent in his drive to generalize relativity and assign an equal status to all coordinate systems (frames of reference), regardless of their motion, insofar as finding the laws of physics goes. In fact, Einstein used the observation that bodies in an accelerated frame of reference behave as though they were subject to a force, which he called the "inertial force," to promulgate his general theory.

In constructing the general theory of relativity, Einstein started from the very general observation first made by Galileo that all bodies falling freely from the same height in the earth's gravitational field fall with the same acceleration regardless of their masses. He also noted that all bodies in an accelerated frame of reference respond to the acceleration exactly the same way regardless of their masses. From these two observations, he proposed one of the most remarkable principles in physics, his famous principle of equivalence—that inertial forces cannot be distinguished from gravitational forces. This principle is the basis of the general theory of relativity, for it denies the possibility of determining one's state of motion (whether our frame of reference is accelerated or not) by observing or detecting inertial forces.

We can best follow Einstein's reasoning by considering briefly his famous elevator thought experiment, in which he pictures an observer in an elevator that at first is suspended above the earth. All the gravitational experiments the observer performs agree exactly with those of an observer outside the elevator on the earth, so that the observer in the elevator, like the external observer, concludes that a downward force, which he calls gravity, is pulling all the bodies in the elevator toward the floor. Compare this situation with the observations and conclusions of this same observer if he and the elevator were suddenly transported far away from the earth or any other massive body and if the elevator were then constantly accelerated in the direction from the floor to the ceiling at 32.2 feet per second squared (the same acceleration as that on the earth's surface). The observer would find all bodies still behaving as they did when his elevator was suspended over the earth, and so, being consistent, he would still conclude that his elevator is fixed and that the bodies in the elevator are being pulled "downward" by a gravitational force. This is the physical significance of the principle of equivalence; it prevents one from concluding that one is in an accelerated frame of reference because all the effects stemming from such an acceleration are exactly the same as those produced by gravity in a frame that is at rest or moving with constant velocity in a gravitational field. The principle of equivalence thus supports Einstein's contention that accelerated motion cannot be distinguished from nonaccelerated motion; the inertial forces produced by acceleration are the same as those produced by gravity, so that an observer does not know from his observations of bodies (at rest or in motion) in his frame of reference (coordinate system) whether he is at rest in a gravitational field or moving with constant acceleration in empty space. Acceleration and rest are indistinguishable. This conclusion holds whether one observes the dynamics or kinematics of material bodies or the propagation of light, which led Einstein to a very important deduction about the behavior of light in a gravitational field.

If a beam of light passes through the accelerated elevator perpendicular to the direction of the acceleration, the beam appears to fall toward the floor of the elevator just as particles of matter do, because the floor is moving with accelerated motion toward the beam. Since the principle of equivalence says that accelerated effects and gravitational effects are indistinguishable, Einstein predicted that a beam of light falls in a gravitational field just as particles of matter do. This prediction was completely confirmed during the 1919 total solar eclipse when a beam of light from a distant star was observed to fall toward the sun as the beam just grazed the sun; the magnitude of this effect was exactly that predicted from Einstein's theory.

Before discussing the other deductions about physical phenomena that stem from Einstein's general theory of relativity, we describe briefly the formal features of the theory that differentiate it from the special theory and Newtonian theory. Although most general relativistic phenomena can be deduced from the principle of equivalence by noting how events unfold in an accelerated frame of reference, the deep insight into the laws of nature and into the behavior and structure of the universe that the theory offers can be obtained only by using the full formalism of the theory.

The general theory encompasses the special theory in that it is based on the four-dimensional space–time (where space and time are merged), but it differs from the special theory in that the geometry of general relativity is non-Euclidean; this non-Euclidean aspect of the theory contains or leads to Einstein's theory of gravity. To see how gravity is related to non-Euclidean space–time, we go back to Einstein's elevator and the principle of equivalence and now picture the elevator falling freely toward the earth. The observer and everything else in the elevator fall at exactly the same speed, and an object thrown across the elevator moves exactly in a straight line as seen by the falling observer; to him, then, there is no gravitational field. But to the fixed observer on the earth, the object thrown across the elevator does not move in a straight line, but in a parabola. To the observer in the elevator, no gravitational force exists, but to the observer outside the elevator, it does exist. How can these two contradictory points of view be accepted? To Einstein, the solution to this paradox was to eliminate the concept of the gravitational force entirely, since it has no absolute meaning, varying from one frame of reference to another, and to recast Newton's laws of motion to incorporate this idea. He did this by reinterpreting Newton's first law for bodies moving in a gravitational field and stating that bodies always move in straight lines whether in a gravitational field or not. But this statement means the redefinition of the concept of straight lines to include lines that are not straight in the Euclidean sense. Einstein did this by stating that space–time geometry determines whether a line is straight or not, and thus geometry can be Euclidean or non-Euclidean depending on the presence or absence of mass in space. If no mass is present, space–time is Euclidean, but the introduction of masses makes the geometry of space–time non-Euclidean. According to Einstein, the concept of a gravitational force is replaced by curved (non-Euclidean) space–time if masses are present. Gravity thus becomes geometry, and bodies move the way they do in a gravitational field because they follow the curvature of space–time in their neighborhood. This motion is straight-line motion because it is the shortest path in the particular non-Euclidean geometry involved.

Einstein's law of gravity, which is a direct consequence of the non-Euclidean geometry of space–time, corrects Newtonian gravity by making a number of remarkable predictions that were verified; it predicts that the path of a light beam passing close to the sun is bent, as previously stated. It states that the orbit of a planet around the sun itself rotates in the direction of the planet's motion. This phenomenon is called the "advance of the planet's perihelion" and has been observed. Finally, it states that light coming from the surface of a star is reddened (the Einstein red shift); this effect is most pronounced for dense, massive stars such as white dwarfs, since the gravitational fields on the surfaces of such stars are very strong.

All the effects stemming from the general theory of relativity can be deduced from the non-Euclidean geometry of space–time produced by massive bodies such as stars. Time is slowed down (clocks run slower) near such bodies; this accounts for the reddening of light, because the atoms that emit light are essentially clocks, and their slowing down shows up in the reddening of the light they emit. Similarly, rods aligned radially in a star's gravitational field shrink, but do not shrink when aligned laterally; this phenomenon accounts for the non-Euclidean geometry of space and leads to the advance of a planet's perihelion.

Since the slowing down of time and the shrinkage of lengths in gravitational fields affect velocity, the speed of light is decreased in a gravitational field. This means that light travels more slowly near a massive star than when far away from it. Essentially, the pull of gravity on the light slows it down, and if the gravity is strong enough, light is unable to move radially (along a line of force), but can move only laterally. This effect is most pronounced on the surface of a massive, compact star; if a star is massive and compact enough, light cannot escape from it at all. Such a star is a black hole, which, in a sense, bends space around itself so that nothing can leave it. Einstein's gravitational theory also predicts that oscillating masses emit gravitational waves just as oscillating electrical charges emit electromagnetic waves. Although gravitational waves have not yet been detected directly, indirect evidence for the emission of such waves has been obtained from the dynamics of a star revolving around what appears to be a black hole.

General relativity has achieved its most dramatic success in cosmology, a subject that can be treated with Newtonian gravitational theory only marginally, if at all. In 1916, Einstein applied his theory of gravity to the universe as a whole and deduced for it a static (neither expanding nor collapsing) model. Other cosmologists, following Einstein, showed that Einstein's theory leads to nonstatic universe models as well, among which is an expanding model. These expanding models predict that the distant galaxies should be receding from us in agreement with astronomical observations. General relativity has contributed greatly to the richness of cosmology, a subject in which observational and theoretical work is proceeding feverishly.

CHAPTER 16

Atomic Theory

The Bohr Atom

Every step of progress that the world has made has been from scaffold to scaffold and from stake to stake.

—WENDELL PHILLIPS

With the discovery of the negatively charged electron and the positively charged proton as the basic constituents of matter, physicists began to construct models of atoms. With experimental evidence indicating that neutral (uncharged) matter contains equal numbers of protons and electrons, it was clear that the simplest atom consists of one proton and one electron; constructing a model of this atom, hydrogen, was the first step in the development of a useful atomic theory. If the structure and dynamics of the hydrogen atom were understood, everything else would fall into place. At first sight, this task seemed fairly simple, because the electron and proton in the hydrogen atom pull on each other with a force that is similar in form to the gravitational pull between two bodies; the difference is that the gravitational pull depends on the masses of the interacting bodies (for example, the sun and a planet), whereas the electrostatic pull of the proton and electron on each other depends on the positive electric charge on the proton and the negative electric charge on the electron. The equal magnitudes of these opposite charges had been measured by Robert Millikan, so that the magnitude of the electrostatic pull of the proton on the electron for a given separation could be written at once, and all the mathematical techniques that had been developed since Newton's time to handle the gravitational two-body problem (sun and planet) could be carried over to the atomic problem (proton and electron). In doing this, one may neglect the gravitational pull between the electron and proton because the masses of these particles are so small (the proton's mass is one trillionth of a trillionth of a gram and the electron's mass is about 1840 times smaller still) that their gravitational pull is smaller than their electrostatic pull by a factor of 1 followed by 39 zeros. According to this reasoning, the atomic problem seemed quite simple: Apply the solution of the two-body gravitational problem of classical physics to the hydrogen atom, replacing masses in the force formula with electric charges.

This plan of attack seemed quite reasonable and attractive, but a number of obstacles stood in the way of carrying it out. First, the evidence for the correctness of the solution of the gravitational problem is directly visible in the orbits of the planets, but the electronic orbits in atoms cannot be seen and must therefore be inferred indirectly from other evidence. Fortunately, such evidence is present in the radiation emitted by atoms when they are excited in some way. This radiation is called the optical "spectrum" of the atom.

Second, no one knew in the early years of the 20th century just how the electron and the proton in hydrogen—or, for that matter, how the electrons and protons in heavier and more complex atoms—moved about or are arranged with respect to each other. The idea of treating atoms like miniature solar (planetary) systems was very appealing, for it indicated a unity of design in nature from the very large to the very small, but this apparent symmetry and simplicity are deceptive, because electrically charged particles behave quite differently from the way masses behave. A planet, such as the earth, revolves around the sun without losing energy according to Newtonian gravity, but according to the laws of electricity and magnetism, a revolving charge must radiate energy continuously. This means that electrons in atoms accelerated as they are would not have stable orbits unless the electromagnetic principles did not apply to such electrons. But physicists at that time did not know how to change the electromagnetic laws to keep the electrons moving in stable orbits, nor did they want to change these laws, which are in such excellent agreement with all observable electromagnetic phenomena, and so the planetary model of the atom was dropped until an important series of experiments forced it on the attention of physicists.

Lord Rutherford, who had become director of the Manchester physics laboratories by 1910, organized a series of experiments that showed conclusively that the protons in an atom are concentrated in a massive tiny central nucleus and the electrons move around this nucleus in some kind of dynamic pattern that was not understood. These experiments forced physicists to consider seriously the planetary model of the atom, even though they did not know how to handle the serious objection raised against it by Maxwell's electromagnetic principles. This impasse remained until 1913, when the Danish physicist Niels Bohr removed the electromagnetic obstacle to the planetary atom by introducing the quantum theory into the atomic model in a way that was very novel and brilliant, but nevertheless objectionable to many physicists. He began by considering the simplest atom (the hydrogen atom), rather than complex atoms. Bohr saw that to construct a stable model of the hydrogen atom, with the electron revolving around the proton in a stable orbit, this orbit would have to remain at a definite distance from the proton, even though such a distance cannot be constructed from the electron's mass and charge alone. It was necessary to introduce another quantity that, together with charge and mass, could give a distance; Bohr found the solution in Planck's constant of action h. This constant had to be introduced into atomic theory in such a way as to permit the electron to move in a stable orbit about the proton. Bohr saw that this could be done if the electron's action in the atom is quantized so that it cannot move in any arbitrary orbit, but instead must move in any one of a discrete set of orbits, each associated with a definite number (integer) of units of action. Bohr pictured these orbits as concentric circles arranged at increasing distances from the proton, with these distances (radii of the circles) determined by the following quantization condition: The lowest orbit (assigned the number 1) represents 1 unit of action when the electron is in it, the second orbit represents 2 units of action (written $2h$), and so on. These orbits were assigned the numbers 1, 2, 3, and so on; these numbers are called the "principal quantum numbers," and the electron's energy in each of these orbits is different, increasing in a definite way when the electron jumps from a lower to a higher orbit. Since the lowest orbit (orbit 1) represents 1 unit of action and no action less than 1 unit of action can exist owing to the quantization of action, the electron cannot get any closer to the proton than this smallest orbit, and so the stability of the atom is assured.

When the electron is in this lowest orbit (called the "Bohr orbit"), it has the lowest possible energy allowed, so that it cannot emit any energy. The quantum theory thus supersedes the electromagnetic theory and permits the electron to revolve around the proton without radiating energy even though it is accelerated. The electron can jump to higher orbits by absorbing energy, one photon at a time, and each such photon absorption produces a transition to a specific higher orbit. As long as the electron is in any one of these discrete orbits, it does not radiate (it does not obey Maxwell's electromagnetic laws); it radiates a single photon when it jumps from a higher to a lower orbit, and the bigger the jump, the bluer the color of the emitted photon.

This idea of discrete orbits in which an electron can revolve without radiating energy, as demanded by Maxwell's theory of electromagnetism, was so strange at the time that Bohr's older contemporaries were either very skeptical of Bohr's atomic model or rejected it outright. Nevertheless, it had enthusiastic supporters who saw in it the only salvation of atomic physics, even though it left many questions unanswered and raised very puzzling new questions of its own. But despite its objectionable features, the Bohr model of the atom had to be accepted because it explained one very important experimental observation that no other model could even begin to explain—the observed spectrum of the hydrogen atom. When atoms are excited in some way, for example, by collisions with other atoms, they radiate electromagnetic energy, which consists of a mixture of different wavelengths (colors) called a ''spectrum''; the typical atomic spectrum does not contain all possible colors, but consists of a discrete number of colored lines that are characteristic of that particular element and no other.

The struggle to understand the hydrogen atom began in 1885 when Jakob Balmer, a young science teacher in a Swiss girls' school, observed four prominent lines in the absorption spectrum of the sun. These lines were soon identified as belonging to the spectrum of the hydrogen atom and became known universally as the ''hydrogen Balmer lines.'' Balmer showed empirically that the frequencies or wavelengths (colors) of these lines can be expressed in terms of the integers 2, 3, 4, 5, . . . by a very simple formula. This remarkable but mysterious formula remained unexplained until Bohr deduced it from his quantum model (discrete electron orbits) of the atom. Everything fell neatly into place, and the discrete electronic orbits of the Bohr hydrogen atom corresponded exactly with the discrete bright lines of the hydrogen spectrum.

Except for Albert Einstein, Niels Bohr was probably the most influential physicist of the 20th century; his replacement of the causality of classical physics with what became known as ''Bohr's complementarity scheme of quantum mechanics'' (which was grounded in statistical probabilities) forms one of the twin pillars of modern physics along with Einstein's theory of relativity. Bohr's conception of nature as a pattern of events that occur based on chance as opposed to the deterministic model

Niels Henrik David Bohr (1885–1962)

offered by Einstein brought these two giants of modern physics into frequent, if friendly, disagreements about the rationality of nature and the intrinsic features of atomic physics.

Niels Henrik David Bohr was born in Copenhagen in 1885, the son of a professor of physiology at the University of Copenhagen. Niels's father was not a man content to immerse himself solely in his own field of expertise; the Bohr home was open to a continual procession of visitors, many of whom were colleagues of his father, specializing in subjects ranging from philosophy to physics. Even as a young boy, Niels listened carefully to their animated and often protracted discussions; the verbal give-and-take stimulated him to think about his own hazy opinions of subjects ranging from theology and science to politics and economics. Niels not only learned a great deal about the world from these evenings of conversation but also heard many things that made him wonder why the physical world was the way it was at a time when Max Planck was announcing his quantum theory and radioactivity was being thoroughly investigated by the Curies and Ernest Rutherford.

Niels had a pleasant childhood, and he and his brother Harald, who later became a distinguished mathematician, spent much of their free time skiing, cycling, and playing soccer. Despite their enjoyment of the outdoors, both boys took their studies very seriously and developed a deep interest in science and mathematics. By the time Niels entered the University of Copenhagen in 1903, he had developed the basic intellectual tools that enabled him to propose a model of the subatomic world that was as revolutionary as the work of Planck or Einstein. From the start, Bohr showed remarkable maturity as a scientist; his first research project, which was designed to measure the surface tension of water, was carried out with such care and thoroughness that Bohr was awarded a gold medal by the Danish Academy of Sciences in 1906, while still an undergraduate student.

Bohr's approach to his work even at this early stage in his career was to consider a problem from every possible point of view and to mull over any inconsistencies for months or even years, making the necessary corrections and polishing the rough edges until he arrived at what he believed to be a satisfactory answer. While Einstein's supreme intellect gave him an intuitive feel for physics that made possible the flashes of insight that led to the theory of relativity, Bohr's approach was more methodical, and he constructed his view of the world from the ground up much in the same way that a bricklayer begins building the wall of a house. Bohr's strength lay in his determination to continue grappling with a given question about physics, to see whether any additional clues might be found, long after he had arrived at what most of his colleagues would regard as a satisfactory solution.

After completing his doctoral dissertation in 1911, Bohr traveled to Cambridge University hoping to do atomic research at the Cavendish Laboratory with J. J. Thomson: "Unfortunately, Thomson had lost interest in the subject, and failed to appreciate the importance of Bohr's dissertation, which the latter showed him in an English translation he had been at great pains to make; this was turned down by the Cambridge

Philosophical Society as too long and too expensive to print, and Bohr's further attempts to get it published were equally abortive.''[1] Disheartened, Bohr cast about for another research director who would be sympathetic to his interest in atomic theory and found one in Ernest Rutherford, who, in 1910, had proposed his nuclear atom consisting of a positively charged nucleus. Bohr soon joined Rutherford at his Manchester laboratory and, in a three-month flurry of activity in 1912, laid the basis for what became known as the ''Bohr theory of atomic constitution.'' Bohr's work helped to eliminate a number of discrepancies that had cropped up in Dimitri Mendeleev's periodic table by admitting ''the occurrence of atomic nuclei of the same charge but different mass, so that there could be more than one species of atom occupying the same place in the periodic table.''[2] The term ''isotope'' was later coined to refer to chemically indistinguishable substances having different atomic weights, but Bohr's discovery attracted little attention at the time, even from his mentor Rutherford, whose innate conservativism prompted him to try to dissuade Bohr from publishing his results. Despite Rutherford's lukewarm enthusiasm, Bohr continued to push ahead with his development of an atomic model that would explain the stability of the electrons orbiting the nucleus. Common sense suggested that the relatively massive nucleus of the atom would gradually pull the much less massive electrons inward until the atom collapsed. Bohr wondered why this phenomenon did not occur, and his search for an explanation of the stability of the hydrogen atom—which is the least complex atom, as it has only one electron orbiting the nucleus—led him to consider Planck's quantum of action as a possible solution.

The success of the quantum theory in the atomic domain prompted physicists to begin applying the Bohr atomic model to complex atoms with moderate success. It was soon obvious that although the Bohr model is basically correct, it has many minor flaws. From 1913 to 1927, the race was on to improve the Bohr atomic model without changing its revolutionary picture of discrete orbits, and a number of improvements suggested themselves. The Bohr model's use of circular electronic orbits was too restrictive to allow the Bohr model to live up to its full potential, and so the first advance in Bohr's atomic theory came with the introduction of noncircular electronic orbits. This change was similar to Kepler's improvement on the Copernican model of the solar system, which replaced the circular orbits of the Copernican system with elliptical orbits; Bohr's circular electronic orbits were replaced with elliptical orbits. Whereas

Bohr's original atomic model dealt only with the sizes of the orbits of an electron as given by the radii of the orbits, the improved model dealt with the shapes of the orbits as well as their sizes; this additional complexity required the introduction of a second quantum number called the "azimuthal quantum number," which is also an integer. Two integers were now attached to an electron in an atom—the principal quantum number and the azimuthal or orbital quantum number—and just as the principal quantum number corresponds to the energy of the electron, the azimuthal quantum number corresponds to its angular momentum (its rotational motion). Thus, the geometry of the electron's orbit is related in a remarkable way to the electron's dynamic properties by two sets of integers.

These alterations did not mark the end of the evolution of the Bohr theory, for as it stood, even with two sets of quantum numbers, it was unable to account for the behavior of an atom in a magnetic field. Owing to the circling of the electrons in an atom around the atom's nucleus, the atom behaves like a rotating magnet and therefore responds, in a very definite way, to a magnetic field. But this response cannot be explained with classical physics; it requires the quantum theory. Thus, a third quantum number, the "magnetic quantum number," had to be introduced; this third number brought the Bohr model into agreement with the observed features of the atom in a magnetic field, which are indicated by changes in the spectrum of the atom. When an atom is not in a magnetic field, the bright lines in its spectrum can be understood in terms of the principal and azimuthal quantum numbers of the electron, but when the atom is in a magnetic field, many more lines appear in its spectrum. This phenomenon is known as the "Zeeman effect," which classical electromagnetic theory can explain only partially, but which the quantum theory explains fully by introducing a third quantum number. We can understand the need for this third quantum number by noting that the atom behaves like a spinning magnet and therefore precesses around a magnetic field just the way a spinning top precesses around the gravitational field when placed on a horizontal surface. This precession of the atom in a magnetic field is a dynamic feature and must therefore be quantized; this procedure is called "space quantization" because it limits the orientation of the atom's axis of rotation with respect to a magnetic field to a discrete number.

The three aforedescribed quantum numbers that define the dynamic properties of an electron in an atom are assigned to the electron according to definite rules that physicists discovered empirically in the decade between Bohr's presentation of his quantum model of the hydrogen atom and the discovery of the quantum mechanics. Even though much of atomic theory after Bohr's initial work was a mixture of classical Newtonian theory and quantum theory, with new rules being added as needed, it worked pretty well, but certain experimental data could still not be accounted for by the theory. For example, the number of spectral lines of an atom as predicted by the theory (even with the three quantum numbers incorporated into the theory) is about half the number actually observed. This flaw in the theory was eliminated with the introduction of a fourth quantum number, which is related to the spin of the electron.

We are all acquainted with spinning bodies from earth down to spinning motors and tiny gyroscopes that are very important in modern technology, but a spinning electron is hard to conceive, and so it was not included as one of the physical properties of the electron in the Bohr theory. In 1925, however, two young Dutch physicists, Samuel Goudsmit and George Uhlenbeck, showed that if the electron does have spin, the difficulty with the spectral lines in the Bohr theory is cleared up. This means that the electron has a fourth quantum number—the "spin quantum number"—that differs from the other three quantum numbers in that it has only two values. The electron's spin axis in a magnetic field can have only two orientations: parallel to the direction of the magnetic field or antiparallel to it (pointing in the opposite direction).

Even before Goudsmit and Uhlenbeck discovered the spin of the electron, the theoretical physicist Wolfgang Pauli had proposed a fourth quantum number for the electron for reasons that had nothing to do with spin. He found the fourth quantum number necessary to account for the way the electrons in an atom with two or more electrons arrange themselves in successive shells, which is the key to chemical valence. Pauli's great contribution to the Bohr model of the atom was his famous exclusion principle, which states that the electrons in an atom must arrange themselves in such a way (in various orbits) that no two electrons can have the same set of quantum numbers. This simple principle accounts for the periodic table of the chemical elements; it is the highest point in the development of the Bohr model, but it could go no further, and a change was needed, which occurred with the discovery of the quantum mechanics. Before we discuss the quantum mechanics, however, we consider the last great contribution of Einstein to the theory of radiation—a contribution that clearly delineates the role of the photon in the Bohr model of the atom and reveals an important dynamic property of the photon that made the transition from the quantum theory to the quantum mechanics inevitable.

Although Einstein fully accepted Planck's derivation of the black-body radiation formula, he felt that it was not general enough, in that it depended on Planck's assumption that the walls of the interior of the furnace, which were in equilibrium with the black-body radiation, consisted of oscillators. Einstein felt that this special assumption is not necessary and showed this to be true in his 1917 radiation paper by deriving Planck's radiation formula using the Bohr model of the atom. He pictured an atom immersed in a sea of black-body (thermal) radiation emitting and absorbing photons of a definite frequency. The electron in the atom is thus constantly jumping from a lower Bohr orbit to a higher one and back again as it absorbs and re-emits a photon of a given frequency. If many such atoms are present, they will be in equilibrium with the radiation in the sense that at any moment, a certain number of atoms have their electrons in the lower orbit (or state) and a different number have their electrons in the higher orbit (state); these numbers are always the same and are related to each other in a way that depends on the temperature of the radiation. As long as the temperature does not change, these numbers do not change.

To obtain the Planck formula, Einstein had to extend or enlarge Bohr's picture of how an excited atom (one with its electron in a higher orbit) loses its energy so that the electron drops to a lower orbit. According to Bohr, this drop happens spontaneously, with the electron jumping from the higher to the lower orbit without any kind of urging and emitting a photon of the given frequency. Einstein accepted this image, but added to it another emission process that he called the "stimulated emission of photons" (radiation). He argued that in addition to energy, the photon has momentum, just like a material particle. If a photon of the given frequency passes the excited atom, the electron in the higher orbit is stimulated (by the passing photon) to emit an identical photon moving parallel to (with the same momentum as) the original photon. Thus, two identical photons moving together are now present. These two photons by the same process quickly become four and the four become eight and so on until (in a very small fraction of a second) an intense beam of identical photons, moving in exactly the same direction, is built up.

Such beams are used routinely in all kinds of current technologies and are produced by devices called "lasers" (the acronym of "light amplification by stimulated emission of radiation"). Einstein had no such idea as the laser in mind when he introduced the concept of the stimulated emission of radiation to deduce the Planck radiation formula. He was concerned only with solving an important problem in radiation theory and not in any kind of optical technology. It is interesting that his two purely theoretical discoveries—the famous equation $E = mc^2$ and stimulated emission of radiation—have spawned two of the most important technologies today—nuclear energy and the laser.

The importance of Einstein's final radiation paper is twofold: It lent great support to the Bohr model of the atom and introduced into physics the important concept of the momentum of the photon, which underlined the particle nature of the photon. This concept later led Werner Heisenberg to his analysis of his uncertainty principle from a physical point of view.

Bohr's explanation for the stability of the atomic nucleus had a tremendous impact on the study of atomic physics, because his adoption of Planck's quantum of action to explain the stability of the atom was simple and elegant, drawing on one of the cornerstones of modern physics—Planck's quantum of action—to usher in one of the most revolutionary changes in the way scientists view the constitution of the atom. Bohr published his conclusions in the Royal Society's *Philosophical Transactions* in 1913, presenting his theory as being based on two postulates: "The first postulate enunciates the existence of stationary states of an atomic system, the behavior of which may be described in terms of classical mechanics; the second postulate states that the transition of the system from one stationary state to another is a nonclassical process, accompanied by the emission of one quantum of homogeneous radiation, whose frequency is connected with its energy by Planck's equation."[3] Bohr's conclusions were gradually accepted, but Bohr was also quite aware of the need for further analysis of the relationship between "the classical and quantal aspects of the atomic phenomena that were embodied in the two postulates."[3] In 1916, Bohr returned to Copenhagen from Manchester to accept a professorship that had been created for him at the University of Copenhagen. Four years later, he became director of the Institute of Theoretical Physics, which had been funded by a group of private individuals anxious to keep Denmark's most distinguished scientist from returning to England to do atomic research. Like the Institute for Advanced Study at Princeton, Bohr's Institute of Theoretical Physics drew many of the world's leading physicists to Copenagen and helped to establish and nurture professional relationships that begat a variety of important discoveries in nuclear physics.

Bohr was awarded the Nobel Prize in Physics in 1922, the year after Einstein had received his award from the Nobel Foundation. Bohr's receipt of the prize preceded a virtual avalanche of honorary degrees and medals. The Nobel Prize made Bohr's name familiar in households throughout Europe and America; his newfound popularity prompted him to give both popular and academic lectures on physics throughout the Continent. His lecture tours also took him to the United States, where he spent much of his time near Princeton and struck up enduring friendships with such physicists as J. Robert Oppenheimer and John Slater. However, Bohr's prestige was greatest in Denmark, where he was regarded as a national hero and his name was known by most schoolchildren. A brewer even donated an opulent mansion for Bohr's lifetime use.

Bohr's satisfaction with his personal affairs and his position as a leader of the quantum physicists led him to consider some of the philosophical aspects of the quantum theory, particularly the question of whether random events on the atomic level invalidate the idea of a deterministic universe on the cosmic level. Bohr's own views were influenced in the 1920s by Werner Heisenberg's development of the matrix mechanics to encompass Bohr's quantum postulates, Paul Dirac's elegant mathematical model of quantum phenomena, and the extension of the quantum theory by Erwin Schrödinger and Louis de Broglie to allow it "to accommodate both discontinuous and continuous aspects of atomic phenomena" based on their conjecture "that the constituents of matter might, like radiation, be governed by a law of propagation of continuous wave fields."[4] The critical factor in the evolution of Bohr's own views about causality, however, was Heisenberg's 1927 discovery of the indeterminacy relations in atomic physics, which caused Heisenberg to conclude that there is a trade-off in the accuracy of the measurements that can be made of atomic phenomena owing to the disturbance introduced into the system by the act of measuring itself. If one tries to specify the position of a particle, for example, the knowledge of the momentum of the particle becomes less precise.

The recognition that it is impossible to pinpoint both the location and the energy of a particle inspired Bohr to formulate his principle of complementarity, which "consecrated the recognition of a statistical form of causality as the only possible link between phenomena presenting quantal individuality, but made it plain that the statistical mode of description of quantum mechanics was perfectly adapted to these phenomena and gave an exhaustive account of all their observable aspects."[5] Bohr's principle of complementarity was accepted by most physicists but steadfastly resisted by Einstein, who offered a variety of reasons to demonstrate the unworkability of a statistical approach toward atomic phenomena only to have Bohr successfully refute his arguments. Although Bohr always regretted that Einstein's preference for rigid causality on the atomic level had caused Einstein to break ranks with his fellow physicists who, by and large, had accepted Bohr's principle of complementarity, it is difficult to deny the extent to which Einstein's objections enabled Bohr to strengthen the cohesiveness of his theory.

Bohr also collaborated with Wolfgang Pauli, Heisenberg, and Dirac to apply quantum mechanics to the electromagnetic field, a sustained effort that culminated with the publication of a famous paper establishing "the basic principles of field quantization" and showing "that the Heisenberg uncertainty relations apply to the measurement of field quantities just as they do to the measurement of dynamical quantities."[6] James Chadwick's discovery of the neutron encouraged Bohr to devise the concepts of the liquid drop model of the nucleus and the compound nucleus, which "made it possible to understand how two nuclei can collide and form a new nucleus with a variety of different particles being emitted."[6]

When World War II erupted, Bohr opened the doors of his institute to scientists fleeing the Nazis from occupied nations. When Denmark was invaded in the spring of 1940, Bohr refused to leave the country, as he believed that his presence in Denmark was needed more than any work he could do in England or the United States. Although the Nazis tried to bribe Bohr to work for the Axis cause, Bohr steadfastly refused to do any research that might contribute to the German war effort. In 1943, Bohr had to flee for his life to Sweden in a small boat because he received advance notice that the Nazis were planning to use him as an example of what would happen to those failing to cooperate with the German government.

Bohr eventually made his way to the United States, where he worked at the Los Alamos laboratory on the Manhattan Project for the duration of the war. Like many of his fellow scientists who were involved in the development of the atomic bomb, Bohr believed that a fission bomb should be developed if only to prevent the Nazis from doing the same and possibly altering the course of the war. However, Bohr was quite aware that once the atomic genie was out of the bottle, it would be possible for any major power to build its own atomic bombs. As a result, the postwar years saw him increasingly obsessed with seeking ways to prevent a nuclear war by creating an international commission to control the development and uses of atomic power. However, Bohr's pleas fell on deaf ears, and his suggestions for concrete actions to both Roosevelt and Truman went unheeded. At the same time, however, Bohr oversaw the expansion of his institute and acted as an advisor to the Danish government's programs to harness nuclear power for peaceful purposes. His death from heart failure on November 18, 1962, ended the life of one of the most versatile physicists who ever lived and in some sense severed the train of epochal developments that had begun six decades earlier with Planck's quantum theory and Einstein's theory of relativity.

CHAPTER 17

Quantum Mechanics

"When I use a word," Humpty Dumpty said, in a scornful tone, "it means just what I choose it to mean—neither more nor less."
"The question is," said Alice, "whether you can make a word mean so many different things."
"The question is," said Humpty, "which is to be master, that's all."

—LEWIS CARROLL

With all its successes and with the various improvements introduced by Bohr's contemporaries, physicists found the Bohr model of the atom basically unsatisfactory for two reasons: First, it is a hybrid compilation of the laws of classical physics and special quantum rules, arbitrarily introduced to bring order out of the vast body of observations about the atom that experimental physicists were rapidly collecting. Second, certain phenomena such as the intensities of atomic spectral lines and the lifetimes of the excited states of atoms cannot be explained at all with the Bohr model. Physicists were therefore convinced that a new theory, replete with its own axioms and laws, would have to be introduced, but nobody before 1924, when Louis de Broglie began the revolution, knew how to construct such a theory. But the direction one had to go to produce such a theory had already been indicated by Einstein in 1909 when he showed that thermal radiation in a cavity has both wave and particle properties and these two aspects occur in tandem so that they cannot be divorced from each other. This feature of radiation was so revolutionary that Einstein felt it could be understood only through a revolution in our thinking about energy and matter and prophesied "that the next phase of theoretical physics will bring us a theory of light that can be interpreted as a kind of fusion of the wave and the emission theory." By "emission theory," Einstein meant the emission of quanta (corpuscles) of light. The revolution that he predicted began in 1924, but it went beyond his surmise that it would deal primarily with radiation, for it involved matter as well, and interestingly enough, its starting point was Einstein's discovery of the equivalence of energy and matter as expressed in his equation that states that energy equals mass times the speed of light squared. This discovery was already a kind of unification of particles (mass) and waves (energy), but more important, it led de Broglie to the first formulation of a wave theory of matter (of particles).

Reasoning that since matter is equivalent to energy, and energy, according to Planck's formula, is a constant (Planck's constant) times a frequency, de Broglie concluded that matter, like energy, has a frequency associated with it and therefore has wavelike properties. This means that a particle like an electron is accompanied by a wave and hence has a wavelength. To find the wavelength (now called the ''de Broglie wavelength'') of a particle, de Broglie started from Einstein's statement that a photon has momentum and reasoned that the same formula that relates the wavelength of a photon to its momentum must relate the momentum of a particle to its wavelength. Einstein showed from his special theory of relativity that the momentum of a photon equals its energy (Planck's constant times its frequency) divided by its speed (the speed of light). From this expression, one finds that the momentum of a photon equals Planck's constant of action divided by the photon's wavelength. Reasoning by analogy, de Broglie proposed the revolutionary hypothesis that a particle's momentum (its mass times its velocity) equals Planck's constant divided by the particle's wavelength (which was at that time purely hypothetical). Expressed differently, the de Broglie wavelength of a particle is Planck's constant divided by the particle's momentum—the faster the particle moves, the shorter is its wavelength.

De Broglie's reasoning was so brazen and unphysical and his particle-wavelength concept so bizarre that his papers aroused little interest until, some three years later, an important observation, made by two American experimental physicists, Davisson and Germer, indicated that de Broglie's formula for the wavelength of an electron might be correct. In 1927, Davisson and Germer bombarded nickel with electrons and discovered that the electrons bounced off the nickel surface similar to the way X rays (which are waves) do. They concluded that electrons interact with the atoms in a crystal of nickel as though they are waves. Though greatly puzzled by this strange behavior of particles (the electrons), they had enough presence of mind to measure the wavelengths of the scattered electrons from their distribution in direction after bouncing off the nickel surface. Their measurements showed that the wavelengths of the electrons agreed with de Broglie's wave theory of particles, although at the time they were not aware of this theory. However, they had heard of the development of a quantum mechanics at Göttingen University in Germany under the direction of the theoretical physicist Max Born, and so they sent their results to him. Born immediately recognized the importance of their work, and so de Broglie's wave theory of particles became the foundation of a new and strange physics—the quantum mechanics.

That de Broglie made such a fundamental contribution to the study of matter or even become a physicist was surprising, as he was a prince, the son of Victor, Duc de Broglie, and Paulini d'Armaillé. Louis-Victor-Pierre-Raymond de Broglie was born in 1892 at Dieppe and enjoyed all the comforts of the royal life, growing up in an ancestral manor. Despite the frivolous behavior often exhibited by modern royalty, the de Broglie household was steeped in tradition, and the elder de Broglie instilled in his children a respect for authority and a determination not to rest on the cushion of the family fortune. Although Louis had originally studied the classics and graduated with a degree in history in 1910, his older brother Maurice, then a respected physicist who had attended the first of the famous Solvay Conferences on physics, gave Louis a detailed account of the proceedings. Although Louis had not up to that point showed a great deal of interest in scientific subjects, especially physics, Maurice's recollections about the discussions among Albert Einstein, Max Planck, Hendrik Antoon Lorentz, and many of the other top scientists about whether the photon is a real particle captured his younger brother's attention and convinced him to switch careers. Almost overnight, Louis gave up his history books for science texts and, after three years of studying mathematics and science, received his degree in physics in 1913.

The outbreak of World War I interrupted his plans to devote his life to studying the nature of matter, but the four years he spent in the military service as a conscript were not entirely wasted, as he was assigned to the French Army's wireless section, which was posted at the Eiffel Tower in Paris. He found the technical problems of the wireless to be of some interest, but his obsession continued to be the behavior of Einstein's particles of light. Fortunately for modern physics, de Broglie spent the entire war in the relative safety of his Paris post, avoiding the blood-drenched plains of northeastern France, where the lives of thousands of his countrymen had been snuffed out.

After the war, de Broglie resumed his photon research on a full-time basis and became interested in the contradictions that arose from the

Louis-Victor-Pierre-Raymond de Broglie (1892–)

quantum theory of radiation, because it seemed that "if the corpuscular theory (Einstein's photons) were accepted, one could not explain interference and diffraction, which are essentially wave phenomena. On the other hand, if the wave picture were accepted, there would be no way to account for black-body radiation."[1] De Broglie's solution was to try not to favor one theory to the exclusion of the other, but to accept both the wave and the corpuscular theories of light as being correct. Light consists of both waves and corpuscles because, as de Broglie stated in his Nobel address, "it is necessary to introduce the corpuscle concept and the wave concept at the same time" and "the existence of corpuscles accompanied by waves has to be assumed in all cases."

De Broglie first suggested that light is both wave and particle in his thesis to the faculty of sciences at the University of Paris in 1924. Because de Broglie's theory lacked the necessary experimental support and because of the seemingly incongruous attempt to marry the conflicting views of the nature of light, de Broglie's ideas were not immediately accepted.

However, his bold idea did attract the attention of a number of physicists, two of whom, Davisson and Germer, while working at the Bell Laboratories in New York, showed that the diffraction of electrons by crystals occurs and follows the laws of wave mechanics exactly. The convincing proof for de Broglie's revolutionary ideas led to his being awarded the 1929 Nobel Prize in Physics, and not surprisingly, his lecture at the awards ceremony in Stockholm discussed the "undulatory aspects of the electron."

De Broglie's academic career as a lecturer at the Sorbonne and then as professor of theoretical physics at the Henri Poincaré Institute preceded his appointment to the chair of theoretical physics at the University of Paris in 1932. Like Einstein, however, de Broglie found that his attempts to derive a causal interpretation of quantum mechanics were ignored by many of his younger colleagues; de Broglie was regarded as something of a dinosaur for his stubborn insistence on the need for determinism in modern physics. Both de Broglie and Einstein greatly disliked the probabilistic interpretation of quantum mechanics offered by Niels Bohr, Max Born, and Werner Heisenberg, but each found it impossible to refute the arguments of those who favored a statistical interpretation.

Although de Broglie was the founder of the new physics, he was not the main figure in its rapid growth. The development of quantum mechanics along two different and apparently disparate paths, the matrix mechanics and the wave mechanics, is credited to Born, Heisenberg, and Jordan (the matrix mechanics) and to Erwin Schrödinger (the wave mechanics). Paul Dirac and Wolfgang Pauli also played a very important and often dominant role in this remarkable work.

The development of the matrix mechanics began in 1925, when Heisenberg, in a seminal paper, initiated the revolt against classical physics and the hybrid classic–quantum theory by arguing that a new quantum theoretical mechanics was needed with its own consistent rules and laws that borrows nothing but the most general concepts from classical mechanics. In particular, he argued that the quantum mechanics should deal only with observable quantities and that such concepts as Bohr orbits, which cannot be observed, should be banished from atomic physics.

One was no longer to speak of the electron as being at some particular point or in some particular orbit in an atom; instead, the position, as a point, was to be replaced by an array of numbers indicating that the electron is to be pictured as spread out over all the Bohr orbits; such an array of numbers is a matrix. The position of the electron is not a numerical quantity but a matrix. Heisenberg also replaced the Newtonian momentum of a particle by a matrix consisting of many possible values of the classical momentum of the particle. This work was the beginning of the matrix mechanics, which was then developed by Born, Heisenberg, and Jordan into a complete self-consistent mathematical technology for solving atomic problems.

Since the order in which two matrices are multiplied makes a difference in the value of the product, the algebra that must be used in matrix mechanics is noncommutative. If the position of an electron is q and its momentum is p, quantities that are precisely determined in Newtonian physics, the product of the two quantities, pq, is the same as the product qp, and the order in which they are multiplied makes no difference. In matrix mechanics, however, p and q are not precise numbers; they are matrices, so that pq does not equal qp. Heisenberg showed that this noncommutativity of multiplication in matrix mechanics means that the quantities involved in the multiplication are governed by an uncertainty principle (Heisenberg's indeterminacy principle) in the following sense: The two terms in the product (q and p in the foregoing example) cannot both be measured with infinite accuracy. The more accurately one measures the position of an electron, the more uncertain is one's knowledge of its momentum, and vice versa. The product of the error in our knowledge of q and the error in our knowledge of p cannot be less than Planck's constant divided by the number 2π.

Heisenberg presented the uncertainty principle from a physical point of view by analyzing the measurement of the position of an electron that is moving in a straight line with constant speed. To locate the electron on the straight line, one must "look at it," which means directing a beam of light at it. When a photon in the beam strikes the electron, the photon is reflected to our eye, and we deduce the position of the electron by applying a simple optical principle (the law of reflection). But the photon, on reflection from the electron, communicates some of its momentum to the electron, which introduces an error in our knowledge of the electron's momentum. The more accurately we wish to determine the electron's position, the greater is the error in our knowledge of its momentum. The reason for this trade-off is that we must irradiate the electron with a very-short-wavelength photon to locate it with only a small error, but that means that a high-energy and therefore a high-momentum photon is needed, which introduces a large error in the electron's momentum. All this stems from the existence of a quantum of action, as defined by Planck's constant of action h, so that we cannot control the disturbance in our measurements produced by a photon.

Heisenberg's uncertainty principle, first applied to position and momentum, applies to other pairs of measurable quantities as well. Pairs of such quantities, which are called "conjugate variables," play an important role in quantum mechanics, for they set limits on the determinism of quantum physics. Another example of an important pair of conjugate variables is time and energy. The uncertainty associated with these two quantities is as follows: The longer the time we have to measure the energy of a particle, the more precise the measurement will be; if we make the measurement in a very short time, the error in the energy measurement is very large.

We can make the uncertainty principle somewhat more comprehensible by deducing it from the existence of the quantity of action h. This means that the action associated with a particle cannot be less than h (the unit of action). If we locate a particle within a small stretch of its path, the length of which is the error in our knowledge of its position, the action of the particle in that stretch is the product of the length of the stretch and the particle's momentum. But this product cannot be less than h. Hence, the particle's momentum must be large enough so that its product with the length does not fall below h. As the error in our knowledge of the particle's position (the length) decreases, the particle's momentum and the error in our knowledge of it increases. This is the uncertainty principle, which stems, as this discussion indicates, from Planck's discovery of the quantum of action.

Werner Karl Heisenberg (1901–1976) was born in Würzburg, Germany, the son of a humanities professor at the University of Munich. Werner grew up in a household that prized the literary classics, and his father's appreciation for the cultural history of Europe undoubtedly influenced Werner's later decision to stay behind in Germany to "carry the torch" of German culture when the Nazis came to power. He received his early education at the Maximilian Gymnasium, where he studied the classics, favoring the scientific works of the early Greek philosophers ranging from Plato and Aristotle to Democritus and Thales. His interest in the relationship of philosophy and science continued throughout his career, and many of his later books are rife with philosophical considerations. His familiarity with the classical works of science enabled him to perceive and understand the uneven manner in which physical theories about the nature of the universe had been conceived, accepted, and discarded and to approach his own work in a patient and methodical manner.

Heisenberg graduated from the Gymnasium in 1920, having studied physics and philosophy. He also had an interest in music, but the lure of

Werner Karl Heisenberg (1901–1976)

theoretical physics was too strong. Heisenberg visited Arthur Sommerfeld, who was then one of the leading atomic theorists in the world, and asked that he be permitted to study theoretical physics. Sommerfeld was somewhat taken aback by the boldness of Heisenberg's request and suggested that he complete the basic physics courses before deciding to specialize in a given area. Heisenberg accepted Sommerfeld's advice and began his studies at the University of Munich, where Sommerfeld taught. He was also permitted to attend Sommerfeld's seminars, where he held his own against the upperclass and graduate students. From the beginning, Heisenberg demonstrated an uncanny mastery of his subjects and passed his doctoral examination in physics in 1923, only six semesters after he had enrolled. He moved to the University of Göttingen, where Max Born was directing a program that transformed that institution's physics department into one of the finest in the world. Heisenberg worked as an assistant to Born and in the next two years developed the basis of what became known as the quantum mechanics. In 1925, he completed a revolutionary paper outlining his ideas. Born recognized the value of Heisenberg's article and forwarded it to the

Zeitschrift für Physik, in which it was published. Heisenberg then collaborated with Born and Pascual Jordan to make additional improvements in his theory, most notably the actual formulation of the noncommutative algebra that describes the matrix mechanics. Heisenberg was gradually led to his uncertainty principle, because it was "clear to him that if the theory was to deal only with physical quantities that can be observed directly, it would have to be cast in a form in which it could not speak of the exact position and momentum of a particle simultaneously, since measuring either one of these quantities would affect the other and hence our knowledge of it."[2] Heisenberg then left Göttingen to work for three years in Copenhagen with Niels Bohr before becoming professor of theoretical physics at the University of Leipzig, where he remained until 1941. This period of Heisenberg's life was extremely productive, as he showed how the quantum mechanics could be applied to the study of magnetic phenomena and collaborated with Wolfgang Pauli "to lay the foundations of quantum electrodynamics and quantum field theory" and introduced the concept of isotopic spin, "in which both the neutron and the proton are considered as two different energy states of the same basic particle, the nucleon."[3] Heisenberg received the Nobel Prize in Physics in 1934. Because Heisenberg was one of the most distinguished physicists in Germany, he was appointed in 1942 to be the director of the Max Planck Institute of Physics in Berlin, where he remained for the duration of the war. Because of the Nazis' persecutions of the intelligentsia in general and the Jews in particular, Heisenberg was sorely tempted to join the exodus of scholars from Germany, but he elected to remain behind because of his own loyalty to the German state, even though he detested Hitler's government. As did Max Planck, Heisenberg believed that he had to remain behind to preserve whatever vestiges of German science remained after the war. In 1945, Heisenberg returned to Göttingen as the director of the Max Planck Institute, where he developed his scattering matrix theory "which attempts to describe events only in terms of what one sees initially before two systems are brought together (two nuclei, for example, or two protons) and what one sees finally, after they have interacted and have separated."[4] Heisenberg spent the remainder of his career trying to derive the properties of the fundamental particles such as electrons and protons and writing about philosophy and science in such books as *Across the Frontiers* and *Physics and Philosophy*.

Although Heisenberg and Wolfgang Pauli solved certain elementary problems with the matrix mechanics, most physicists found its mathematical structure too difficult to work with. Quantum mechanics would have evolved very slowly had Erwin Schrödinger not discovered the wave mechanics. The wave property of particles (the wave–particle dualism), as revealed in de Broglie's paper, inspired Schrödinger to a high degree of creativity that culminated in his production of what is now called the "wave equation of the electron," although it has a much more general character than was first attributed to it, for it applies to any particle, not just the electron.

Erwin Schrödinger (1887–1961) was born in Vienna at a time when it was the cultural center of Europe and the capital city of the decaying Austro-Hungarian Empire. His father, Rudolf Schrödinger, despite having been trained as a chemist, had devoted himself to Italian painting and botany. As the family was very well off financially, Erwin's parents had the time and energy to introduce him to the cultural and historical riches of Vienna. Both parents taught their son to appreciate the arts and to enjoy the pursuit of knowledge itself. Against the backdrop of beautiful Vienna, it was not long before Erwin developed a keen love of life and an interest in the biological processes of living organisms.

Erwin received his early education at the Vienna Gymnasium, where he studied the sciences and mathematics. He was also very fond of literature and enjoyed poetry and languages. Like Einstein, Erwin did not enjoy rote learning, but he did not develop the hatred for authority that Einstein did during his school years, even though Erwin would have preferred to structure his courses more to his own tastes. He received high marks in his subjects and in 1906 entered the University of Vienna, where he studied classical physics. One of his areas of study at Vienna was the physics of continuous media, which gave Schrödinger the intellectual tools to understand the wave theory of light. This understanding was of crucial importance to his later formulation of the wave equation for particles.

After graduating from the University of Vienna in 1910, Schrödinger was appointed as an assistant professor of physics there. Despite his theoretical interests, he was hired to set up laboratory experiments for the students. Schrödinger did not really enjoy his work in the laboratory, as he considered himself to be a theoretical physicist. Finding no such job available, however, he was forced to swallow his pride and make the best of his work in the laboratory, even though he never considered himself to be anything more than a passable experimentalist.

Schrödinger spent World War I as an artillery officer, then resumed his academic career in 1920 as an assistant to Wilhelm Wien, the farsighted physicist who, as the editor of *Annalen der Physik* in 1905, had been the first scientist to recognize the brilliance of the three papers by

Erwin Schrödinger (1887–1961)

Einstein on the photoelectric effect, Brownian motion, and the special theory of relativity. Schrödinger then accepted successive appointments at Stuttgart and Breslau before succeeding Max von Laue as professor of physics at the University of Zurich. Schrödinger's years at Zurich were the most productive of his academic career, because he published technical papers on thermodynamics, statistical mechanics, the specific heats of solids, and atomic spectra. He also formulated his wave equation because he disliked the idea of the quantum jumps: "He therefore sought to return to some kind of continuous classical description by treating the spectrum as the solution of an *eigenvalue* problem. His reasoning was that if the discrete modes of vibration of a classical system like a violin string could be obtained as the solution of an *eigenvalue* problem, so could the Bohr stationary states. One could thus eliminate, so he thought, the idea of quantum jumps and replace it by the concept of transitions from one mode of vibration (*eigenvalue*) to another.''[5]

Schrödinger's discovery of the wave equation followed his merger of the work de Broglie had done on the wave nature of the electron with the mathematical framework that William Hamilton had devised for Newtonian mechanics.[5] His skill in reconciling the works of the two physicists to fashion an equation of great value to physicists in the modern era was recognized in 1933 when he shared the Nobel Prize in Physics with Paul Dirac. Because Schrödinger regarded the electron not as a particle but as a real wave that was spread out through space in different concentrations, he rejected Born's statistical interpretation of quantum mechanics or the wave-particle duality of matter.[6] His opposition to Born's quantum mechanical scheme sparked a friendly but protracted debate with Born that lasted for the rest of their careers.

On the retirement of Max Planck in 1927, Schrödinger was invited to Berlin to become Planck's successor, a position he held for six years. He found his daily contacts with the distinguished faculty there very stimulating, but he decided to resign his post and leave the country when the Nazis came to power in 1933. Schrödinger was not Jewish, and his Austrian citizenship would not have prevented him from enjoying a very comfortable life had he chosen to collaborate with the government, but he doubted that he could live in Hitler's Germany, especially when so many colleagues, including Born, were being forced to leave owing to the Nazi racial laws. Schrödinger accepted a fellowship at Oxford, where he taught for two years before moving on to the University of Graz in 1936.

The annexation of Austria by Germany forced Schrödinger to flee to Italy in 1938 and then to Princeton, where he settled temporarily. He then became the director of the school of theoretical physics at the Institute for Advanced Studies in Dublin, Ireland, where he remained until his retirement in 1955. Although he continued to do theoretical research in physics, Schrödinger's best-known work while at Dublin was a little book that he published in 1944 called *What is Life?,* in which he tried to determine whether biological phenomena such as heredity could be explained by quantum jumps. Although Schrödinger's biological views have been replaced by more recent developments in that field, such as the discovery of the DNA molecule, his book enjoyed great popularity and encouraged a number of physicists to study molecular biology.

After his retirement from Dublin in 1955, Schrödinger returned to his beloved Vienna, where he was given many honors. He continued to think about questions of physics and biology until his death in 1961, but he also spent much of his time walking around the streets of Vienna and the surrounding countryside, enjoying the sights and sounds of a city whose cultural and artistic soul had so heavily influenced his own education as a young man and nurtured his own multifaceted genius.

Schrödinger was a brilliant theoretician who had distanced himself from the Bohr theory of the atom because he abhorred the concept of discrete orbits and discontinuous jumps of the electron from one orbit to another. Steeped in classical physics, he was a master of classical wave equations and the solutions of problems stemming from vibrating systems such as plucked strings and oscillating plates. With his orientation toward classical physics and his bias against the Bohr model of the atom, he was very receptive to de Broglie's wave theory of matter, for he saw in it a way to replace the discontinuous Bohr model and the cumbersome matrix mechanics by a single wave equation of the electron from which all the desirable features of the Bohr model and the matrix mechanics could be deduced. Moreover, such a wave equation would be in the tradition of classical physics, in which wave equations abound. Physicists knew how to handle and solve such equations and could therefore work with quantum mechanics in wave form.

To obtain his electron wave equation, Schrödinger went back to the classical (Newtonian) description of an electron moving in an electrostatic field (in the field of the proton in the hydrogen atom). He saw that if he replaced the momentum in the classical expression for the electron's energy by a mathematical operator (the operator being the rate of change of the electron's wave with respect to the change in the electron's position) and the energy itself by another operator (the rate of change of the electron's wave with respect to time), he would obtain the wave equation he wanted. He was convinced that in doing this operation he had brought continuity back into physics and banished the repugnant discrete electronic orbits and their attendant discontinuities, but he really had not done so. The discontinuities were camouflaged and hidden in the wave itself.

Schrödinger immediately applied his prescription for obtaining a wave equation to the electron moving in the electrostatic field of the proton. He wrote the classical (Newtonian) equations for the electron's energy and then replaced the momentum in this expression and the energy itself with his operators. He then applied the entire expression, as an operator, to a function of space and time (a quantity that varies with the position of the electron and with time). This is the initial Schrödinger wave equation. An expert in handling such equations, Schrödinger showed that this wave equation breaks up into three distinct equations, one of which gives the Bohr orbits with their principal quantum numbers (the electron's energy states), another the shapes of the orbits (the azimuthal quantum numbers), and the third the discrete orientations of the atom's axis of rotation in a magnetic field (the discrete magnetic quantum numbers). By using a single equation, Schrödinger produced in a matter of hours everything that physicists had struggled for more than a decade to obtain with a series of arbitrary, *ad hoc* rules. This success made the wave mechanics the primary mathematical tool of quantum mechanics for solving atomic problems. The matrix mechanics played only a minor role in the rapid growth of the quantum mechanics that followed.

But even with its great success, Schrödinger's wave equation raised many questions and was surrounded by impenetrable mysteries that still have not been cleared up. The most pressing of these questions was and is the nature of the wave that is associated with the electron. In all classical wave phenomena, the wave is a real, physical entity that can be observed and the intensity of which can be measured with physical instruments. Thus, the intensity of a light source or a sound source is given by the square of the amplitude (the size of the vibration) of the light wave or sound wave emanating from the source. But the Schrödinger wave is not a real wave, so that its intensity cannot be measured; the expression for the Schrödinger wave contains the imaginary number $\sqrt{-1}$ and is therefore a complex quantity.

This unphysical feature of the wave associated with a particle predisposed the matrix mechanics supporters to accept the wave mechanics with some strong reservations, but these were finally swept away when Max Born, one of the cofounders of the matrix mechanics, proposed an entirely new and radical interpretation of Schrödinger's electron wave: It is a measure of the probability of finding the electron in a given region of space. To be specific and explain this more fully, we consider an electron moving about in a box. We know nothing about its position at any moment, but we can describe its motion by the Schrödinger equation, and if we solve this equation, we obtain the electron wave function, which tells us much more about where we can find the electron. We first replace $\sqrt{-1}$ wherever it appears in the wave function with $-\sqrt{-1}$, which gives us the complex conjugate wave function. If we now multiply the wave function by this complex conjugate, the quantity we obtain (the absolute value of the wave function) gives us the probability for finding the electron at any desired point in the box. In accordance with Born's interpretation, the Schrödinger wave is a probability wave.

This picture of the wave function was universally accepted, for it permits one to calculate all kinds of probabilities for the electron. One can thus calculate the probability for an electron in an excited atom to jump down to any given lower level and emit a photon of a particular frequency. In this way, the lifetimes of excited atoms (the times spent in excited states) and the intensities of their spectral lines can be calculated, which cannot be done with the Bohr theory.

As more and more atomic problems yielded to the Schrödinger equation, the equation dominated atomic physics and became synonymous with quantum mechanics. However, difficulties with it began to appear, some of which were overcome but some of which were basic flaws in the theory. The equation is so constructed that it treats time in a special way as compared to the way it treats space; this difference is objectionable, because it conflicts with the requirements of relativity theory, which demands that space and time be treated on an equal footing. This means that the Schrödinger equation is not relativistically invariant (it changes when transformed from one space–time coordinate system to another), as all correct theories must be. Schrödinger and all the other top physicists of that day knew this was true, but they did not know how to obtain a relativistically correct electron wave equation.

The flaw of nonrelativism in Schrödinger's wave equation arises because it is derived from the Newtonian relationship between the energy and the momentum of a particle, instead of from the Einstein relationship. Noting this, Schrödinger did obtain a relativistically correct wave equation by starting from Einstein's nonquantum energy–momentum equation and replacing the energy and momentum with the appropriate mathematical operators (a time operator for energy and space operators for the momentum) to obtain a relativistically correct wave equation. But this equation led to meaningless physical phenomena, because the solutions of this equation (the wave functions) give negative as well as positive probabilities for events, and negative probabilities have no physical interpretation. These difficulties persisted until Paul Dirac, the great British theoretical physicist, derived what is now known as the "Dirac relativistic equation of the electron."

Paul Adrien Maurice Dirac (1902–1984) was born in Bristol, England, the son of a Swiss father and an English mother. As a child, he was shy and withdrawn; even though he could speak both English and French, he rarely said very much in any language. Although his reluctance to carry on conversations gave his parents some cause for worry, Dirac soon showed that he was not mentally deficient by his fine academic work in the primary and secondary schools he attended in Bristol. He favored

Paul Adrien Maurice Dirac (1902–1984)

mathematics, but he sought a vocation that would enable him to use mathematics in a practical manner, and so he settled on engineering. Dirac studied electrical engineering at the University of Bristol and received his baccalaureate degree in 1921. However, he could not find an engineering job and, perhaps because of his ease with numbers, became interested in physics. Feeling that his becoming a physicist would be aided by further study of mathematics, Dirac spent the next two years immersing himself in that subject at Bristol, even though he was not in a formal degree program. Once he had sharpened his analytical skills to his satisfaction, Dirac left Bristol to work as a research assistant in mathematics at St. John's College, Cambridge. During this time, he continued his mathematical studies and received his Ph.D in mathematics in 1926. He remained as a fellow at St. John's College for several years, until he was appointed in 1932 to what is perhaps the most famous chair in the world—

the Lucasian professorship of mathematics at Cambridge, the position formerly held by Isaac Newton. That Dirac should receive such an honor at such a young age showed that he had already been recognized as one of Europe's outstanding mathematical physicists for his work in formulating a general noncommutative algebra for quantum mechanics not dependent on Schrödinger's wave equation or the matrices of Heisenberg and Born as well as his development of the relativistic theory of the electron, the latter of which earned him half of the 1933 Nobel Prize in Physics.

Before assuming Newton's old post, Dirac had already earned his substantial reputation as a theoretical quantum physicist by his work in two different areas: He had demonstrated (as Schrödinger himself had done independently) that the matrix mechanics and the wave mechanics are equivalent and had taken the first step in the development of quantum electrodynamics by showing that Maxwell's electromagnetic theory (Maxwell's equations of the electromagnetic field) can be cast into a quantum mechanical form—the quantization of the electromagnetic field. In Dirac's version of the quantum mechanics, the physical state of any particle or system of particles is represented by a function of space and time (the state function) that contains all the information that one can learn about the system by observing it. Each observation is a physical operation (for example, observing an electron's position or momentum) that is represented by a mathematical operator that is applied to the state function, and these mathematical operators are governed by a noncommutative multiplication. In the wave mechanics, the state function is Schrödinger's wave function; in the matrix mechanics, it is a matrix.

Dirac treated the electromagnetic field as a collection of oscillators, for each of which the quantum mechanical behavior was well known and each of which represents a photon. In this way, Dirac replaced the classical Maxwellian electromagnetic field by quantum mechanical oscillators, each of which obeys its own Schrödinger wave equation. This breakthrough marked the beginning of what is now called "quantum field theory," which is used extensively by high-energy particle physicists today. By introducing distinct oscillators to represent the electromagnetic field, Dirac replaced continuous quantities (the electric and magnetic field intensities), which are difficult to handle in quantum mechanics, with discrete quantities (the oscillators).

In 1927, when Dirac's contributions to quantum mechanics were universally recognized, he was, as he mentioned to Niels Bohr, "trying to get a relativistic theory of the electron." Recognizing that the relativistic wave equation obtained by Schrödinger and other physicists is unsatisfactory, since it gives negative and therefore incorrect probabilities for finding the electron in any given region of space, Dirac saw that the difficulty arose because the Einstein relationship between energy and momentum, from which one must start, relates the square of the energy to the square of the momentum, rather than the energy to the momentum. Wave mechanics requires that the energy rather than the square of the energy be used in setting up the wave equation. Dirac therefore had to take the square root of the Einstein energy expression, which he did in a very ingenious way that gave him just what he wanted, but it was at a certain price. His procedure for obtaining the energy and working with it instead of with the square of the energy introduced an unexpected complication: The Schrödinger nonrelativistic wave equation for the electron was replaced by four distinct equations. Dirac's entire procedure and his fourfold equation were so distasteful to Heisenberg and Pauli that they rejected the whole idea initially. Not until the solutions of Dirac's four equations gave results for atomic dynamics (spectral lines and others) considerably superior to those obtained from the nonrelativistic equation did Heisenberg and Pauli grudgingly accept Dirac's equation. They were particularly prompted to do so because the "Dirac equation" (actually the four equations) gives spin to the electron, just as Goudsmit and Uhlenbeck had proposed, which the Schrödinger equation does not. But Dirac's equation was still repugnant to Pauli because, in addition to endowing the electron with spin, it has what at first appeared to be a very undesirable feature—it predicts electrons with negative energy.

To see how this result came about and how Dirac interpreted negative-energy electrons to make them acceptable to his contemporaries, we note that negative energies are present in the special theory of relativity, in which the square of the energy of a particle is related to the square of the particle's momentum plus its mass, so that the square root of this sum must be taken to obtain the energy itself. But the square root of a quantity may be either positive or negative, so that negative energy is inherent in the special theory of relativity. But negative energies were simply disregarded in relativity theory before quantum mechanics was discovered; negative energies cannot be disregarded in quantum mechanics, however, because quantum mechanics predicts that if negative energy states exist, electrons will jump down to fill them (unless they are prevented from doing so) and all matter will disappear into negative states in one vast explosion, leaving behind a radiation-filled universe. Since this does not happen, some mechanism must prevent such a catastrophe, and Dirac proposed such a mechanism; he suggested that each negative energy state in the vacuum already has one electron (a negative-energy electron) in it, so that, by the Pauli exclusion principle, it cannot accommodate another electron. Thus, the Pauli exclusion principle saves all the matter in the universe from disappearing into the infinitude of negative energy states in the vacuum. The vacuum is not empty, but is infinitely filled (tightly packed) with negative-energy particles, which we cannot detect precisely because their energies are negative.

We now see why the Dirac relativistic wave equation of the electron assigns four equations to a moving electron; the electron can be in a state of either positive or negative energy (two equations are required), and it may spin either clockwise or counterclockwise about its spin axis, which requires two more equations (one for each state of spin). This accounts for the four wave equations that Dirac deduced from the theory of relativity. Although the concept of negative energy states was very bothersome at first and appeared to be a definite drawback to the theory, it was gradually accepted because it predicted the spin of the electron and gave the correct magnetic properties of the electron (it behaves like a very tiny magnet).

But the negative energy states soon proved to be a great asset of the theory, for under Dirac's analysis, they led to a remarkable prediction. Dirac showed that if a photon with enough energy (equivalent to at least twice the mass of the electron) were absorbed by a negative-energy electron, the electron would become a positive-energy electron and leave a hole in the vacuum in its place. This hole, being the absence of negative charge and negative energy, would behave like an electron with positive charge and positive energy. This theory is known as "Dirac's theory of holes" and was merely a fanciful curiosity until Carl Anderson discovered such positive electrons, which he called "positrons," among cosmic rays. These positrons are indeed the Dirac holes (also called "antielectrons"), for when a positron and an electron meet, the electron disappears in the hole and the hole, being filled with the electron, also disappears. Thus, the positron and electron annihilate each other, so that each is the antiparticle of the other. This event is a remarkable example of the predictive power of pure theory.

With the Dirac theory of the electron accepted, the wave mechanics began to permeate all of physics, and chemistry as well. Even without the Dirac theory, physicists and chemists began to apply the Schrödinger equation to molecular dynamics, and the various molecular bonds (ionic and homopolar bonds) were analyzed and explained satisfactorily. Such chemical properties of atoms as their valences were also deduced with the wave mechanics. The application of the wave mechanics to ensembles of particles such as the molecules of a gas led to the discovery of certain important symmetry properties of the wave function of such an ensemble in relation to the statistics of the particles in the ensemble. The first important discovery was that the statistics of the ensemble must be so designed that interchanging any two identical particles (atoms, electrons, or molecules) makes no difference to the results deduced from the statistics. This is different from classical statistics, which assumes that one can distinguish between two identical particles. The next symmetry feature of the wave function that affects quantum statistics stems from the algebraic change in the wave function when two identical particles of the ensemble are interchanged. Since the interchange of the positions of two identical particles cannot be detected in quantum mechanics, such an interchange cannot alter any dynamic property of the ensemble such as its energy, but it can alter the algebraic sign of the wave function from plus to minus or vice versa without affecting the dynamics of the ensemble. The reason is that the wave function itself is not the determining factor in the dynamics, but rather the product of the wave function and its complex conjugate, and this product is the same whether the wave function is positive or negative.

But the statistics that one must apply to an ensemble do depend on whether the sign of the wave function changes or remains unaltered on the interchange of two identical particles. If the sign does not change, the wave function is said to be symmetric; if the sign does change, the wave function is antisymmetric. The importance of this kind of symmetry, in addition to its relationship to the kind of statistics that must be used in analyzing an ensemble of identical particles (for example, electrons), is that antisymmetric wave functions (those that change sign) must be used to describe an ensemble of particles that obey the Pauli exclusion principle, whereas symmetric wave functions must be used for particles that are not governed by the exclusion principle.

Another important physical property of the particles related to this symmetry of the wave function is their spin. The basic unit of spin (rotation) in nature is Planck's constant h divided by 2π, or $h/2\pi$, which is written as $\hbar$. The fundamental particles in nature, such as electrons, are placed in one of two different categories depending on whether they have either half a unit of spin ($\hbar/2$) or zero or one unit ($\hbar$) of spin. Particles such as electrons and protons with a half unit of spin are called "fermions" (named after the great Italian physicist Enrico Fermi); they must be described by an antisymmetric wave function, because they obey the Pauli exclusion principle. Particles like photons (one unit of spin) that have either zero spin or one unit of spin must be described by a symmetric wave function, because they do not obey the Pauli exclusion principle; such particles are called "bosons" (named after the Indian physicist Sir Jagadis Chandra Bose). In considering the symmetry properties of the wave function, one must take into account the dependence of the wave function not only on the positions of the particles (their spatial coordinates) but also on their spins. Thus, interchanging any two particles in an ensemble means interchanging not only their positions but also their spins. The relationship between the spins of particles and their statistics was discovered by Fermi for spin $\frac{1}{2}$ particles and by Bose for spin 0 or spin 1 particles. We thus speak of fermions obeying Fermi statistics and bosons obeying Bose statistics. Since electrons obey Fermi statistics, many phenomena of our daily world, in which electrons play a vital role, are governed by Fermi statistics. The electrical conductivity of metals, superconductivity, and many other aspects of solid-state physics are governed by Fermi statistics.

These statistics also play an important role in the structures of stars like the sun near the end of their evolution when they become white dwarfs and in the structure of neutron stars (pulsars). White dwarfs are maintained in equilibrium against gravitational collapse by the outward pressure of their free electrons, in accordance with the Fermi statistics. Free neutrons, also governed by Fermi statistics, do the same thing for neutron stars. Bose showed that free photons in a container (a photon gas) do not obey the Pauli exclusion principle and are described by a symmetric wave function because a photon has one unit of spin ($\hbar$). They are therefore governed by the Bose statistics, from which Bose deduced the Planck radiation formula. Because the nuclei of helium-4 (ordinary helium) have zero spin, they obey Bose statistics; all the remarkable properties of helium at temperatures near absolute zero stem from its Bose statistics.

QUANTUM ELECTRODYNAMICS

With the great success that physicists had achieved in applying quantum mechanics to atomic dynamics, they began treating the electromagnetic field and its interaction with charged particles quantum mechanically also. This general and broad field of physics is called "quantum electrodynamics"; it is at the very heart of the relationship between matter (for example, electrons) and pure energy (radiation or photons). This branch of physics consists of the quantum mechanics of the radiation field (quantizing Maxwell's electromagnetic equations) and the quantum mechanics of the interaction between charged particles (for example, electrons) and the electromagnetic field. This branch of quantum mechanics was started by Dirac in 1927 and continued by Heisenberg, Pauli, and, particularly, by Fermi, whose treatment was the simplest and most direct.

The quantum mechanics of the radiation field itself is quite straightforward and fairly easy to grasp; the field is represented dynamically by a collection of harmonic oscillators in various states of excitation that can emit and absorb photons, thus changing the states of the radiation field, which is pictured as creating or annihilating photons via these oscillators.

The difficulty in quantum electrodynamics arises in problems involving the interaction of charged particles (for example, electrons) with the electromagnetic field. This is a fairly simple problem in classical electrodynamics, but a complex one in quantum electrodynamics, since one cannot solve exactly the wave equation of the field interacting with the electric charge; instead, one must use a perturbation procedure (a series of successive approximations), with terms of ever-increasing complexity appearing as one goes to higher approximations. The reason for this is that the interaction between charge and field is explained as arising from the emission and absorption of virtual photons by the charge. A virtual photon has no observable existence, but is created and reabsorbed in a very short time, which is allowed by the uncertainty principle without violating the conservation of energy. Two charges interact by tossing virtual photons between themselves. A charge also interacts with its own electromagnetic field by emitting and reabsorbing its own virtual photons. Calculating electromagnetic interactions involves an infinitude of virtual photon emissions and absorptions; hence, its complexity is such that the calculation must be done step by step, and when such calculations are done in the standard way, they blow up at each step and become infinite. The problem of these infinities was intractable until the theoretical investigations by Shin'ichirō Tomonaga in Japan and by Julian Schwinger and Richard Feynman independently traced the difficulty to the mass (the self-energy) and the electric charge of the electron. Since the interactions of the electron, via its charge, with its own electromagnetic field contribute to the electron's mass, such interactions must be calculated in determining the electron's mass, but they give infinite answers. In an analysis similar to that of Tomonaga, Schwinger and Feynman showed how to sweep these infinities under the rug by subtracting them from the calculations to obtain finite results. This kind of subtraction physics is called "mass and charge renormalization"; in all calculations, one replaces the theoretical mass of the electron, as it would be if the electron had no charge, with the experimental mass and neglects the interaction of the charge with the field. One does the same thing with the electron charge, so that the charge and mass are just numbers that one does not worry about any more. This procedure gives incredibly accurate theoretical results for most experiments involving electrons.

Schwinger developed his renormalization theory using a step-by-step mathematical analysis of the interaction of the electron's charge and the electromagnetic field (its own or an external one) to all orders of approximation, making sure that the analysis at each step was in accord with the theory of relativity (relativistically invariant). Feynman, however, did the same thing from a global point of view, but without Schwinger's complicated and difficult mathematics. He pictured all interactions between charged particles (or a particle and a field) by diagrams (known as Feynman diagrams) in space–time in which the world line of a particle is represented by a straight solid line and that of a photon by a wavy line. All possible interactions can be represented by sets of straight and wavy lines; by introducing enough such lines, interactions can be represented pictorially to any desired approximation. Since these lines are world-lines in space–time, the Feynman diagrams are always and automatically in accord with relativity theory. Feynman completed his representation of events involving electrons and photons (interaction of charges) by assigning to each line and to each intersection of two lines in any diagram a definite mathematical expression from which the wave function along that line can be deduced. By properly combining all such expressions for a given set of events, one can trace the evolution of the wave function from the beginning to the end of the set of events and thus calculate the probability for the events to unfold. Since Feynman's diagrams can be applied to any kind of interaction involving the emission or absorption of particles, they are used extensively in such branches of physics as general relativity, nuclear physics, and high-energy particle physics.

Richard Phillips Feynman was born in New York City in 1918 and received his early education in the New York public school system. He first demonstrated an unusual talent for mathematics in high school, and his physics teacher was so impressed with Feynman's abilities that he permitted him to sit in the back of the class and use advanced calculus to solve the assigned problems while the rest of the class worked through them using basic algebra.[7]

After graduating from high school, Feynman enrolled at the Massachusetts Institute of Technology, perhaps the most prestigious scientific

Richard Phillips Feynman (1918–1988)

university in North America, and began a rigorous curriculum in mathematics and physics. He quickly mastered the intricacies of quantum physics and received his degree in 1939. With his interest in the probabilistic nature of atomic physics whetted, Feynman went to Princeton on a fellowship to do his graduate work with John Archibald Wheeler, then one of the foremost authorities on nuclear physics. Since Wheeler has been one of the most versatile physicists of the 20th century, making theoretical contributions to the study of the atomic nucleus and black holes, he was the perfect mentor for the brilliant Feynman, who was interested in electrodynamics and the ''fundamental problem of the interaction between charged particles and whether such an interaction is best treated as 'action at a distance' or the action of a field.''[7] Feynman completed his doctoral program in 1942 and joined the westward trek of many of the finest scientific minds to the secret government laboratory at Los Alamos, where he did research on the Manhattan Project.

Feynman remained at Los Alamos until the end of the war. In 1945, he accepted a position as an associate professor at Cornell University, where he developed his "Feynman graphs" to illustrate the various processes that occur when one charged particle interacts with another.[8] Within a few years, Feynman had become a full professor at Cornell, and his work in quantum mechanics was becoming well known among his colleagues. His papers on quantum electrodynamics and his Feynman graphs (which illustrate the overall behavior of a system of particles instead of trying to track the behavior of particles from moment to moment) were typical of Feynman's efforts to understand the processes of nature in as simple a manner as possible with a minimum of mathematical formalism.

Feynman's work in quantum electrodynamics (along with that of Schwinger and Tomonaga) "led to a reconstruction of the fundamentals of quantum electrodynamics, which was accompanied by a great advance in the accuracy with which the behavior of the electron could be computed."[9] Feynman left Cornell for the California Institute of Technology in 1950 to become a professor of theoretical physics; he remained there until his death. In the early 1950s, Feynman continued to devote most of his efforts to fine-tuning his theory of quantum electrodynamics and developing new mathematical techniques. Satisfied with the theoretical edifice he had painstakingly created over the previous two decades, Feynman began to apply some of the methods he had developed to high-energy particle physics and collaborated with Murray Gell-Mann, who formulated the theory of quarks, on an important theory of beta decay. Feynman also did important work in low-temperature physics, especially the phenomenon of superconductivity. The fundamental importance of Feynman's work, particularly his theory of quantum electrodynamics, was recognized when he was awarded a share of the 1965 Nobel Prize in Physics with Tomonaga and Schwinger.

By the time he died in 1988 after a long battle with cancer, Feynman had become quite well known to the general public and had written a best-selling autobiography, *Surely You're Joking, Mr. Feynman,* and had been named by the readers of *Esquire* as the most intelligent man in America. He also wrote a number of more technical works, the best known of which is the *Feynman Lectures on Physics*.

The renormalization technique of Schwinger, Feynman, and Tomonaga for solving quantum electrodynamical problems is only a mathematical procedure and not new physics, for it is based entirely on all the well-known laws of physics. However, this does not detract from its usefulness and its great success in producing answers to problems to an incredibly high degree of accuracy (in some instances to one part in billions). But it also underscores a serious flaw in quantum mechanics: Applied conventionally to electrodynamics, it gives nonsensical results. In any event, the work of Feynman, Schwinger, and Tomonaga marks the end of the "golden age of physics," which saw the emergence and brilliant development of the two great theories, relativity and quantum theory, that guide us in our search for an understanding of nature.

CHAPTER 18

Nuclear Physics

It was a saying of the ancients, that "truth lies in a well"; and to carry out the metaphor, we may justly say that logic supplies us with steps whereby we go down to reach the water.

—ISAAC WATTS

The development of physics may be presented as a series of stages that progressed from domains of low-energy interactions to domains of ever-increasing energy. Newtonian physics dealt primarily with the vast regions of space (for example, interplanetary, interstellar, intergalactic), in which the relatively weak gravitational interactions dominated. In the next stage, physics entered the molecular domain (for example, the law of gases), where the energies that bind molecule to molecule to form solids are larger by many orders of magnitude (powers of ten) than the gravitational interaction energies between molecules. The forces that operate in this domain—the domain of chemistry and the chemist—which are called "van der Waals forces," are complex forms of the electromagnetic force and are generally subsumed under the latter. The discovery of the electron and the proton led physics into the atomic world and the relatively strong electromagnetic interaction energies between the atomic electrons and the atomic nucleus. This progression represents an inward journey into ever-decreasing spatial dimensions that correspond to the increasing interaction energies. The smaller the spatial domain, the larger the interaction energies involved.

This hierarchy of interaction energies may be considered from a different point of view that stems from the concept of the binding energy of a structure, which is the energy required to separate from each other the particles (electrons, protons, atoms, or molecules) that constitute the structure. In general, the smaller the structure, the more tightly bound it is and the greater is its binding energy. Thus, much less energy is required to separate an electron from the earth (gravitational binding energy) than to tear it away from a molecule or an atom.

Owing to this relationship between the size of a structure and its binding energy, molecular and atomic structures could not be studied experimentally by bombarding them with sufficiently energetic particles (for example, photons or electrons) or having them collide with each other violently enough to induce an observable response until physicists had developed ways of accelerating charged particles to sufficiently high speeds. The solution to this problem came with the development of electromagnetic technology, particularly discharge tubes. But the energies of the accelerated particles thus produced were far too small to probe the atomic nucleus, which is far smaller than the atom. A brief survey of the energies involved in these various structural domains of matter elucidate this point. In atomic and nuclear physics, the electron volt (eV) is the unit of energy; it is approximately one trillionth of an erg, which is the energy (kinetic) that a 2-gram mass has when moving at a speed of 1 centimeter per second. An electron acquires a kinetic energy of 1 eV when, starting from rest, it moves unhindered through the vacuum from the negative to the positive plate of a condenser, if the potential difference between the two plates is 1 volt—hence the name electron volt for this energy.

The binding energy of molecules is of the order of a few electron volts, as indicated by the ease with which these molecules can be broken up or dissociated. The electron in the hydrogen atom is bound to the proton by about 14 eV, and the electrons in the helium atom are bound to the helium nucleus by some 50 eV. Thus, the energies involved in molecular and atomic physics are a few hundred electron volts, at most.

Before discussing the atomic nucleus and its binding energy, we present a few more numbers in the energy hierarchy to make the concept of the electron volt more understandable in terms of other physical concepts. If the average kinetic energy of the molecules in a gas is 1 eV, the absolute temperature of the gas is of the order of 5000°K; this statement relates the electron volt to temperature. If an electron's mass is completely changed into energy, the energy generated is 0.5 million (mega) electron volts, written as 0.5 MeV. If a proton's mass is transformed into energy, the energy generated is about one billion (giga) electron volts, written as 1 GeV. These energy units, Mev and Gev, are used extensively in nuclear and in high-energy particle physics.

Although nuclear physics developed most dramatically in the 1930s and 1940s, it really began experimentally with the discovery of radioactivity, for it is clear from the very large energies of the alpha, beta, and gamma particles emitted from radioactive nuclei that they cannot be emitted from the low-energy outer regions of the atom, but must originate in the very tightly bound nucleus. But the study of these particles, particularly by Ernest Rutherford and the Curies, did not reveal much about the structure of the nucleus. Indeed, in those early years, the concept of the nuclear atom was not favored by most physicists; Joseph John Thomson, the head of the Cavendish laboratory at Cambridge and the discoverer of the electron, had proposed the so-called "raisin-bun" model of the atom, in which the atom's positive charge is distributed uniformly throughout the atom's volume and the negative charges (electrons) are distinct points within the positive charge. Things stood this way until Rutherford's decisive experiments, in which he bombarded strips of thin gold foil with alpha particles (the positively charged particles emitted by radioactive uranium) and noted that the way the alpha particles were scattered by (bounced off) the gold atoms indicates quite clearly that they were violently repelled by a heavy concentration of positive charge deep within the gold atoms. The concept of the nuclear atom was thus born, but at its birth, it presented physicists with what appeared to be insoluble problems.

Before considering these problems and how they were solved, we introduce two important numbers that characterize the nucleus and help us describe it—the atomic weight, A, and the atomic number, Z. When 19th-century chemists began to study the chemical properties of known elements, they found it convenient to assign masses (atomic weights) on some arbitrary scale to the smallest, chemically indivisible unit of an element (an atom). Since hydrogen is the lightest element, chemists naturally assigned to it the atomic weight 1, and they discovered that the atomic weights of all the other elements on this mass scale are very nearly integers. Chemists, particularly William Prout, thus inferred that hydrogen is the building block of all the elements, so that the elements are "multiples of hydrogen." This poorly defined idea has no meaning when applied to atoms as a whole, but is meaningful for atomic nuclei; if the atomic weight is the mass of the nucleus of an atom on a scale on which the mass of the proton (the nucleus of the hydrogen atom) is 1, the masses of all nuclei are very nearly integers, as though they consisted of undivided protons. Leaving aside, for the moment, the question as to the number of protons in a nucleus, we now turn to the atomic number, Z.

The atomic number was not clearly defined until Dimitri Mendeleev discovered the periodic table of the chemical elements; if the chemical elements are arranged in this table in order of increasing atomic weights, an atom's atomic number is its position in this table. This observation led to the important discovery that an atom's atomic number is approximately half its atomic weight. This tells us something important about the nucleus of the atom rather than about the atom itself: The atomic number is the total positive electric charge on the nucleus and hence equals the number of electrons in an electrically neutral (un-ionized) atom. The atomic number—not the atomic weight—determines the chemical properties of the atom. This discovery led to another important discovery: the existence of chemically identical elements with the same atomic number but different atomic weights. Such elements, called "isotopes," were first discovered by Frederick Soddy in 1911. The study of the atomic nucleus has revealed that all elements exist in different isotopic forms.

The properties of the atomic nucleus revealed by these early investigations and discoveries presented physicists with the following problems: determining the size of the nucleus, which, from all indications, is thousands of times smaller than the atom as a whole; determining the nuclear constituents, which cannot be only protons, for the nuclear positive charge would then be more than twice its measured value; and determining the nature of the nuclear force (the force that keeps the constituent nuclear particles in the nucleus). From the energies of the emitted alpha and beta particles in the radioactive decays of various heavy elements (for example, uranium) and from the way alpha particles are scattered by heavy nuclei, the nuclear diameter is extremely small—on the order of one tenth of a trillionth of a centimeter. From this investigation, we conclude that the binding energy of a nuclear particle is about one million times as large as the binding energy of an electron in an atom. This finding posed an immediate problem: How can the protons, all with the same electric charge, remain confined in so small a region as the nuclear volume? The electrostatic repulsive force between any two protons separated by the dimensions of the nucleus is so strong that the nucleus would blow up unless some unknown very strong force prevents such a nuclear explosion. From the early years of the 20th century until its third decade, the stability of the nucleus, despite the enormous electrostatic repulsive forces among its constituent protons, was the principal nuclear problem that preoccupied most physicists.

The nuclear stability problem is closely related to the nature of the particles, other than protons, within the nucleus, and almost all the top physicists of those early years speculated about it. The most obvious guess, which appealed to many physicists, was that the nucleus contains electrons in addition to protons, with one electron for each proton in excess of the atomic number. This suspicion seemed to be confirmed by the identification of beta rays with electrons, which was taken as very strong evidence for the presence of electrons (beta rays) in nuclei, for, it was argued, if electrons are emitted by nuclei, they must be in those nuclei to begin with. This "self-evident" proposition was generally accepted even by Rutherford, who supported it until the beginning of the 1930s.

With the development of the quantum mechanics and the discovery of the uncertainty principle, the assumption of electrons in the nucleus became completely untenable for a number of convincing reasons. To begin with, if an electron were placed in a nucleus, its momentum in so small a region would be so large, owing to the uncertainty principle, that it could not be confined; its kinetic energy would far exceed nuclear binding energies. Put differently, the wavelength of the electron, except for speeds close to the speed of light, is so large that its wave would extend far beyond the nuclear dimensions. It is clear that the electrostatic attraction between electrons and protons in a nucleus would hardly be strong enough to keep the nucleus from exploding. Some very strong attractive force between particles in the nucleus is required to give the nucleus its remarkable stability. Finally, electrons in the nucleus would give wrong values for the spins of nuclei, some of which had been measured by 1926. If both protons and electrons, each with a spin of $\frac{1}{2}\hbar$ (where $\hbar$ is equal to the unit of spin), were in nuclei, the total spin of the nucleus would be an odd number of $\frac{1}{2}$ spin units if the sum of the number of protons and electrons were odd and an even number of such $\frac{1}{2}$ units otherwise. This does not agree with the measured spins of such nuclei as the nitrogen-14 isotope (atomic weight 14—the ordinary nitrogen). If this nucleus had 14 protons and 7 electrons (as required to give it its atomic number of 7), its spin would be an odd number of $\frac{1}{2}$ units, but its measured spin is 1 (two $\frac{1}{2}$ units).

All these arguments convinced physicists that free electrons are not present in nuclei, so that the proton–electron model of the nucleus was discarded. But Rutherford did not discard entirely the idea of electrons in the nucleus; he replaced the idea of free nuclear electrons by bound ones, insisting that under the appropriate circumstances, an electron and a proton can form a very tightly bound structure, much smaller than the hydrogen atom. He considered such a neutral structure, which he called a "neutron," to be absolutely essential for the buildup of the nuclei of heavy elements and was convinced that a careful search among the heavy atoms, in which the inner electrons are close to the nucleus, would reveal it. The search for Rutherford's neutron was begun in 1924 by James Chadwick, a close colleague and co-worker of Rutherford, but most of his eight-year search was fruitless because he was looking for the wrong thing—the capture of an electron by a proton in a nucleus with the atomic number of the new nucleus reduced by 1. But in 1932, Chadwick discovered that if beryllium-9 (^{9}Be) is bombarded with alpha particles (from radioactive polonium), carbon-12 (^{12}C) nuclei are produced with the emission of very energetic neutral particles that Chadwick identified with the Rutherford neutron—that is, a composite proton–electron structure. It soon became clear with the very rapid development of neutron physics that Chadwick's neutral particle is the neutron we know today as the electrically neutral constituent of nuclei. With a mass slightly larger than that of the proton and with $\frac{1}{2}$ unit of spin, it is the neutral counterpart of the proton. This discovery was really the beginning of nuclear physics, the rapid development of which in the 1930s and 1940s has influenced our lives and society in such an important and dramatic manner. In any event, the discovery of the neutron opened the floodgates of nuclear research; the physics of the outer atom was almost entirely abandoned as physicists of all ages and abilities contributed to theoretical and experimental nuclear physics. To be called a nuclear physicist was to have conferred on one a title of great esteem.

An electrically neutral particle, with a mass slightly larger (about 3 electron masses larger) than that of the proton, the neutron has just the right properties to account for the structure of the nucleus and for its stability; with a spin of $\frac{1}{2}$ unit, it is a fermion like the proton, with which it forms what physicists call an "isotopic doublet": the "nucleon." This designation was first applied to the proton–neutron pair by Werner Heisenberg, who argued that the proton and neutron are two different aspects of the same basic particle—the nucleon—in the sense that either can be transformed into the other under appropriate conditions.

Since the neutron's mass is about 1840 times that of the electron, its wavelength is correspondingly smaller than the electron's, and it can therefore reside comfortably within the nucleus without violating the uncertainty principle; its large mass allows it to move around rather sluggishly inside the nucleus, rather than too violently to be confined, as the electron would. The atomic number of a nucleus is just the sum of its neutrons and protons; two nuclei with equal numbers of protons but different numbers of neutrons are isotopes of the same element. The stable nuclei of small atomic weight, such as helium, carbon, and oxygen, consist of equal numbers of protons and neutrons, but the number of neutrons becomes progressively larger than the number of protons for the heavy nuclei. The nucleus of the ordinary uranium-238 isotope contains

96 protons (its atomic number) and 142 neutrons. The nuclear isotopes that have too large an imbalance between their protons and neutrons (either too many protons or too many neutrons) are unstable, which explains why the heavy nuclei such as uranium and radium are unstable (that is, radioactive).

While theoretical physicists were trying to construct nuclear models that can explain the nuclear properties such as size, mass, spin, and the nature of the very strong nuclear force that makes tampering with the nucleus very difficult, the experimentalists were collecting data at a dizzying pace. A very important breakthrough occurred shortly after the discovery of the neutron when the French husband and wife team Frederic and Irène Joliot-Curie produced radioactive isotopes of ordinary light nuclei by bombarding such nuclei with alpha particles. This team just missed discovering the neutron because, while working with high-energy alpha rays, they detected the neutral rays from the alpha–beryllium reaction, but interpreted these rays incorrectly as gamma rays (very penetrating photons) and so reported them in a paper they published in January 1932. Chadwick read this paper and, rejecting the Joliot-Curie interpretation, announced his neutron discovery a month later. But the Joliot-Curies did not lose out entirely, for they continued their alpha-particle research, bombarding ordinary aluminum nuclei to produce beta-radioactive phosphorus (emission of positrons) and doing the same with boron to produce beta-radioactive nitrogen-13. This is the first example of artificially produced beta radioactivity.

But this was just the beginning of a vast area of nuclear research for the great Italian theoretician and experimentalist Enrico Fermi (1901–1954). Fermi noted at once that slow neutrons are much more effective in producing new isotopes than are alpha particles because, unlike the latter, which, owing to their positive charge, are strongly repelled by nuclei, the uncharged neutrons are not repelled and enter nuclei very easily. With this perception, Fermi began in 1934 a systematic study of the absorption of slow neutrons by all the known nuclei and showed that in almost all cases, radioactive isotopes are produced. On subjecting uranium nuclei to slow neutrons, he obtained what he thought were the nuclei of a transuranic element of atomic number 93. This conclusion was one of the few mistakes that Fermi made in his scientific career, but it was fateful, for the products he had obtained in his neutron–uranium experiment were not the nuclei of a new heavy element, but those of barium and iodine, the individual masses of which are about half those of the uranium nuclei. Without knowing it, Fermi had produced nuclear fission, which some

Enrico Fermi (1901–1954)

eight years later led to the first nuclear bombs. One may speculate about the historical developments had Fermi discovered his error and remained in Italy instead of emigrating to the United States; the secret of nuclear fission would have been in the hands of the Axis powers.

Fermi was the last of the great physicists who were equally comfortable with pencil and paper and with laboratory equipment. With his superb intelligence and all-consuming interest in all branches of physics and his mastery of neutron physics, Fermi became the undisputed leader among nuclear physicists. Most of the outstanding physicists in Italy were attracted to him and came to his laboratory in Rome to make it the center of European nuclear research.

Enrico Fermi was the greatest Italian scientist since Galileo and one of the most influential physicists of the 20th century. He grew up in a comfortable home; his father was a civil servant who worked for the Italian railroads, and his mother was a schoolteacher. He spent his early years in the public schools in Rome, where he first became interested in science. He was a bright boy who learned more by reading on his own than he did in his classes. His aptitude for science, especially physics, carried over into mathematics, and Fermi later recalled that "when he was about ten years old he tried to understand why a circle is represented by the equation $x^2 + y^2 = r^2$."[1] He first became deeply interested in physics when he studied a book on mathematical physics written in Latin because there were no Italian physics textbooks at that time. Although Enrico was not yet 13 years old, he spent many evenings making careful notes in the margins of the book's 900 pages until he had thoroughly mastered the subject. A colleague of his father, A. Amidei, noticed Enrico's scholastic tenacity and lent him a book on projective geometry: "[A]fter a few days Fermi had read the introduction and the first three lessons, and after two months he had mastered the text, demonstrated all the theorems, and quickly solved the more than two hundred problems at the end of the book."[2]

Enrico continued to study physics on his own, reading the important works of Planck, Poisson, and Poincaré, as he completed his secondary school studies, excelling in all subjects including Latin and Greek. He also enjoyed poetry and committed many poems to memory; this exercise in memorization helped Enrico to keep abreast of the major theoretical and experimental developments in physics once he began his career. Enrico spent much of his spare time during these years with a schoolmate, Enrico Perisco, building apparatus to conduct experiments in physics. Not only did they both become proficient experimentalists, but also they "determined precisely the value of the acceleration of gravity at Rome, the density of Rome tap water, and the earth's magnetic field."[3] Fermi took the entrance examination for admission to the University of Pisa at the age of 17: "His examination paper must certainly have amazed the examiners, for instead of giving the usual high-school level solutions to the problems, he applied the most advanced mathematical techniques, such as partial differential equations and Fourier analysis, to problems in sound."[4] His essay not only convinced his examiners of his genius but also earned him a fellowship, making it possible for him to complete his education without expense to his family.[5]

Fermi received his doctorate from the University of Pisa in 1922 and, recognizing the need to study at the premier educational institutes, competed successfully for a foreign fellowship that financed his studies with Max Born in 1923 in Göttingen and Paul Ehrenfest in Leiden.[5] Ehrenfest recognized the extraordinary abilities of his young assistant and taking a great interest in his career, introduced him to many other physicists. In

1924, Fermi returned to Italy to become a lecturer at the University of Florence. During the next two years, he developed what is now known as Fermi statistics, which are applicable to particles obeying Pauli's exclusion principle, which "prevents more than one electron from occupying an orbit completely defined by its quantum number."[6] Since Fermi statistics is applicable to all particles having half integral spin, it is of fundamental importance in atomic and nuclear physics; this work established Fermi as one of the preeminent theoreticians in Europe. The value of Fermi's work was officially recognized when Fermi was appointed to the first chair of theoretical physics at the University of Rome.

Fermi's appointment to Rome was instrumental in reviving Italian physics, which was not very highly regarded in the 1920s. Although he did not enjoy administrative duties, Fermi devoted much of his time to recruiting outstanding students, many of whom later became respected scientists in their own right. His work encouraged many foreign students, including Hans Bethe and Edward Teller, to come study in Rome. The overhauling of the school of physics in Rome also helped to revive other university physics departments in Italy.

While in Rome, Fermi developed his theory of beta decay in response to the neutrino hypothesis that had been offered by Wolfgang Pauli owing to the apparent nonconservation of energy and momentum in beta decay: "Pauli sought a way out of the apparent paradoxes by postulating the simultaneous emission of the electron and of a practically undetectable particle, later named 'neutrino' by Fermi."[7] To solve the problems posed by Pauli's neutrino, Fermi proposed a new force—the so-called "weak force," which was given its name because it is much weaker than the nuclear (strong) force. This weak force is present in all particle interactions involving neutrinos, and physicists believe that it and the gravitational, electromagnetic, and strong forces are all the forces that are needed to construct the universe.

In 1934, Frederic and Irène Joliot-Curie bombarded boron and aluminum with alpha particles and discovered artificial radioactive isotopes, receiving the Nobel Prize for their work. Fermi believed that the uncharged neutron would be a better "bullet" for bombarding the elements because it would not be repelled by the nuclear charge and would have a much greater probability of penetrating the target nuclei.[8] Fermi bombarded a number of elements with neutrons and discovered what are known as "slow neutrons" after interposing paraffin between the source of the neutrons and the target element. These slow neutrons were able to penetrate the nuclei of the target elements even more readily than ordinary neutrons because they collided with the hydrocarbon atoms of the paraffin, which slowed them down and enabled them to remain "in the vicinity of the target nucleus sufficiently long to increase their chance of absorption."[9] Fermi used the slow neutrons to bombard many different elements; he investigated the properties of the radioactive isotopes created by these experiments. When he bombarded uranium with slow neutrons, which leads to nuclear fission, however, Fermi erroneously believed that he had produced two transuranic elements that he named "ausonium" and "hesperium." The true results of this experiment were not uncovered until 1938, when Otto Frisch and Lise Meitner discovered that nuclear fission was taking place. Despite this oversight, however, the value of Fermi's experimental work with slow neutrons was recognized worldwide, and he received the 1938 Nobel Prize in Physics.

Fermi's work in Rome took place under the lengthening shadow of Mussolini's Fascist dictatorship. Although Fermi opposed the militarism of the Italian government and was particularly concerned about the growing German influence in Italy, which was mirrored in increasingly restrictive laws dealing with Jews, he did not criticize the government openly. His wife, Laura, whom he had married in 1928, was a member of a prominent Jewish family, and Enrico feared for her safety. That she was the daughter of an admiral in the Italian navy did not allay his concerns. When Fermi went to Stockholm in 1938 to receive the Nobel Prize, he had already decided to leave Italy for good, so after the awards ceremony, he proceeded directly to New York, where he accepted a professorship at Columbia University.

After learning of the discovery of nuclear fission, Fermi began to conduct experiments at Columbia to see whether a nuclear chain reaction could be sustained. The key was to slow the neutrons down, but not so much that too many of the neutrons were captured without producing fission. After extensive experiments, Fermi finally settled on graphite as a moderating substance.

The military potential of a sustainable chain reaction was not lost on Fermi or his collaborators, and they tried to convince the United States government to fund nuclear fission research programs. In the beginning, however, owing in part to the newness of the field, adequate support was not forthcoming. Part of the problem stemmed from the manpower and material requirements for building a successful nuclear reactor, which were far greater than Fermi or anyone else anticipated at the time. Fermi's own efforts were hampered by his status as an enemy alien and his unwillingness to take a more active role in the administration of an atomic research project. His research at Columbia centered around efforts to obtain a chain reaction using ordinary uranium, though it soon became clear that only the uranium isotope ^{235}U is fissionable.

By the time the United States entered World War II at the end of 1941, the bureaucracy that was to direct the development of the atomic bomb had been established. The defeats initially suffered by the Allied forces in 1941–1942 gave added urgency to exploiting the military applications of atomic fission. In 1942, Fermi went to Chicago, where he directed efforts to construct the first self-sustaining fission reaction at a secret laboratory beneath Stagg Field at the University of Chicago: "Using pure graphite as a moderator to slow the neutrons, and enriched uranium as the fissile material, Fermi and his colleagues began the construction of the pile. It consisted of some 40,000 graphite blocks, specially produced to exclude impurities, in which some 22,000 holes were drilled to permit the insertion of several tons of uranium."[9] On December 2, 1942, almost a year after the Japanese attack on Pearl Harbor, Fermi's atomic pile went critical and sustained a chain reaction. In 1944, Fermi left Chicago to go to Los Alamos, New Mexico, to work as a general consultant to J. Robert Oppenheimer, the head of the Manhattan Project. While at Los Alamos, Fermi witnessed the first successful test of the atomic bomb near Alamogordo, New Mexico, on July 16, 1945.

Fermi returned in 1946 to the University of Chicago, where he remained for the rest of his life. Since he was now a professor at the newly created Institute for Nuclear Studies, he was spared the administrative duties he disliked, so that he had the time to instruct an outstanding group of graduate students, several of whom, such as Murray Gell-Mann, Tsung-Dao Lee, and Chen Ning Yang, went on to become Nobel laureates. Fermi gradually shifted his own focus from nuclear to particle physics, as he felt that the most interesting discoveries had already been made in the former. Fermi's interest in particle physics was whetted by the chaotic proliferation of particles in the 1930s, 1940s, and 1950s. Like many physicists, Fermi searched for some sort of underlying pattern in this particle menagerie, but this effort was not very successful, even though Fermi did useful experimental work in pion-scattering and theoretical work on the origin of cosmic rays. Fermi did play an indirect role in the research that culminated in the proposal by Murray Gell-Mann, his former student, that all matter consists of fundamental particles that he called "quarks."

Although Fermi continued to be active in the scientific community, dividing his time between his own work and collaborations with other physicists, he began to slow down noticeably in 1954, when an exploratory operation revealed stomach cancer. Although Fermi knew that he had little time left, he tried to continue his research as best he could and refused to let his illness stand in the way of his daily routine until his health had so deteriorated that he was hospitalized in September 1954. Fermi fought the debilitating effects of the cancer for two months, but finally succumbed in November and was buried in Chicago. After his death, the element 100 was named fermium, and the unit length of 10^{-13} cm was named the fermi in his honor. In recognition of Fermi's unique excellence as both a theoretical and an experimental physicist, the National Accelerator Laboratory at Batavia, Illinois, was renamed the Fermilab, and a special national award—the Fermi prize—was established in his honor.

Fermi's work in nuclear physics made it possible to measure the sizes of nuclei using scattering experiments in which high-energy protons were bounced off nuclei; at the same time, these experiments led to the determination of the strength and the range of the nuclear force. It was found to be some hundreds of times stronger than the electromagnetic force and to have an extremely short range—about one tenth of a trillionth of a centimeter; owing to its great strength, the nuclear force is also called the "strong force" or the "strong interaction." The scattering of high-energy neutrons from nuclei shows that the nuclear force, which is always attractive, is the same for both neutrons and protons; the strong force is thus charge-independent, so that protons attract protons, neutrons attract neutrons, and protons attract neutrons with equal strengths. The very short range of the nuclear force means that a proton must first penetrate the repulsive, Coulomb electrostatic barrier produced by the positive charge on the nucleus before it feels the strong nuclear attractive force, and is thus pulled into the nucleus to form a new, heavier nucleus. Indeed, two different nuclei, if they are traveling fast enough, can interpenetrate each other's Coulomb barriers and stick together, owing to their nuclear mutual attraction, to form a compound nucleus. This kind of nuclear chemistry plays a very important role in the structure and evolution of stars, for stars generate their radiation by this process (thermonuclear fusion).

As physicists began to probe the structure of nuclei in ever-increasing detail, they had to develop special electromagnetic machines that could accelerate the probing particles (electrons and protons) to ever-increasing speeds. There are two kinds of such particle accelerators, linear and circular ones, the former primarily to accelerate electrons and the latter to accelerate protons. In a linear accelerator, or linac, the electrons move through a series of identical cylinders with a voltage difference maintained between each pair of cylinders so that the speed of each electron is given a boost as it moves from one cylinder to the next. The modern linac may contain thousands of such cylinders and may be several kilometers long. With the voltage difference between two adjacent cylinders about 100,000 volts, such a modern linac can accelerate an electron to a speed very close to the speed of light and an energy of about 50 billion electron volts (50 GeV).

The circular accelerator is the modern version of the cyclotron, the first of which (a few centimeters in diameter) was built in the early 1930s by Ernest Orlando Lawrence. Today's circular accelerator is a ring of successive very powerful magnets arranged along the ring's inner circumference. The diameter of the vacuum pipe through which the protons travel is about a centimeter, and the protons are kept moving along the pipe without spreading by two sets of magnets, one of which directs them along the circular orbit and the other of which (the focusing magnets) keeps the protons in the beam together. The protons are accelerated step by step to the desired speed (energy) by an array of increasing voltages. The larger the perimeter of the vacuum pipe, the greater the increase of the proton's speed each time the proton traverses the pipe. To obtain very large proton energies, circular accelerators with circumferences of miles or more have been built. The largest one, with a diameter of 1.5 miles, is in operation at CERN (the acronym for the French name of the European Center of Nuclear Research). With this device, energies of the order of 540 GeV have been obtained, but this threshold has been reached by a colliding technique in which protons move through the vacuum pipe in one direction and a stream of antiprotons moves in the opposite direction. In a collision between a proton and an antiproton, the total kinetic energy of both particles is tranformed into particles of all kinds. This is twice as much collision energy as one obtains by having just protons collide with a stationary target. The circular accelerators in which particles and antiparticles collide are called "colliders." At the moment, colliders are being planned with circumferences of 30 or 40 miles. If these gigantic colliders are ever built, they should produce collision energies of trillions (tera) of electron volts (TeV).

The measurement of the masses of nuclei is very important, for the Einstein relationship between mass and energy ($E = mc^2$) indicates that in nuclear reactions, an appreciable fraction of the masses of the interacting particles is transformed into energy. Measuring the masses of nuclei can thus be used to identify new nuclei that may be formed in nuclear reactions and to test Einstein's mass–energy relationship; thus far, no exceptions to it have been found in nuclear reactions.

Spin is another important nuclear property that has been measured with great precision. This measuring is done by the "molecular beam" procedure, which was developed with great ingenuity and success by the American physicist Isidor Isaac Rabi. A spinning nucleus is a very tiny magnet owing to its electric charge; the strength of such a nuclear magnet depends on the magnitude of its spin (the number of units of spin, $\hbar$, it has) and the magnitude of its electric charge. Hence, if a beam of nuclei is sent through a magnetic field, the nuclei interact with the field (acquire rotational energy from it), and the strength of this interaction depends on the spin of the nuclei. If the strength of the magnetic field varies across the beam of nuclei, the beam is broken up into a number of distinct beams, which depends on the number of spin units the nuclei in the original beam have. If each nucleus has a spin of $\frac{1}{2}$ $(h/2\pi)$—$\frac{1}{2}$ of a spin unit—the original beam is broken up into two beams; if the nuclear spin is 1 (one spin unit), three distinct beams are produced by the magnetic field; and so on. This technique gives nuclear spins at a glance. Later, it played an important role in various practical applications and was the basis for nuclear magnetic resonance (NMR).

A series of theoretical questions concerning nuclear physics immediately arose as the experimentalists began collecting more detailed nuclear data. Most of these questions dealt with the nature of the nuclear (strong) force that keeps neutrons and protons locked in the tiny nucleus despite the strong electrostatic repulsive force between protons. To invent a model of the nucleus (as Bohr had done for the atom), the theorist had to know the mathematical nature of the nuclear force. He also had to know whether or not the laws of quantum mechanics apply to the nucleus. This question was answered in the affirmative by two separate theoretical investigations. In 1928, the Russian physicist George Gamow explained the alpha-particle decay of heavy nuclei like uranium-238 by applying the Schrödinger equation to alpha particles inside a heavy nucleus and showing that such alpha particles, owing to their wave properties, "tunnel through" the potential barrier of the nucleus. From the rate of such tunneling, as given by the alpha-particle wave function, Gamow successfully calculated the lifetime (the half-life) of the uranium nucleus. This kind of calculation later played an important part in the calculation of thermonuclear fusion in stellar interiors.

Gamow did his work before the discovery of the neutron, but with the neutron's advent, the first theoretical model of a simple nucleus—the deuteron—was constructed by Hans Albrecht Bethe and Rudolf Peierls. This model was suggested by Harold Clayton Urey's discovery in 1931 of heavy water, a compound that consists of an atom of heavy hydrogen (the deuteron) and the usual two oxygen atoms. The deuteron, consisting of a single proton and neutron held together by the nuclear force, is the simplest nucleus and therefore is to nuclear physics what the hydrogen atom is to atomic physics. Bethe and Peierls felt that solving the deuteron problem quantum mechanically would lead to a quantum mechanical model of the nucleus, just as the Schrödinger equation for atomic hydrogen led to the development of the quantum mechanics in general. They were successful to some extent, but the quantum mechanical theory of the deuteron is still incomplete, since the mathematical nature of the force between the proton and neutron is still unknown, so that a correct Schrödinger equation for the deuteron cannot be written. Bethe and Peierls skirted this difficulty by assuming an interaction between the proton and neutron of the right strength and correct range and showed that within these constraints, the shape (mathematical form) of the interaction does not matter. In any case, they showed that quantum mechanics is valid in treating nuclear interactions and setting up models of nuclei. However, the quantum mechanical treatment of nuclear structure is far more difficult than that of the outer atom, because the protons and neutrons within the nucleus must be treated equally; the nucleus—unlike the atom itself—is not a planetary system with a massive center surrounded by a swarm of very light particles moving in well-defined orbits. All in all, the quantum mechanical treatment of the nucleus has been fairly successful.

Although the strong (nuclear) force determines the structure of the nucleus and dominates the dynamics of the nuclear particles, another force—the weak interaction—comes into play in radioactive nuclei that emit beta particles (electrons). These are nuclei (isotopes) that have too large an excess of neutrons compared to protons, as was demonstrated by Fermi in his famous experimental production of artificial radioactivity by bombarding stable nuclei with slow neutrons to produce neutron-rich unstable isotopes. In their study of natural beta-radioactive atoms in the 1920s, physicists discovered what appeared to be a spin and energy imbalance between a radioactive nucleus that emits a beta particle and the resulting nucleus, the mass of which is always smaller than that of the emitting nucleus. According to Einstein's mass–energy relationship, this mass difference must equal the mass of the emitted electron plus the mass equivalent of its kinetic energy, but this is not always so because the electrons emitted from a large number of identical beta-radioactive nuclei have a spread of velocities that range from zero to a maximum value. This continuous velocity spectrum led to consternation among physicists, for it was in serious conflict with the sacrosanct principle of the conservation of energy. So severe was the crisis produced by this seemingly unavoidable energy imbalance that some prominent physicists, including Niels Bohr, proposed abandoning energy conservation in nuclear processes. But this drastic move would not have eliminated the difficulty altogether, for the spin imbalance would still have remained; when a beta ray is emitted from a nucleus, the spin of the residual nucleus remains the same or changes by just one unit. But this cannot be so if only an electron is emitted, because its spin is $\frac{1}{2}$ a unit, so that the spins of the original and final nuclei should always differ by half a unit in all beta-radioactive processes. Clearly, nuclear beta decay must involve something more than the emission of an electron, and so Pauli, in 1930, very reluctantly proposed the concept of a new particle to accompany the electron as a "desperate way out" of the dilemma presented by the emission of an electron only in beta decay. He called this particle the "neutron," but this name was later changed to "neutrino" (the little neutral one) by Fermi. This proposal was considered both outrageous and courageous by most physicists, since the only basic particles known at the time were the photon, the electron, and the proton, and these three particles were thought to be adequate to account for all known properties of energy and matter. But something drastic had to be done to save the principle of the conservation of energy, so the neutrino slowly began to gain acceptance among physicists.

To bring beta decay into line with accepted physical laws, the neutrino was assigned a spin of $\frac{1}{2}$ unit, zero electric and magnetic charge, and, as far as could be observed, zero mass. In other words, the new particle—the neutrino—appears to have been designed to be unobservable, except that it can transport energy so that neutrinos of different energies can exist. This multiplicity of neutrinos is essential to keep the correct energy balance in beta decay, and the $\frac{1}{2}$ spin of the neutrino gives the correct spin balance; in beta decay, the energy of the neutrino that is emitted with the electron makes up for the energy gap that is associated with the electron alone. Although the neutrino is very elusive (it interacts so weakly with matter—hence the weak interaction—that it can pass through the entire galaxy without being deflected), it was finally observed experimentally in 1956 by C. Cowan and F. Reimes using a very ingenious experiment.

The beta-decay process can now be understood completely in terms of the neutron, which had not been discovered when the neutrino was proposed. When it is outside the nucleus, the neutron is unstable; it was originally believed that it decays into a proton, an electron, and a neutrino in about 12 minutes on the average. Actually, the neutron decays into a proton, an electron, and an antineutrino, as we now know, instead of a neutrino. Just as the electron has its antielectron (the positron), the proton its antiproton, and the neutron its antineutron, so the neutrino has its antineutrino, but since the neutrino has no charge, the antineutrino can be distinguished from the neutrino only by the way it is spinning (clockwise or counterclockwise) relative to its velocity. But all these facts do not alter the role that the decay of the neutron plays in the beta decay of neutron-rich nuclear isotopes. In stable isotopes, the neutron does not decay, but in unstable isotopes, which have an excess of neutrons, one of the neutrons changes into a proton, and the electron and antineutrino are emitted from the nucleus to give the beta-decay process: The formation of a neutron occurs by the fusion of a proton, an electron, and an antineutrino. This process is the first step in the thermonuclear fusion of four protons to form a helium nucleus in the deep interior (the core) of stars like the sun. Such stars produce their energy by the thermonuclear burning of hydrogen.

Although much of this beta-decay process was known empirically, no quantum mechanical theory of beta decay had been produced until Fermi presented one in 1934 in one of the most famous and important papers in physics. In this paper, he developed the theory of the emission and absorption of electrons and neutrinos (or antineutrinos) by neutrons and protons in beta decay in analogy with the emission and absorption of photons by electrons in electromagnetic interactions. To carry out his analysis, he introduced a new force field—the weak force—to which the neutrino is coupled by a coupling constant (the Fermi constant), similar to the coupling of the electron to the electromagnetic field through the charge of the electron. With this field theory, Fermi correctly reproduced the velocity spectrum of the electrons in beta decay. From the empirical data of beta decay, he also computed the magnitude of the neutrino coupling constant and showed that it is many powers of ten smaller than the electromagnetic coupling constant of the electron. In other words, the weak interaction is much weaker than the electromagnetic force, but stronger than the gravitational force.

With beta decay taken care of and the role of the neutrino understood, we now turn to some of the models of the nucleus that have been proposed. Since the mathematical form of the nuclear interaction is not known, an exact quantum mechanical treatment of a complex nucleus is impossible, but one can construct a reasonable model that gives good results in analyzing the interactions of nuclei such as those that occur in stellar interiors. To take into account both the strong, short-range attractive nuclear force and the long-range, repulsive Coulomb (electrostatic) force, we represent a nucleus, metaphorically, as a deep, narrow well (the strong attractive force) surrounded by a high, steep rim (the electrostatic repulsion or Coulomb's barrier) that slopes down to ground level on all sides (like the crater of an extinct volcano); the greater the atomic weight of the nucleus, the deeper the well, and the higher the atomic number, the higher the rim. A compound nucleus can be formed from two nuclei if they are moving fast enough to penetrate each other's Coulomb barrier, for then they fall into the common nuclear well that they form. Despite the crudity of this model, it has been applied with great success to the generation of energy and the buildup of heavy nuclei from light ones inside stars. This type of analysis was initiated by Hans Bethe in the late 1930s when he showed that in stars as massive as or less massive than the sun, the dominant thermonuclear process that governs their structure, their energy generation, and their slow evolution is the proton–proton chain of reactions in which four protons, in a series of steps, are fused to form the ^{4}He (helium) nucleus. In stars more massive than the sun, the dominant thermonuclear process is also the fusion of four protons to form the ^{4}He nucleus, but instead of being a direct process, it is an indirect one involving the carbon nucleus, which begins the process by absorbing a proton. Three other protons are then absorbed, one at a time, with a new (heavier) nucleus appearing at each absorption until four protons have been absorbed. But after the last such absorption, the original carbon nucleus reappears together with an alpha particle (a helium nucleus). Since the carbon nucleus is a nuclear catalyst in the process, the process is called the "carbon cycle."

Two important technologies were created by nuclear physics, one destructive (nuclear bombs) and one constructive (nuclear energy). Both were born together owing to the military urgencies of World War II. As previously noted, Fermi had unknowingly produced nuclear fission by bombarding uranium with slow neutrons, but the official discovery of nuclear fission is credited to Otto Hahn, Fritz Strassman, and Lise Meitner, who, in 1938, repeated Fermi's experiment and correctly concluded that a slow neutron, on entering the uranium nucleus, causes it to explode into halves. This process is best understood with the aid of a nuclear model proposed by Niels Bohr—the liquid drop model. The molecules in a drop of water are held together by a relatively strong, short-range force (the van der Waals force) between neighboring molecules, but the drop can be broken into two equal droplets by an external radial force. One sees this when a liquid at the end of a vertical glass rod forms successive drops that fall off; here, gravity is the force that keeps pulling drops off the end of the rod. A heavy nucleus that absorbs a slow neutron is set oscillating by the neutron's energy, and as the two halves of the oscillating nucleus are pulled away from each other, the nuclear attraction between the two quickly weakens, owing to the short range of the nuclear force, until the Coulomb force of repulsion separates the two completely and fission occurs. The two nuclear fragments then rush away from each other, owing to their mutual electrostatic repulsion, with a total kinetic energy of the order of 200 MeV. This vast amount of energy does not come from the slow neutron, which merely initiates the whole process, but from the electrostatic repulsion.

Physicists quickly discovered that the nucleus of the ordinary uranium isotope of atomic weight 238 (^{238}U) does not fission on interacting with a neutron, but the uranium isotope of atomic weight 235 (^{235}U) does. Since the proportion of the 235 isotope in natural uranium deposits is less than 1%, one of the most important projects in World War II was the isolation of enough pure fissionable uranium-235 (about 15 pounds) to construct the first nuclear bomb. During this project, physicists found that although the abundant uranium-238 isotope is not fissionable, it captures a fast neutron to form the isotope of uranium (^{239}U) that quickly emits a beta ray (an electron and antineutrino) to become the transuranic nucleus of the element neptunium. The new element in turn emits another beta ray to become the nucleus of plutonium, which becomes highly fissionable after absorbing a slow neutron.

The discovery of neutron-induced nuclear fission was the first step in the development of useful nuclear energy, which culminated in the construction of the first nuclear reactor (known as an atomic pile) under the direction of Fermi. The theoretical basis on which the concept of the pile rests is quite simple. If a large enough cube (the pile) of pure uranium slabs is built, a self-sustaining fission process, with the constant production of energy, occurs automatically in it. The reason is that any stray external neutron entering the pile carries out a random walk process, owing to its collision with uranium nuclei, and ultimately is absorbed by a ^{235}U nucleus, which fissions and produces additional neutrons. These neutrons are slowed down by adding slabs of graphite to the slabs of uranium in the pile. The fast neutrons are absorbed by ^{238}U to produce plutonium, and the slow ones are absorbed by ^{235}U to keep the pile going. The modern nuclear reactor uses cylinders packed with rods of ^{235}U instead of slabs of natural uranium, which greatly increases the rate at which the energy is produced.

CHAPTER 19

Particle Physics

I have found you an argument: but I am not obliged to find you an understanding.

—SAMUEL JOHNSON

When the electron and proton were discovered in the late 19th and early 20th centuries, physicists viewed these particles with a sense of fulfillment, because they had been striving for many years to discover these two kinds of particles with opposite but equal electric charges and because these particles appeared to be all that was needed to construct a correct theory of matter. These discoveries were all the more gratifying in that Michael Faraday's earlier electrochemical experiments quite clearly pointed to the existence of a basic unit of charge and Hendrik Antoon Lorentz's accepted theoretical work on an "electron" theory of matter required a real electron to validate it. But physicists soon found that the discoveries of the electron and proton marked the beginning of a new chapter of physics known as "particle physics," rather than the end of the search for the fundamental building blocks of nature, a goal that had persisted in science since the time of the Greek atomists.

In a sense, particle physics began with the speculations of Democritus as to the nature of matter and his atomic theory. But the discoveries of the electron and proton changed speculation to reality. That this reality, however, was just the beginning of particle physics is indicated by the profound questions that the electron and proton raised, such as whether they were the basic particles physicists were seeking or whether there were still more basic particles that make up electrons and protons. If the electron and proton are basic and equally important in the structure of matter, physicists wondered why the proton is so much more massive than the electron. That the electron and proton are structurally different is indicated by the failure of the Dirac relativistic wave equation to describe the proton correctly despite its success in describing the electron; it gives the correct value of the magnetic moment of the electron, but not of the proton. Since the Dirac equation applies to a point (zero extension) electric charge, this difference seems to tell us that whereas the electron may be treated as a point, the proton may not. This situation is already very unsatisfactory, because it presents us with an unacceptable asymmetry in nature. Why should the two basic particles that play such important roles in the structure of matter be so different structurally? Is it not aesthetically more satisfying to consider both the proton and electron not as elementary, but as consisting of still more basic particles? But all attempts to construct an acceptable model of the electron failed, and so it is at present thought to be a structureless pointlike particle, whereas the proton is not.

During the early part of this century, when physicists were concerned with constructing a correct model of the atom, little attention was paid to the nature of either the electron or the proton. But as more kinds of particles were revealed experimentally, questions about their structure began to absorb the attention of theoreticians. Planck's discovery of the quantum theory (the quantum of action) introduced the photon as the first massless (zero rest mass) particle in nature and with it the concept of duality (the simultaneous existence of particle and wave properties), which the quantum mechanics extended to the electron and the proton. Treating the photon as a particle and as the quantum of the electromagnetic field was the beginning of what is called "field theory," which now permeates all particle physics. That the photon has no rest mass or charge distinguishes it in a very important way from the electron or proton, but it is different from the other two in its spin, which is a unit of spin ($h/2\pi$) or $\hbar$ instead of half a unit of spin. The photon is a boson (which obeys Bose statistics) instead of a fermion (which obeys Fermi statistics); the Planck radiation formula stems from this property of the photon.

The existence of a massless, chargeless particle with unit spin, the photon, should have suggested to physicists, just from considerations of symmetry, the existence of a massless, chargeless particle of spin $\frac{1}{2}$, but in the early years of the 20th century, the introduction of a new particle was highly repugnant to most physicists and was proposed only under the greatest duress. This same situation existed when Wolfgang Pauli proposed the neutrino to save the principle of conservation of energy from Niels Bohr's powerful attacks on it. Even so, Pauli was very tentative and even apologetic for his temerity in proposing such an outlandish idea as a neutral (no charge) massless, spin $\frac{1}{2}$ particle to account for the continuous spectrum of electron velocities in beta decay. He began a letter to participants in a conference on radioactivity in 1930 as follows: "I have come upon a desperate way out regarding the 'wrong' statistics of the N^{14} and Li^6 nuclei as well as the continuous beta-spectrum in order to save the 'alternation law' of statistics and the energy law. . . . For the time being I dare not publish anything about this idea. . . .''

The neutrino was projected into the arena of particles without a very well-defined role. But it did save the day for the conservation of energy, and it does give the correct statistics for beta-decay processes. Since the neutrino's spin is $\frac{1}{2}$, it may well be that nature uses it to alter the statistics of a system without altering the electric charge or the rest mass of the system. With zero rest mass, the neutrino must travel at the speed of light to conform to the special theory of relativity. The neutrino is the most mysterious of all the known particles, and nothing at all is known about its structure. Since it has energy, it must also have frequency, because the quantum theory requires that its energy be equal to Planck's constant times its frequency. According to this line of thought, neutrinos with an unlimited range of frequencies exist, and since they interact with matter very weakly, most of the neutrinos that existed when the universe began (the Big Bang) are still here.

The positron (the antielectron) predicted by Dirac's relativistic wave theory of the electron and observed in cosmic rays by Carl Anderson was the fourth particle to enter the particle zoo. Its discovery altered drastically the previously held concepts about elementary particles, because it convinced physicists that every particle has its antiparticle, so that the number of elementary particles was immediately doubled. A particle and its antiparticle have the same rest mass and the same spin, but opposite electric charges. These qualities mean that any particle can suddenly be created in any neighborhood of a point in space, provided that its antiparticle is also created and that enough energy is available for this miracle or that it is a sudden, very short fluctuation of the vacuum. The photon is its own antiparticle, but the electron, the proton, and the neutrino have their own distinct (from themselves) antiparticles.

The neutron, a particle slightly more massive than the proton, was discovered almost simultaneously with the positron, but unlike the latter, it is a relatively stable particle without which nuclei could not exist. The term ''relatively stable'' means that it is completely stable inside nuclei that are not beta-radioactive. Outside the nucleus, as already noted, it decays into a proton, an electron, and an antineutrino in about $12\frac{1}{2}$ minutes, on the average; a period of time defined as its half-life [half of a given number of neutrons decay (beta decay) in this way every $12\frac{1}{2}$ minutes]. Just as the electron and proton have their antiparticles, the neutron has its antineutron. If a positron is in completely empty space, it is as stable as an electron, but in passing through matter, it and an electron annihilate each other in a burst of energy (two gamma rays are emitted). It is thus technically not correct to think of the positron as unstable in the sense that the neutron is unstable, that is, the positron does not decay spontaneously.

Following the discovery of the neutron, physicists knew that they had all the basic particles they needed to develop a complete theory of the structure of matter from nuclei to molecules; everything seemed to fall into place quite nicely, so that the search for new particles in the decade before World War II slowed down considerably. Cosmic ray physics accelerated steadily, however, and during the years from 1933 to 1936, cosmic ray physicists reported some strange findings in cosmic rays, the very high-energy particles that enter the earth's atmosphere from every spatial direction. The solar system is bathed in a sea of rapidly moving particles (charged and uncharged), the nature of which was not fully understood until these rays were first studied photographically and with ion-detecting instruments sent up in balloons and rockets into the outer regions of the earth's atmosphere. Before these balloon and rocket experiments were performed, there was vigorous argument among physicists as to whether these energetic particles emanated from the earth or originated in interstellar space. This argument was settled in favor of origin in interstellar space when the Austrian physicist Victor Hess went up in a balloon himself and demonstrated conclusively that the intensity of cosmic rays increases with increasing height above the earth's surface, which is just the opposite of what would happen if the earth were the source of the rays. Based on these observations, the American physicist Robert Millikan called these rays "cosmic rays," and that very appropriate name has persisted.

Millikan, who did a great deal of cosmic-ray research, proposed his theory that cosmic rays are electromagnetic waves (gamma rays) that are generated in distant galactic regions by some unexplained processes involving the transformation of matter into energy. But this idea was never accepted and was discarded, with great reluctance, by Millikan himself, when the study of the paths of cosmic rays in cloud chambers immersed in magnetic fields clearly showed that these rays consist of very high-energy electrically charged particles. More detailed studies revealed that cosmic rays have two components: the primaries, which originate in interstellar or intergalactic space and consist of very energetic protons (with energies up to trillions of eVs) and heavy nuclei, and the secondaries, which are produced in the earth's atmosphere by the primaries. A single energetic primary, colliding with an atmospheric nucleus, can produce a shower of thousands of secondaries, among which different kinds of short-lived particles are found. These secondaries opened up a whole new field of research and discovery in particle physics.

The first important discovery of a new particle (other than the positron) in cosmic rays was that of Carl David Anderson (the discoverer of the positron) and S. H. Neddermeyer, who noted in 1934–1935 that certain of the charged cosmic ray secondaries have very high penetrating powers, from which their rest masses were estimated to be more than 100 MeV. This discovery was extremely startling, because physicists could find no way of fitting such a particle into their scheme of things and their model of matter. The exact rest mass of this particle, as measured later, is 105.57 MeV (about 200 times that of the electron); it is negatively charged, has a spin of $\frac{1}{2}$ (a fermion), and decays into an electron, a neutrino, and an antineutrino after a lifetime of two millionths of a second. This particle was the first in a series of very short-lived particles found in cosmic rays or produced in high-energy accelerators. This particular particle was called the "mu-meson," the word "meson" indicating that its mass lies between that of the electron and that of the proton.

The mass of the "mu-meson" was particularly exciting to the particle theoreticians, because they were looking for an explanation of the nuclear (strong) force in terms of a force field. Just as the electromagnetic force is carried by the quanta of the electromagnetic field (photons) and the gravitational force is carried by gravitons (never-observed quanta of the gravitational field), so, it was argued, is the strong force carried by the quanta of its field, which they assumed to be the "mu-mesons."

The reason for this assumption goes back to the work of the Japanese theoretician Hideki Yukawa, who, in 1935, used general quantum mechanical arguments to show that if a force field has a short range, its quanta must have a nonzero rest mass, and the shorter the range, the greater this mass. Infinite-range forces like gravity and the electromagnetic force have quanta, gravitons and photons, of zero rest mass. Using the measured range of the force between two nucleons (from scattering experiments), Yukawa deduced the mass of a strong force quantum to be about 200 times that of the electron. It was no wonder, then, that particle physicists, in the early 1930s, eagerly accepted the "mu-meson" as Yukawa's strong-force quantum. But a careful analysis of the properties of the "mu-meson" soon showed that it is not a force field quantum at all, but a fermion in its own right, like the electron.

To see why the ''mu-meson'' particle is not the carrier of the strong force, we describe Yukawa's picture of the interaction between two nucleons (proton–proton, proton–neutron, and neutron–neutron). According to the Yukawa theory, in each of these interactions, massive field quanta are exchanged between the two interacting particles, and these quanta must have exactly the right properties to satisfy the experimental observations: (1) The forces in the three cases above are equal; (2) the electric charges of the two interacting nuclei may be interchanged, but the total charge must not change; and (3) the quantum of the strong field must be strongly absorbed by nucleons.

The interactions between two nucleons via the strong field are now pictured as follows when the nucleons are within each other's range: Each nucleon momentarily creates virtual quanta, one of which is quickly absorbed by the other nucleon. Thus, two nuclei interact strongly by tossing the quanta of the strong field between them. This means that three distinct types of such quanta—with very nearly the same rest masses but with electric charges of +1, −1, and 0—must exist to account for the equality of the strong force regardless of the charges of the interacting nuclei (charge independence of the strong force). Moreover, these quanta must be strongly absorbed by nuclei and their spins must be integers (for example, 0, 1, not half integers—that is, they must be strongly absorbed bosons and not weakly absorbed fermions. The ''mu-meson'' has none of these characteristics: It passes right through matter, interacting only weakly with nuclei; only positively and negatively charged ''mu-mesons'' exist; it is a fermion with spin $\frac{1}{2}$. For these reasons, the ''mu-meson'' is now called the ''muon,'' and the ''meson'' designation is reserved for particles, later discovered, that have the properties required of the Yukawa quantum. The muon is now classified as an unstable, heavy electron, and its existence is as much of a mystery to most particle physicists now as it was when it was discovered, for it seems to play no role in the particle physicist's scheme of things; if it did not exist, the universe could go on as it has before, so many physicists have wondered why nature created what appears to be an unnecessary particle. This view misses the point, because the existence of the muon is demanded by the existence of the electron, *if we picture the muon speculatively as an excited state of the electron.* Just as the existence of the hydrogen atom leads to excited states of this atom via photon absorption, so the electron, as the unexcited, ground state of a composite structure, implies the existence of an excited state of this structure, the muon, produced by neutrino absorption.

Convinced that the Yukawa theory of the nuclear force is correct, experimental physicists continued their search for a set of three particles in cosmic rays that have the right spin, masses, and charges expected of the Yukawa quantum and that interact with nuclei; in 1946–1947, Cecil Frank Powell and his assistants found such a triplet with electric charges of $+1$, 0, and -1 and with zero spin in cosmic rays. These three bosons, called the "pi-mesons" or "pions," have very nearly equal masses (about 140 MeV) and are strongly absorbed by nuclei; particle physicists have therefore accepted them as the carriers of the nuclear force. Using these bosons, Yukawa deduced a mathematical formula for the dependence on distance of the interaction between two nucleons that is like the Coulomb electrostatic interaction multiplied by an exponential distance factor to give this interaction a very short range. This formula for the interaction gives no better results than the various empirical formulas that were used before Yukawa's work. It therefore cannot be said that the pion theory of the nuclear force has given us any deeper insight into the nature of this force than we had previously.

A charged pion (π^+ or π^-), which has a very short-life (a half-life of about one hundredth of a millionth of a second), decays into a muon with the emission of a neutrino; occasionally, the pion decays into an electron (or a positron, depending on its charge) with the emission of a neutrino or an antineutrino. The neutral pion (π°) decays into two gamma rays in a much shorter time than the charged pions. Pions are produced copiously in collisions between two nucleons; this was taken as additional evidence in support of the Yukawa proposal that pions are the carriers of the strong (nuclear) force.

The lifetimes of the pions and the muon can be used to give very strong experimental support to Einstein's deduction (from his special theory) that the faster a clock moves relative to a given observer, the slower it runs as a clock, relative to that observer. This was verified when it was noted that the faster a pion or muon moves relative to an observer, the longer it lives as measured by the observer's clock. The assigned pion and muon lifetimes are those measured by an observer in whose frame of reference (coordinate system) the pion and muon are at rest.

The discovery of the pion and its acceptance as the true meson led to the first step in the classification of particles into distinct categories defined by special characteristics. Even before the discovery of the muon and the pion, three particle categories were known: the spin $\frac{1}{2}$ heavy particles (the neutron and proton), the spin $\frac{1}{2}$ light particles (the electron, positron, and neutrino), and the spin 1 photon with zero rest mass. These particles were not divided into different categories at the time, because doing so led to no real simplification of the physics or no deeper understanding of it. With the discoveries of the pions and muons, however, arranging these particles into categories suggested itself as a way of establishing some kind of order in the "particle zoo," as the totality of the ever-increasing number of newly discovered particles was dubbed.

The spins and masses were the first distinguishing characteristics that were used in setting up groups; all spin $\frac{1}{2}$ (odd multiples of the $\frac{1}{2}$ unit of spin) particles are grouped together as fermions (electrons, neutrinos, muons, protons, neutrons, and their antiparticles), and all spin 1, 0, and 2 particles (photons, mesons, and gravitons, respectively) are grouped together as bosons. The fermions are the building blocks of all matter (atoms and molecules), whereas the bosons are supposed to be the carriers of the fields of force between fermions.

The fermions are divided into two distinct groups, the heavy fermions (nucleons and other heavier ones that were discovered later), called "baryons," and the light fermions, called "leptons." As more and more baryons popped up in high-energy accelerators, some of them were found to have a spin of $\frac{3}{2}$, so that it was convenient to subdivide the baryons into spin $\frac{1}{2}$ and spin $\frac{3}{2}$ baryons (both of which groups are fermions, of course). Another complication appeared in the baryon electrical charges: All the spin $\frac{1}{2}$ baryons have electric charges of 0, +1, or −1, but the spin $\frac{3}{2}$ baryons have electric charges that range from +2 to −1 in steps of 1. Such an array of particles (each with its own antiparticle) made no sense at first, but in time, some semblance of order was produced with the introduction of the "quark," which we discuss later in this chapter.

The leptons, which are all spin $\frac{1}{2}$ particles and their antiparticles, consist of two groups: those that are charged and, except for their masses, are like the electron, and the uncharged neutrinos, which have zero rest mass. The first of these groups contains the electron, the muon (mass 105 MeV), and the tau lepton, the mass of which is 1784.2 MeV, about twice that of the proton. These masses show that it is wrong to think of all the leptons as "light," so that the name "lepton" is misleading. Three kinds of neutrinos have been detected by particle physicists: the ordinary neutrino that appears in beta decay, which is called an "electron neutrino"; the neutrino that the pion emits when it decays into a muon, which is called the "muon neutrino"; and the neutrino that the tau lepton emits when it decays. The existence of six and only six distinct leptons is accepted as one of the basic features of the world of particles. The assignment of its own neutrino to each lepton is one of the mysterious curiosities of modern particle physics; since all neutrinos behave exactly the same way as far as their spin, their zero rest mass (speed of light velocity), and the relationship of their spin direction to their velocity direction are concerned, it is difficult to assign any physical meaning to this distinction between an electron neutrino and the other two neutrinos. This distinction was introduced when the experimentalists discovered that the muon interacts with neutrinos emitted by pions, but not with the beta-decay neutrinos; however, it has never been experimentally verified that the tau lepton has its own neutrino. The experimental distinction between the muon neutrino and the electron neutrino dates from the 1950s when L. Lederman, M. Schwartz, and J. Steinberger (1988 Nobel Laureates) showed that the neutrinos emitted by pions combine with protons to produce neutrons and muons, but never to produce neutrons and positrons.

The mesons are integer spin bosons with charges of +1, 0, or −1 and with masses ranging from those of the pions (140 MeV) to more than 10,000 MeV (10 GeV). These very massive mesons are produced in the very powerful accelerators that have been operating in various countries. The existence of this vast array of massive mesons raises a very important question as to their role in the universe: If the pions, the least massive of the mesons, carry or mediate the nuclear force, why are the other mesons around? One answer may be that they are all excited states of the pions; but if this is so, why are the pions not excited states of a particle that is in a still lower energy state (the lowest possible energy state of matter) than the pions? This question is important, since in all atomic and nuclear phenomena, excited systems decay to their lowest possible energy state.

With the construction of very high-energy accelerators that can accelerate charged particles up to energies of hundreds of billions of electron volts, cosmic rays have taken second place as a source of new particles, although no accelerator has yet been constructed that even comes near (in the energies of its particles) the highest energies found in cosmic rays—10 billion GeV—so cosmic rays are still where we must look to find the very, very high-energy (extremely large rest mass) new particles, if they exist at all. The disadvantage of scouring cosmic rays for such particles is that hundreds of thousands of photographs of cosmic-ray tracks must be carefully examined to pick up a rare event. Accelerators, on the other hand, can be designed to produce numerous particles of just the energy in which one is interested. In any case, in cosmic rays and in the products of accelerators resulting from the collisions of high-energy primary beams of particles with fixed or moving targets, a bewildering array (in number and kind) of new particles has been discovered since the 1950s, and theoretical particle physicists have brought some order out of this chaos by arranging them not only into the three groups described above but also into subgroups of these major divisions. The idea was to try to construct a table of such particles similar, in a sense, to the Mendeleev table of the chemical elements (that is, of the nuclei and their isotopes). There was immediately revealed, however, a difference between the Mendeleev table and the particle table that would have to be constructed to encompass all the new high-energy particles that were rapidly popping up in the new accelerators: All the nuclear isotopes in the Mendeleev table consist of only two kinds of basic particles, neutrons and protons, but building up all the baryons (the two nucleons and all the heavier ones) and all the mesons (the pions and all the heavier ones) with just two kinds of more elementary constituent particles cannot be done; three or more such particles are needed, and we describe below how such basic particles were introduced. Before doing that, we point out certain empirical rules that were revealed in the behavior of the newly discovered high-energy particles. These rules appear as additional conservation principles that must be added to the dynamic conservation principles, such as the conservation of energy–momentum–mass and the conservation of angular momentum, that have played such an important role in our understanding of natural phenomena.

The most important of these additional conservation principles are the conservation of electric charge, the conservation of baryon number, and the conservation of lepton number. In all particle interactions, regardless of whether these particles are baryons, mesons, leptons, or any mixture of these particles, the total electric charge before the interaction must equal the total charge after the interaction. Electric charge can be neither created nor destroyed; if a new particle with a positive charge appears after the interaction, which gives a charge excess of $+1$ over the initial charge, the antiparticle of this particle with a charge of -1 must also appear after the interaction.

The conservation of baryon number stems from the observation that baryons decay only into baryons plus other particles. A baryon number is assigned to every baryon, which is just the number of nucleons (neutrons or protons) it can decay into spontaneously; all known baryons have the baryon number $+1$ because they ultimately decay into a neutron or a proton. The neutron itself has the baryon number $+1$ because it decays (very slowly) into a proton. The proton itself also has the baryon number $+1$; the conservation of baryon number means that the proton is absolutely stable, for if it were to disappear, the baryon number would go from $+1$ to 0, which is forbidden. It may be that this conservation principle is obeyed only to a very high approximation, but not absolutely, and that the proton has a very long, but not infinite, lifetime, as deduced from one of the current Grand Unification Theories. This is not supported by any evidence. The baryon number of an antibaryon is -1. The conservation of baryon number thus states that the total baryon number of a group of interacting particles (the sum of the baryon numbers of all the baryons and antibaryons in the group) must, after the interaction, equal the baryon number before the interaction. Baryons can be neither created nor destroyed; if an additional baryon appears after an interaction, an antibaryon must appear simultaneously to keep the baryon number the same. Mesons and photons are not conserved in such interactions; thus, a baryon more massive than the proton may decay into a pion (meson) and a proton or a neutron without violating any conservation principle. Baryons may also decay into other baryons plus photons or into baryons plus photons and mesons.

The conservation of leptons (that is, electrons, neutrinos, muons, and other particles) or lepton number is also maintained in all interactions. The lepton number of a particle is the number of electrons or neutrinos into which the particle can decay. The lepton number of the electron and the neutrino is $+1$ (their baryon number is 0). The lepton number of any meson is 0, but the lepton number of the muon is $+1$. The lepton number of the positron is -1, and the lepton number of any antiparticle that decays into a positron or an antineutrino is -1. The principle of the conservation of lepton number states that the total lepton number of a group of interacting particles and antiparticles must be the same after the interaction as it was before the interaction. A few examples of the operations of these conservation principles illustrate their usefulness.

Before a neutron decays, its baryon number is 1 and its lepton number is 0. Conservation of baryon number means that it must decay into a proton, and conservation of electric charge means that the proton must be accompanied by an electron. But the appearance of the proton and the electron alone would violate conservation of lepton number, which is why an antilepton (an antineutrino) also appears in this decay. The decay of the charged pion is another simple but clear example of this conservation principle. The π^- and π^+ (lepton numbers 0) decay into the μ^- and μ^+ (lepton numbers $+1$), respectively, but in each such decay, an antineutrino also appears to conserve lepton number (the lepton number of the antinuetrino is -1).

Before leaving the conservation principles, we consider in some detail the conservation of mass–energy in particle interactions and particle decays. First, though, we note and discuss the fall of (that is, the nonapplicability of) a certain conservation principle—the conservation of parity—that, until the discovery of the neutrino, was thought to be universally valid. The neutrino shows, however, that conservation of parity does not apply to processes in which neutrinos are either absorbed or emitted (that is, processes involving the ''weak interaction''). The parity concept stems from the analysis of how we are to compare the description of events in the real world with how they appear in a mirror image of the real world. At first sight, this notion does not appear to present any difficulty or raise any questions, because all a mirror does is change all left-handed phenomena to right-handed ones and vice versa, so that we ought to expect the laws of nature to be the same for the image universe as for the actual universe; according to this point of view, no event as seen in the mirror should enable us to know that we are looking at the reflected image of the universe. This concept is called the ''principle of conservation of parity,'' which seems quite reasonable, since we see no reason for the laws of nature to be different for left-handed and right-handed phenomena. In modern physics, the parity of a system or a phenomenon can be defined in terms of the wave function that describes the system or the phenomenon. This wave function is a mathematical expression that depends on the frame of reference (the coordinate system) we use to describe events. If we now change our coordinate system by reversing the direction of one of the three axes of our coordinate system, we obtain the equivalent of viewing the events we are dealing with in a mirror. Such a coordinate transformation is called a ''reflection.'' The question that arises from a quantum mechanical point of view is as follows: What happens to the wave function that describes the events when we change our coordinate system in this way?

General quantum mechanical arguments show that either the wave function remains unaltered, in which case we call the parity of the system or phenomenon ''positive'' (even), or the wave function changes sign, in which case the parity is called ''negative'' (odd). Physicists discovered that both kinds of parity occur in nature quite naturally, but until neutrino phenomena and weak interactions, in general, were discovered, no change in parity from even to odd or vice versa had ever been observed. For that reason, physicists proposed the principle of conservation of parity, which was universally accepted until certain phenomena in the behavior of a group of mesons called K mesons raised serious questions about this conservation principle. These mesons decay into either two or three pions, and it is clear from the definite parity of the K meson and that of the individual pions that parity is not always conserved in this decay, since the parity of two pions is even and that of three pions is odd.

Nobody knew what to do about this parity difficulty until the physicists Tsung-Dao Lee and Chen Ning Yang boldly proposed that parity is conserved in strong, electromagnetic, and gravitational interactions, but not in weak interactions (interactions involving mesons or neutrinos). This hypothesis was later verified experimentally by S. Wu in her careful analysis of the beta decay of the cobalt nucleus, which her experiment showed to emit its beta rays preferentially in a direction opposite to its spin direction, rather than equally in that direction and in the spin direction. This result is clearly a violation of the principle of parity, for this asymmetry tells us that the mirror image of the emission of beta rays from the spinning cobalt nucleus is different from the real emission, for, depending on how the spin of the nucleus is aligned with the mirror (either parallel or perpendicular to it), the mirror reverses the direction of the spin without reversing the direction of the emitted beta rays or vice versa.

The neutrino itself shows very simply that the conservation of parity does not apply to it because of the fixed direction of its spin with respect to its velocity; an observer from whom the neutrino is receding always sees it spinning counterclockwise, so that the observer whom it is approaching always sees it spinning clockwise. This tells us that the neutrino moves at the speed of light and has zero rest mass; otherwise, an observer could move fast enough to overtake it, and for that observer it would be receding but, as he views it, spinning clockwise, contrary to the way a receding neutrino should behave. The antineutrino's spin and velocity have just the opposite alignment of that of the neutrino.

Now consider a neutrino that is moving away from you (spinning counterclockwise) and approaching a mirror. Its image in the mirror is approaching you, but the spin of the image neutrino is also counterclockwise; therefore, the image of the neutrino in a mirror is not a neutrino, but an antineutrino. The mirror-image universe therefore does not obey the same laws as the real universe does; parity is not conserved in transforming our universe into its mirror image. This idea does not mean that a mirror-image universe cannot exist; it can, indeed, exist, but it is a universe in which all particles are replaced by antiparticles. Otherwise, all the laws are the same, but on an atomic scale, time appears to be reversed, for a positron behaves like an electron going from the future to the past.

We turn now to the conservation of mass–energy in particle interactions and particle decays. Even before high-energy physics developed into its present complexity, particle interactions, particularly in nuclear physics, had been studied extensively, and the basic conservation principles such as conservation of charge, spin, and energy that govern such interactions were well known, but the energy interchanges in such phenomena are relatively small compared to the rest masses of the systems involved. But in high-energy particle physics, the energy changes are themselves as large as the rest masses involved. How, then, is the mass–energy conservation principle to be applied as a guide to tell us in which direction a phenomenon involving interactions of particles or the decay of a single particle will go? To be specific, suppose that A represents the state of a single particle (a baryon at rest) and B represents the state of the decay products of A—that is, B consists of a collection of particles (baryons, photons, mesons, and leptons) the total energy of which (kinetic plus rest mass) equals the mass of A, as required by energy conservation. Since the energy of the initial mass exactly equals the energy of the final state, why does the process go at all, since energy is conserved whether it occurs in the direction A→B or in the reverse direction A←B? The determining factor is the rest mass of A compared to the rest mass of the baryon in the state B; the baryon A will always decay into some other baryon with a smaller rest mass, and the energy thus released (the difference in the two baryon masses) will appear as the kinetic energy of the final baryon plus the kinetic energies and masses of the mesons, leptons, and antileptons, and the energy of the gamma rays produced. The entropy of the final B state is thus larger than the entropy of the A state, so that the second law of thermodynamics is the determining factor: For a process to go, entropy must increase. Since a neutron is slightly more massive than a proton, it decays spontaneously into a proton, electron, and neutrino, and this decay represents an increase of entropy.

This state of affairs was all well understood in the early 1950s when certain events were observed in high-energy cosmic rays and in large accelerators. In addition to pions and muons (mesons and leptons), which had become well-accepted ingredients of cosmic-ray beams and products of collisions in accelerators, new families of very massive baryons were discovered that were something of a puzzle, owing to some unusual properties. When families of unfamiliar particles are discovered, physicists try to arrange them into groups in accordance with the properties that are understood, for example, mass, spin, charge, and parity. The archetypal new massive baryons discovered in the late 1950s and early 1960s (masses ranging from that of the proton to about three times the proton's mass) arranged themselves quite naturally into a group of eight spin $\frac{1}{2}$ baryons (the octet) and (the more massive ones) a group of ten spin $\frac{3}{2}$ baryons (the decimet); each of these "supermultiplets," as they are called, consists of subgroups arranged according to electric charge, mass, and a designation called "strangeness." Thus, the octet consists of the charge nucleon doublet (proton, neutron), the neutral singlet Λ° (lambda), the charge triplet Σ^+, Σ°, Σ^- (sigma), and the charge doublet Ξ°, Ξ^- (xi). Each of the subgroups is characterized by an integer value of "strangeness," which is 0, −1, −1, −2 for each of the four subgroups, respectively; the masses of the members of a given subgroup are about equal, but the average mass increases from about 940 MeV for the nucleon doublet to 1321 MeV for the xi doublet. Particle physicists have introduced still another numerical designation (integer or half odd integer), the so-called "isotopic spin," which gives the number of members in a subgroup; it has nothing to do with actual spin, but rather has to do with electric charge. The number of different charged members (0 charge included) in a subgroup is twice the isotopic spin plus 1. Thus, the isotopic spin of a singlet is 0, that of a doublet is $\frac{1}{2}$, that of a triplet is 1, and so on.

The decimet consists of the subgroups Δ^{++}, Δ^+, Δ°, Δ^- (delta quartet, isotopic spin $\frac{3}{2}$, strangeness 0); Σ^+, Σ°, Σ^- (sigma triplet, isotopic spin 1, strangeness −1); Ξ°, Ξ^- (xi doublet, isotopic spin $\frac{1}{2}$, strangeness −2); and Ω^- (omega singlet, isotopic spin 0, strangeness −3). The story of the discovery of these groups and how they were named gives an interesting insight into the thinking of particle physicists during the last three decades. The first experimental evidence, outside cosmic rays, for the existence of baryons other than nucleons was found in high-energy accelerators that produced violent collisions of very energetic (high kinetic energy) negatively charged pions (π^-) with stationary protons in a bubble chamber resulting in the appearance of zero charged particles. The sudden end of the π^- path indicates that the pion was captured by a proton (demanded by conservation of electric charge), but the conservation of momentum (the pion's momentum) required the emergence of two other uncharged particles moving away from the end of the pion path, each with just the right momentum, so that the sum of their two momenta equaled the momentum of the captured pion. Being uncharged, the two newly emitted particles left no tracks in the bubble chamber, but two new tracks appeared farther on, from which a new kind of baryon, the lambda zero, Λ°, and a new kind of meson, K°, were inferred to have been created at the point where the original π^- was absorbed by a proton. The process is written symbolically as $p^+ + \pi^- \rightarrow \Lambda^\circ + K^\circ$.

The Λ° possessed certain characteristics that were quite puzzling vis-à-vis the force (the strong force) that had created it. It is created very quickly—in a trillionth of a trillionth of a second (and very abundantly)—as one would expect from the action of the strong force, but it has a very long lifetime—a tenth of a billionth of a second—compared to its creation time, before it decays into a proton and a π^- or a neutron and a π°. For this reason, particle physicists call the Λ° a "strange" particle and explain its slow decay (long lifetime) by saying that it possesses a kind of "charge" called "strangeness" that is conserved in strong interactions. It therefore cannot decay via the strong interaction route—a very rapid decay—and must decay via the weak interaction route with strangeness not conserved. The strong interaction conserves "strangeness," but the weak does not!

But in view of conservation of "strangeness" in strong interactions, how can we account for the appearance of Λ° at all as a strong force product of two zero-strangeness particles? This question is explained by the idea that strange particles are produced in pairs (called "associated production"), one member of which, the Λ°, has negative "strangeness," -1, and the other member of which, the K°, has positive "strangeness," $+1$, so that strangeness is conserved in the creation process. Thus, the creation of Λ° and K° may be compared to the creation of an electron–positron pair. The positron is thus in a state of existence that it cannot leave unless it meets and is annihilated by an electron (antipositron). A positron thus takes a much longer time to disappear than to be created. In the same way, the Λ° can decay rapidly (strong force decay) with conservation of "strangeness" only if it does so together with a K°; lacking that, it can decay only weakly, with "strangeness" not conserved. All these concepts—associated production, "strangeness," its conservation and nonconservation—were introduced to account for the long life of Λ° and all the other strange particles created in the high-energy accelerators that were constructed and operated in the 1960s, 1970s, and 1980s.

The question that arose with the proliferation of all the heavy baryons was whether their arrangement into supermultiplets and subgroups of these supermultiplets can be understood in terms of more basic (elementary) particles in the same way that the structures of atomic nuclei and isotopes can be understood in terms of neutrons and protons. That the baryons are not elementary particles but consist of particles more elementary than neutrons and protons (nucleons) had already been demonstrated by the basic experimental work of Robert Hofstadter and his co-workers. Using very energetic (high-speed) electrons (essentially a very powerful electron microscope), Hofstadter probed the structure of protons and concluded, from the paths of the electrons scattered from the protons, that the proton is not a point charge, with no internal structure, but behaves like a charge distributed over a finite volume, with a complex structure. Subsequent experiments with very high-energy electrons (very deep probes) showed that the charge in a nucleon is not uniformly distributed throughout the nucleon's volume, but is concentrated in discrete nodules of charge. Additional experiments indicated further that each of the baryons that constitute the octet and decimet supermultiplets contains just three distinct electric charges; these charges were then accepted as the fundamental particles of which all matter is constituted.

Robert Hofstadter was born in New York City in 1915, the son of Louis Hofstadter and the former Henrietta Koenigsberg. Like the parents of many other outstanding physicists, Robert's parents instilled in their son an appreciation of culture and learning that was not limited to any particular discipline.[1] He received his primary and secondary education in the New York City public school system and, not surprisingly, earned high marks in his subjects. Hofstadter received his undergraduate education at the College of the City of New York, where he majored in mathematics and physics, graduating magna cum laude in 1935. That same year, he received a fellowship from the General Electric Company that enabled him to study at Princeton University, where he received his M.A. and Ph.D degrees in 1938.[1] His doctoral work dealt with the infrared spectra of organic molecules, but after receiving his doctorate, he became interested in the use of crystals as electron detectors, which proved to be extremely valuable to his later work in high-energy electron-scattering experiments.[1] Hofstadter left Princeton in 1939 to go to the University of Pennsylvania, where he became interested in nuclear physics, particularly the structure of the proton and neutron.

The Japanese attack on Pearl Harbor brought the United States into World War II and forced many physicists, including Hofstadter, to put their personal research projects on hold and contribute their talents to the war effort. Hofstadter worked at the Bureau of Standards and then at the Norden Corporation, where he remained until the end of the war.[2] After the war, Hofstadter returned as an assistant professor of physics to Princeton, where he developed counters made of crystals of sodium iodide doped with thallium.[2] These crystals were soon found to be excellent measuring devices for gamma rays and charged particles such as electrons.[2]

Hofstadter joined the faculty at Stanford in 1950 as an associate professor. Since a high-energy linear (straight-line) accelerator was being built near the campus when Hofstadter arrived there, he used the experimental skills he had developed at Princeton to help design and build equipment that could be used in the scattering experiments at Stanford.[2]

Robert Hofstadter (1915–)

Once the accelerator became operational, Hofstadter devoted himself fully to high-energy particle physics and in the 1950s conducted a series of experiments that culminated in his measurements of the distribution of charge and magnetic moments in nucleons.[2] ''More precisely, he determined their four electromagnetic 'form factors,' '' each of which ''is a technical quantity that describes how the particle interacts with other particles and fields,'' so that ''its behavior is a more general way of describing size and shape than can be done with a model.''[3] Hofstadter received the 1961 Nobel Prize in Physics for his work with nucleons, especially his discovery that ''protons and neutrons were made up of a central core of positively charged matter, about which were two shells of mesonic material.''[4] This discovery showed that ''the proton and neutron are bodies of considerable complexity, not pointlike or 'elementary' as previously supposed.''[5]

Hofstadter's experimental techniques still guide particle physicists, because the use of very energetic charged particles (i.e., electrons) remains one of the most useful ways to probe the individual protons and neutrons of the nucleus. The more energetic the charged particle, the more deeply it can penetrate into the nucleus before veering off, so there is a direct relationship between the energy of the charged particles and the extent to which the nucleus can be accurately described by electron-scattering experiments. Hofstadter himself deduced the possible existence of "mesons more massive than those already known; these included what he called the rho-meson and the omega-meson."[6] These mesons play very important roles in the interactions between nucleons.[7] Hofstadter also measured the dimensions of many nuclei and discovered how nuclear matter arranges itself in its most stable state; he also found that "nuclei form around themselves a surface region of constant thickness in which their density gradually falls to zero and that in the centers of nuclei the nuclear density is approximately constant."[7]

Hofstadter's experiments initiated the vast theoretical development of high-energy particle physics that is now called the "quark (q) theory" or the "quark model" of hadrons (baryons and mesons). It was begun by Murray Gell-Mann, who showed, from very general symmetry considerations, that the arrangement of baryons into supermultiplets of eight (octet) and ten (decimet) can be understood in terms of what he whimsically called "quark" triplets; three distinct kinds of quarks are required, however, to explain the subgroups (octet and decimet) in each supermultiplet.

The chronology of the events that led to the introduction of "strangeness" and the concept of "associated production" is interesting. In 1952 physicists at the Brookhaven National Laboratory began to operate their cosmotron accelerator which produced billion-electron volt (GeV) collisions between the colliding particles; as predicted, new particles, in the mass range from 500 to 1700 MeV, were created. These particles were called hyperons as a group, but they were soon divided into a subgroup of "heavy" mesons (the K mesons or kaons) with spin 0 and masses of about 500 MeV, that like pions, interact strongly with nucleons, and a subgroup of six "heavy" baryons (named and described above) in the mass range from 1100 to 1400 MeV. How were these six baryons to be divided into still smaller subgroups? As we have seen, these baryons all have the same baryon number (+1), the same spin ($\frac{1}{2}$), parity + or − and their electric charges are −1, 0, +1. But spin, baryon number, parity, and charge alone cannot account for the four mass subgroups into which the two nucleons and six ''heavy'' baryons break up naturally; another physical quantity (a quantum number) was required to label these four mass subgroups uniquely. Such a quantity, ''strangeness,'' was proposed almost simultaneously in 1955 (but independently) by the Japanese physicist K. Nishijima, who thought of it as a kind of charge, and Murray Gell-Mann. That ''strangeness'' is governed by a conservation principle was implied in A. Pais's observation in 1954 that strange hyperons are always produced in pairs so that each member of the pair shares a common property (strangeness), with its total value equal to zero. This type of pair production was later called ''associated production.'' With the introduction of quarks in 1964, a strange quark, to carry strangeness, was introduced to complete the quark trinity of the octet as described in detail below.

Murray Gell-Mann was the son of an Austrian immigrant and was born in New York City in 1929. Like many other scientists, Murray demonstrated his knack for mathematics and physics at an early age and was able to enroll as an undergraduate at Yale University at the age of 15. After receiving his baccalaureate degree in 1948, he left New Haven for Cambridge, Massachusetts, and enrolled as a graduate student at the Massachusetts Institute of Technology, where he earned his Ph.D after three years of study. In 1952, Gell-Mann left MIT to study under Enrico Fermi at the University of Chicago, where he remained until 1955, the year after Fermi's death. That Gell-Mann had been admitted to Fermi's research group was one indication of his promise as a theoretical physicist, but even more striking was his meteoric rise after he joined the faculty at the California Institute of Technology at Pasadena in 1956. Within one year, he was promoted to full professor in recognition of the important work he had begun in particle physics: ''In 1953, Gell-Mann proposed that certain subatomic particles possessed an invariant quality, which he called 'strangeness,' that was conserved in strong and electromagnetic interactions, but not in 'weak' interactions.''[8]

Gell-Mann's proposal, one of the earliest symmetry schemes, brought some coherence to the chaotic proliferation of particles in the 1940s and 1950s, and helped explain ''a number of peculiarities in the behavior of the short-lived, heavy, artificially produced 'strange particles.' ''[9] This proposal was a precursor to later schemes offered by Gell-Mann and other physicists to explain the behavior of strongly interacting particles: ''In 1961 Gell-Mann as well as the Israeli physicist Ne'eman announced a new system of united classification of strongly interacting particles, which Gell-Mann called the 'eightfold way.' In this scheme, the varied strange particles, and others, are expressed as 'recurrences' of a few ground states.''[10]

The utility of Gell-Mann's eightfold way was soon demonstrated when the Ω^- particle (which he had predicted) was detected by experimental physicists in 1964. One unconventional feature of Gell-Mann's classification scheme was his postulation of pointlike particles having fractional charges; he called these particles ''quarks'' after a line in James Joyce's *Finnegans Wake* (''Three quarks for Muster Mark.''). Gell-Mann suggested that there are three quarks in nature, which he called ''up,'' ''down,'' and ''strange'' quarks. These quarks have $\frac{1}{3}$ or $-\frac{2}{3}$ the charge of the electron, which, despite the arbitrariness of this value, has been accepted as the unit electrical charge. Because quarks have no known internal structure (like leptons such as electrons and neutrinos), all matter in the universe is believed to consist of combinations of quarks and leptons. Since there can be no further division of such particles, several physicists have argued that the search for the fundamental building blocks of nature will end with the anticipated discovery of the top (t) quark.

Although quarks have not yet been directly detected, Gell-Mann's scheme has been accepted by physicists and, to some extent, embellished by others who have posited additional quarks such as ''beauty,'' ''charm,'' ''top,'' and ''bottom.'' This proliferation of quarks has made many physicists uneasy that physics may be returning to the chaos that beset particle physics in the 1940s, but Gell-Mann's original eightfold way has continued to remain a cornerstone of particle physics. For his work in organizing particle families, Gell-Mann was elected to the National Academy of Sciences in 1960 and received the 1969 Nobel Prize in Physics.

Initially, Gell-Mann (and, independently, G. Zweig) proposed just two kinds (also called ''flavors'') of quarks, an ''up'' (u) quark, with an electric charge of $\frac{2}{3}e$ (e = the numerical value of the charge on the electron), and a ''down'' (d) quark, with an electric charge of $-\frac{1}{3}e$. The proton is thus pictured as the composite structure udu (2 ups and 1 down) and the neutron as the structure dud (2 downs and 1 up). The quark hypothesis, with its assignment of different fractional electron charges to the u and d quarks, was very distasteful to physicists initially, for they had been nurtured on the idea that the smallest charge in nature (the ''unit'' charge) is the charge e on the electron and that all other charges must be positive or negative integral multiples of this basic charge. But this idea is clearly untenable if baryons consist of three charged quarks, and the very idea of a unit charge loses its meaning because it is no more sensible to call the charge on the electron the unit charge than it is to call that on the u quark or on the d quark the unit charge.

In the Gell-Mann quark theory, the three quarks in a nucleon, or in any baryon, are held together by the strong force, but since nothing is known about the mathematical form of this force, the theory does not give a dynamical picture of the quarks inside a baryon. Even the masses of the constituent quarks are not known, except that the d quark is assumed to be more massive than the u quark, which is itself puzzling, since the electrical charge on d is smaller than that on u, and one expects the larger mass to go with the larger charge because charge itself contributes to mass.

With the discovery of the Λ° baryon and the heavier "strange" baryons in the octet, it was immediately clear that the u and d quarks alone cannot account for the structure of the strange baryons; a third quark is required. Such a quark—the s (for strangeness) quark—was proposed; its charge is $-\frac{1}{3}e$, and its mass is greater than that of the d quark. A single s quark in a baryon means that the baryon has "strangeness" of -1, and with two s quarks, the baryon has "strangeness" of -2. Thus, Λ° with the quark configuration dus and Σ^- with the configuration dsd have "strangeness" of -1, whereas the negative xi, Ξ^- with the configuration sds has "strangeness" of -2, and Ω^- (in the decimet) with the configuration sss has "strangeness" of -3.

Like every other known particle, each quark (q) has its antiquark ($\bar{q}$) that has an electrical charge opposite in sign to that of the quark and strangeness opposite to that of the quark. Thus, the configuration $\bar{u}\bar{d}\bar{u}$ is the antiproton $\bar{p}$, and the configuration $\bar{d}\bar{u}\bar{d}$ is the antineutron $\bar{n}$. All mesons consist of a quark and an antiquark, so that the mesons π^-, π°, and π^+ are the quark–antiquark combinations ($\bar{u}$d), ($\bar{u}$u or $\bar{d}$d), and ($\bar{d}$u), respectively; π° is thus its own antimeson, π^- is the antimeson of π^+, and vice versa. To account for the fermion (spin $\frac{1}{2}$ or spin $\frac{3}{2}$) character of the baryons, all the quarks must have a spin of $\frac{1}{2}$ (that is, the quarks are themselves fermions); this means that the mesons are bosons with spins of 0 or 1.

The simple quark theory, as described above, encountered certain difficulties that required the introduction of additional quark features that increased the complexities of the theory enormously. The quark model, in its simple form, is in conflict with the Pauli exclusion principle, which states that two identical fermions cannot exist in the same quantum state.

The simple quark model treats all u quarks as identical fermions, as it does all d and s quarks. This approach means that baryons consisting of three u quarks in the same state or three d quarks in the same state cannot exist, but in the decimet, the uuu baryon (Δ^{++}) does exist, as does the ddd baryon (Δ^{-}). The conflict between these baryons and the Pauli exclusion principle arises because in the quark theory, as developed by Gell-Mann and others, the three quarks in a baryon are all placed in the same spatial ground state. But if the three quarks are identical, as in Δ^{++} and Δ^{-}, at least two of the three must also be in the same spin state, since the quark spins can be only parallel or antiparallel to each other; thus, only two can have antiparallel spins and therefore be permitted in the same spatial ground state. Since the spin of an identical third quark would have to be parallel to the spin of one of the other two identical quarks, having three identical quarks in the same ground state conflicts with the Pauli exclusion principle and therefore is inadmissable. This conflict was eliminated or sidetracked by Gell-Mann and his co-workers by the suggestion that each quark comes in three varieties, which, for want of a better name, are called "colors," so that "red," "yellow," and "blue" u, d, and s quarks are assumed to exist. The conflict with the Pauli exclusion principle is thus eliminated by picturing the Δ^{++} baryon as consisting, not of three identical quarks, but of a "red," a "yellow," and a "blue" u quark, so that Δ^{++} itself is "colorless."

In analogy with the electromagnetic force between electric charges, which is mediated by the emission and absorption of photons by the charges, the strong force between quarks is described as carried by photonlike (0 rest mass) spin 1 bosons called "gluons," which themselves are "colored," so that they interact with themselves. In carrying the "color" charge (the source of the strong force), gluons differ dramatically from photons, which carry the electromagnetic force but carry no electric charge (the source of the electromagnetic force), so that photons do not interact with themselves. In this picture of the strong force, the gluons can change the "colors" of quarks when they are emitted or absorbed by quarks; a gluon does this because it simultaneously carries one of the three "colors," r (red), y (yellow), or b (blue), and one anticolor $\bar{r}$, $\bar{y}$, or $\bar{b}$. Thus, if a gluon carries the color r and the anticolor $\bar{b}$ (specified as $G_{r\bar{b}}$) and is absorbed by a "blue" quark, the quark's color is changed to "red," and the color of the quark that emits this gluon changes from "red" to "blue"; a gluon that carries a color and the same anticolor does not change the color of a baryon. According to the gluon model of the strong force, six color-changing gluons and two color-preserving ones (eight in all) exist.*

As the QCD model now stands, it is burdened by a number of assumptions that are unwarranted and for which no observational evidence exists. Most particle physicists assume that the strong force has the peculiar property of increasing in strength as the distance between quarks increases, becoming infinite very quickly and decreasing just as quickly to zero as quarks approach each other. The rapid increase of the "color" force with increasing separation from the quarks is called "infinite quark confinement"; according to this concept, free quarks cannot exist, so that "color" can never be observed. This means that quarks must always occur in "colorless" combinations (either as ryb triplets—baryons—or as "color"–"anticolor" doublets—mesons). This assumption of infinite confinement is not based on any concrete evidence, but is a theoretical extrapolation stemming from the fact that free quarks have not been observed in cosmic rays or in any high-energy accelerators. But this may mean, not that quarks are totally confined, but that their binding energies in baryons and mesons are extremely large—on the order of ten million trillion GeV—as would be the case if the mass of a free quark were the Planck mass (one hundredth of a thousandth of a gram), as proposed by one of the authors (Motz).

The assumption that the three quarks within a baryon move around freely because the "color" force vanishes when quarks get very close to each other (called "asymptotic freedom") is also an unwarranted interpretation of scattering experiments. In these experiments, high-energy leptons (electrons or muons) are sent into nucleons; after they leave the

*Just as the photon theory of the electromagnetic force is called quantum electrodynamics (QED), the color gluon–quark theory of the strong force is called quantum chromodynamics (QCD), but despite all the efforts that have gone into the theoretical development of this theory, very little in the way of numerical results has been achieved. The theory has predicted only very general features of baryons and mesons, which can be done just as easily with a simpler theory and with far less theoretical scaffolding than now encumbers QCD. Thus, if quarks are very massive particles (mass equal to 10^{-5} grams—the Planck mass) that lose most of their mass when they combine gravitationally in triplets to form baryons, the baryons are linear rotators with one quark at each end and the third at the center. This model eliminates the need for "colored" quarks, since the three quarks are not in the same ground state and the strong force is gravity. This model also accounts correctly for the magnetic moments of the baryons in the octet, which the QCD model cannot do. The most unsatisfactory aspect of QCD is the numerous arbitrary parameters that burden it whose values must be inserted manually. As long as QCD offers no dynamical model of a baryon, construction of competing models of baryons remains an essential part of physics today.

nucleons, their behaviors (paths) are carefully studied to glean any information about the dynamic state of the quarks within the baryons. The conclusion drawn from these paths is that the quarks seem to move about as though they were uncorrelated—that is, as though they were free. But a lepton probe of a quark structure would behave the same way after being scattered from the composite structure if the quarks were gravitationally bound.

The number of quarks now accepted has increased by two since the introduction of the s quark. A fourth quark, called "charm" (c) (in three colors and with a $\frac{2}{3}e$ charge), was proposed in 1970 by a Harvard group of theoretical physicists to account for certain types of hadronic reactions that occur very rarely as compared to their expected abundance. In 1974, a group of experimentalists at the Brookhaven National Laboratory discovered a meson they called J and another group at the Stanford Linear Acceleration discovered a meson they called psi; these two particles were later found to be the same meson consisting of a quark more massive than the s quark and an antiquark. The existence of the J/psi meson is taken as evidence for the discovery and existence of the "charm" quark, since the J/psi quark mass (3.7 GeV) is too large to be accounted for by any combination of an s quark and a u or d antiquark.

After the discovery of "charm," a group of physicists at Fermilab, led by L. Lederman, in 1977 discovered a new meson with a mass of 9.46 GeV and a charge of $-\frac{1}{3}e$; they accepted this meson as evidence for the existence of a fifth quark, which they called b (bottom). But this discovery has not ended the search for quarks, since, particle physicists argue, symmetry between leptons and quarks requires the existence of a sixth quark. The symmetry argument goes somewhat as follows: Three so-called "generations" (families) of leptons exist: (e, ν_e), (μ, ν_μ), (τ, ν_τ), that is, the three electronlike particles, the electron, the muon, the tau, and the three kinds of neutrinos, ν_e, ν_μ, ν_τ, a different kind for each kind of electron; therefore three generations or families of quarks must exist: (d, u), (s, c) and (b, t), t standing for the undiscovered "top" quark. This reasoning is very tenuous and untrustworthy. To begin with, a τ-type neutrino has never been observed, and the observational evidence for a top quark is extremely weak. But even if a very massive meson is discovered, the reasoning outlined above is faulty, because the leptons and quarks are basically different kinds of particles, as most drastically emphasized by the zero rest mass of the neutrino.

We close this chapter on elementary high-energy particles with a brief discussion of the intermediate bosons $W^{\pm}$ and Z°, which are assumed to be the carriers or mediators of the weak interactions, the force that produces the interactions between neutrinos and baryons and mesons. Since this interaction occurs only when the neutrino is extremely close to the baryon or meson (the range of the weak force is 10^{-15} cm), the masses of the intermediate bosons must be very large; certain theoretical considerations stemming from an attempt to unify the electromagnetic and weak interactions (the so-called "electroweak theory") indicated a mass of about 80 GeV for the $W^{\pm}$ and 92 GeV for the Z°. The need for two charged intermediate bosons W^{+} and W^{-} arises because neutrinos can interact with baryons and mesons by changing their electric charges; thus, the weak interaction changes a neutron into a proton (increase of charge by +1) with the emission of an electron and an antineutrino. This beta-decay process is produced by the emission of a W^{-} from the d quark in the neutron (the d quark becomes a u quark), and the W^{-} then decays very quickly into an electron and an antineutrino. If a muon–neutrino (ν_{μ}) interacts with a neutron, it does so (according to the theory) by emitting a W^{+} that is absorbed by the neutron, which then becomes a negative muon ($\mu-$). In accordance with the theory of the intermediate bosons, a Z° is interchanged between a proton and a high-energy neutrino when the neutrino is scattered by the proton without acquiring charge and with the proton's electric charge unchanged. The W^{+} and Z° are vector bosons because their spins are 1 (that is, $\hbar$).

In 1981, experimental physicists at CERN, under the direction of Carlos Rubbia, turned on their super proton synchrotron collider, in which a beam of protons streaming in one direction collides with a beam of antiprotons streaming in the opposite direction. The collider is designed so that a total energy of 540 GeV is released when a proton and an antiproton in the oppositely moving beams collide and annihilate each other. On release, after the annihilation, this amount of energy produces large numbers of particles, among which, it was hoped, a W or a Z would appear, since the energy of 540 GeV was chosen to give particles of the theoretically predicted masses of the W and Z. In July 1984, the CERN group announced that they had found six events among a billion or more events recorded on their photographic plates. In no case is the boson itself observed, since it decays in 10^{-20} seconds (one hundredth of a millionth of a trillionth of a second), but it is supposed to decay into a lepton (or an antilepton) and an antineutrino (or a neutrino). Thus, a track was found transverse in direction to the beam direction that is interpreted as that of a very high-energy electron. The absence of a track opposite to the electron's track on the photographic plate, as required by conservation of momentum, is taken as evidence that an antineutrino really appeared with the electron, but left no track because it has no charge.

One cannot avoid a feeling of uneasiness about this kind of physics, since so much of it is based on the assumed existence of particles that cannot be observed. Thus, in the announced discovery of the W^-, as described above, it is argued that the existence of an electron track with no track to balance it is sufficient evidence for the appearance of the W^- as its precursor. But this conclusion raises two questions: First, why could not the electron and its antineutrino have been produced directly instead of via an intermediate particle state? Second, why single out the electron track from the dozens of others on the plate for special treatment, since nothing in the appearance of that electron track points to precursors any more than do the others? We mention these points to indicate that many unanswered questions still remain in connection with experiments such as the CERN experiment, and one should subject every feature of it to intense analysis.

CHAPTER 20

Cosmology

And God said, Let there be light: and there was light.

—GENESIS 1:3

Cosmology is as old as civilization itself because every society, however ancient, was deeply concerned about its place in the universe, and the awesome sight of the clear, moonless night sky inevitably stimulated speculation about the nature of the planets, the stars, and the Milky Way. Although the ancient Sumerians, the Babylonians, the Phoenicians, the Egyptians, and the Chaldeans developed an accurate positional astronomy as a guide to their navigation, their agriculture (knowledge of the variations of climate with changes in the position of the sun was absolutely essential), and their calendar making, they never developed a rational cosmology. Their picture of the universe stemmed from their mythology, their astrology, and their theology; thus, it varied from group to group and was primitive, anthropomorphic, and without any rational base.

The ancient Greeks were the first to try to develop a rational cosmology—a cosmology based on careful observations of the heavens and thus amenable to measurement and testing. Pythagoras probably began Greek cosmology by trying to explain the motions of the planets in terms of simple numerical relationships, and this idea was pursued by Plato. Nothing much came of all this work, however, since the thinking of all these early Greek philosophers was bogged down by their belief in a geocentric universe. Aristarchus of Samos broke with this anthropomorphic belief and, using impeccable logic and careful observations of the positions of the sun and moon relative to the earth at different lunar phases, demonstrated that a geocentric solar system leads to ridiculous, untenable conclusions. Unfortunately, Aristarchus wrote about his work in only one book, which was neglected for centuries, so that his heliocentric theory had no Greek followers. Some 1800 years later, however, Nicolaus Copernicus partially justified his own acceptance of the heliocentric theory by pointing out that Aristarchus had already set a precedent for such a theory.

Though Aristarchus's Greek contemporaries and those who followed him did not accept his cosmology, they continued with their careful astronomical observations and their use of mathematics wherever possible. This work culminated in the incredibly accurate, naked-eye celestial observations of Hipparchus, and Ptolemy's theory of epicycles, but cosmology was not much further advanced as a model of the structure of the universe at the end of the ancient Greek era than it was at the beginning of that era. All that was changed with the publication of Copernicus's great work, *On the Revolutions of Celestial Bodies,* but not until Johannes Kepler discovered and published his three laws of planetary motion was astronomy launched on a rational path that ultimately led to a complete cosmology. But Kepler's laws alone do not provide a sufficient base on which to construct a cosmology, since they are empirical and do not have the generality to permit us to make any deductions about objects, such as galaxies, beyond our solar system. The development of a general cosmology based on fundamental universal principles had to wait for Isaac Newton and his laws of motion and of gravity, but before that occurred, Galileo's telescopic observations gave some indication of the vastness of the universe and demonstrated that the scale on which a cosmology would have to be developed is far grander than anything that the Greeks had envisioned. The nature of the Milky Way had been a mystery for thousands of years and had remained so until Galileo examined it through his telescope and discovered that it consists of thousands of discrete points of light, which he correctly identified as individual stars. From the faintness of these stars and his knowledge of the intrinsic luminosity of stars (solar luminosity), he concluded that the Milky Way is at a vast distance from us, a distance much greater than the distances of the naked-eye stars. We know today that the Milky Way consists of the spiral arms (accumulations of star clouds) of our galaxy that lie between us and the center of our galaxy.

Newton's laws of motion and his law of gravity opened the door to the development of a rational cosmology based on established physical principles and amenable to rigorous mathematical analysis. To Newton, it was clear that solving the cosmological problem was no more than solving a gravitational problem involving the motions of many bodies like the sun. But in contemplating the problem, he encountered what appeared to be an insurmountable difficulty; neither an infinite universe with an infinite number of stars, distributed uniformly throughout space, nor an infinite space containing a finite number of gravitationally interacting stars can possibly give the kind of universe we see. If the universe were populated by an infinite but uniform distribution of point masses (stars), the gravitational field at a point would not be uniquely determined; it could take on any value we please. To understand this idea, we consider any point in such a model of the universe and pass through that point the surface of a sphere of any given radius R. The intensity of the gravitational field at the given point depends only on the total mass (total number of stars) within the sphere; the stars outside the sphere contribute nothing to the gravitational field at the point. Since the mass within the sphere increases with the volume of the sphere, that is, with the cube of the radius R^3, and the gravitational field on the surface of the sphere (at the given point) decreases with the square of the radius R^2 (that is, varies inversely as the square of the radius), the intensity of the field at the point goes as R^3/R^2, or increases with R. Since we may choose a sphere of any size we please in this analysis, we can obtain any value we please for the field; this conclusion, of course, is nonsensical, so that an infinite distribution of stars is physically untenable.

A similar objection, based on the dark appearance of the evening sky, was voiced against an infinite stellar distribution by H. W. Oblers in 1890. The intensity of the light from the stars at any point would be so great that the night sky would be as bright as the surface of any star, so that the sky should never be dark. Certain legitimate objections have been raised against Oblers's analysis on the basis of the finite lifetimes of stars—the very distant stars will become cold and dark long before their light reaches us, so that they should not be counted in an Oblers-type analysis. But this fact does not affect the gravitational field argument against the infinite star distribution.

Newton's objection to a finite number of stars in an infinite space was that such a distribution could not possibly be the uniform distribution of the stars that we actually observe; the stars would be either infinitely dispersed, so that the sky would appear to be essentially empty, or distributed around a concentrated nucleus, but thinning out uniformly in all directions away from the nucleus. Newton knew nothing about galaxies or clusters of galaxies, but the argument against a finite distribution of stars in an infinite space applies to a finite distribution of galaxies or clusters of galaxies. These difficulties associated with the Newtonian cosmological problem delayed cosmological modeling until Albert Einstein applied his general theory of relativity (his gravitational theory) to the universe as a whole and obtained a set of cosmological equations that are the starting point of all cosmologists. But before we consider these equations, we describe the universe as it has been revealed to us by the various types of telescopes, ranging from radio telescopes to gamma-ray and neutrino telescopes, that modern technology has produced for the astronomer. With these telescopes, we can study the universe through many radiation and particle windows that were beyond the wildest dreams or imaginations of astronomers just half a century ago. Astronomers today have the additional advantage of contributions from orbiting telescopes, so that the distortion of the earth's atmosphere is eliminated.

The observational data collected through all these various telescopes reveal that however far we look out into space, the building blocks of the universe are aggregates of galaxies in which the individual members are held together by their mutual gravitational interactions. Thus, the Milky Way, the galaxy in which we live, belongs to the local cluster of galaxies to which the great spiral nebula in Andromeda and some 20 other galaxies belong. The Andromeda galaxy is about 2.5 million light-years away from us, so that our local cluster has a diameter of about 4 million light-years (1 light-year is the distance light travels in 1 year—about 6.5 trillion miles). Many clusters contain hundreds and even thousands of galaxies; thus, the Virgo cluster, at a distance of about 50 million light-years, contains 2500 visible galaxies, and the Hydra cluster, at about 2 billion light-years, consists of a few hundred galaxies.

An interesting and puzzling feature of such clusters is that the individual galaxies are moving about so rapidly that the observed number of the individual galaxies in a cluster is too small, by a factor of almost 100, to give the gravitational force that is required to keep the cluster from dispersing. This puzzle is known as the "mystery of the missing mass" (actually the hidden mass); only about 1% of the total mass in the universe can be observed or can be accounted for by ordinary baryons (nuclei). The nature of this "hidden mass" has puzzled physicists and astronomers for the last 50 or so years, but we do know that it must be of a very esoteric nature to have escaped detection by the many observational astronomers who have been searching for it so assiduously. It may be that this hidden mass accounts for the vast energetic events associated with certain galaxies that radio telescopes have detected.

Galaxies themselves are aggregates of anywhere from tens of billions to hundreds of billions of stars that are arranged either in fairly uniform, undifferentiated ellipsoidal structures (elliptical galaxies) or in structures consisting of smooth cores from which two spiral arms that start from two diametrically opposite points wind around the cores to form up to five distinct spiral arms (spiral galaxies), as in the Andromeda galaxy, or three arms, as in the Milky Way. Most spiral galaxies have no more than three such arms, which can be traced out from the core with radio telescopes because neutral hydrogen, which emits a very definite radio signal that can be easily detected even with small radio telescopes, is concentrated in the spiral arms of such galaxies. The spiral arms are also marked by vast clouds of dust, which, together with hydrogen, is the raw material from which new stars are constantly being created; the spiral arms are thus the star nurseries. The cores of spiral galaxies are uniform, with no dust and very little neutral hydrogen, so that no new stars are born in the cores. Astronomers have accordingly divided stars in a galaxy into two categories: the very oldest stars, which are found, generally, in the cores of spirals and throughout ellipticals, and younger stars like the sun, which are found in the spiral arms of spiral galaxies, but not in elliptical ones. The oldest stars, which are metal-poor, are called "population II" stars, whereas the younger, metal-rich spiral-arm stars are called "population I" stars. The term "metal-rich" (heavy elements) is a relative one: The atmospheres of population I stars have a total proportion of heavy elements (carbon, oxygen, iron, and others) of about 3%, whereas the proportion of such elements in the atmospheres of population II stars is very close to zero. Since both stellar populations consist mostly of hydrogen and helium (73% hydrogen and 24% helium in the sun), the absence of the heavy elements from the population II stars is taken as evidence that these stars were born first from the primordial hydrogen and helium (no heavy elements present) in the early universe. Most population I stars were born anywhere from hundreds of millions of years to 5–6 billion years after the population II stars from material that had been cooked at very high temperatures (millions of degrees) in the furnaces of the cores of population II stars and then spewed into space in vast explosions (supernovae). The "cooking" had fused about 3% of the original hydrogen and helium in the cores of the population II stars into the nuclei of the heavy elements, starting with carbon, so that population I stars were born from enriched material.

Population II stars are also the only constituents of the large spherical aggregates of stars called "globular clusters," which form halos around the cores of galaxies. About 100 globulars, each containing anywhere from 100,000 to 1 million population II stars, revolve around the core of the Milky Way; these globular stellar aggregates, free of dust and gas, are probably relics of the birth of our galaxy, as they are of the birth of any galaxy to which they are gravitationally attached. Another important feature of a galaxy is its amorphous halo, which is very nearly spherical, with a radius about twice that of the galaxy's diameter. The material nature of this halo is unknown, and so astronomers have speculated that galactic halos may account for some of the "missing mass."

Galaxies like the Milky Way have diameters of about 100,000 light-years and spherical cores with diameters of about 15,000 light-years; they contain about 200 billion stars. The Andromeda spiral, about twice the size of the Milky Way, contains some 400 billion stars in various stages of aging and evolution like the stars in the Milky Way.

With this brief description of galaxy types, we consider again the clustering of galaxies. A careful analysis of the distribution of galaxies in space has revealed not only that galaxies are organized into individual clusters, but also that these clusters are themselves organized into superclusters. Our local cluster and the Virgo cluster are two members of a vast supercluster. But the surprising property of superclusters is that they are not necessarily spherical in shape, but seem to be linear. They appear to form lacy patterns in space, like cobwebs on a lawn. Thus, large filaments of clusters of galaxies seem to be strung through space, with large spherical gaps of what appears to be empty space. But the evidence for emptiness in these gaps is very tentative, and it may well be that they contain large quantities of massive dark particles.

The life histories of stars are intimately related to the cosmological problem. The formation of a star from an amorphous distribution of dust and gas is essentially a gravitational phenomenon, but all the natural forces play a more or less important role at different times in directing the life of a star. The shapes of stars clearly indicate that a spherically symmetric force (gravity) produced all the stars, but since large variations in stellar properties exist from star to star, we consider why such variations exist. These variations are determined by two parameters: the mass of the gas–dust cloud that contracts gravitationally to become a star and its chemical composition (its mean molecular weight, which is essentially determined by its hydrogen–helium content). It is intuitively clear that if an initial amorphous configuration of gas and dust with a given chemical composition and a definite mass contracts down to an equilibrium state, it can do so in only one way; that is, the final gravitational configuration is unique. Thus, only the mass and the chemical composition of the initial raw material from which the star emerges determine the life history of the star.

To see how the mass and the chemical composition play their separate roles in this stellar drama, we picture a cluster of stars of various masses formed by the gravitational fragmentation of a vast distribution of gas and dust. All the stars in the final cluster are chemically identical when they are born, but differ in mass, so that the observed variations among them are due entirely to mass differences. To see the role that the mass plays, we consider a massive fragment of the initial cloud breaking off and beginning to contract gravitationally. Clearly, the speed of its contraction depends on its mass and the basic laws of motion and gravity; they, together with the laws of gases and thermodynamics, enable us to follow the collapse and see how the internal configuration of the contracting gas and dust changes. The principle of conservation of energy and the second law of thermodynamics tell us at once that the in-falling atoms and molecules can form a stable sphere of gas only if the entire configuration loses energy so that the total entropy increases (release of energy means entropy increase). Energy is released in a very simple way: As the force of gravity pulls the atoms and molecules together, they move with ever-increasing speeds (the gravitational potential energy changes to kinetic energy) and the temperature of the entire configuration increases, with radiation released from the collapsing configuration at a rate proportional to the 4th power of its absolute temperature. But the temperature does not tell us the origin of this electromagnetic radiation, which can come only from accelerated electric charges; the radiation is released by electrons in the atoms, which are set vibrating when the atoms collide with each other. As the contraction of the configuration continues, its energy diminishes, and it must go on contracting, while its temperature rises because only half the released gravitational potential energy is changed into electromagnetic radiation; the other half remains in the configuration as kinetic energy (internal energy) and thus raises the temperature continuously. This process goes on until the central temperature reaches 10 million degrees K; the contraction then ceases because thermonuclear fusion begins with the transformation of protons (in groups of four) into helium nuclei. At this point, the configuration becomes a star, and its radius, luminosity, and surface temperature are determined primarily by its mass, which also determines the rate at which it evolves. Its chemical composition plays an essential but minor role in this evolutionary process; the hydrogen content, together with the temperature, determines how rapidly it generates energy by fusing protons to form helium nuclei, and the heavy elements determine the rate at which the radiation, generated in the star's core, gets to the surface. Thus, in a star like the sun, the radiation generated by nuclear fusion at any given moment reaches the surface some 30 million years later. In that time, its character has changed drastically from very energetic gamma rays to the life-giving radiation we receive from the sun now.

Since we cannot penetrate into a star's interior to measure the various parameters (for example, pressure, temperature, density) that determine the physical conditions from point to point, we must deduce these conditions from the basic physical laws that govern the stellar interiors. Radiation cannot reach us directly from stellar interiors, but neutrinos do; however, the information they convey is ambiguous. Fortunately, the stellar material is in a gaseous state (actually a perfect gas), so that we may apply to it the laws of thermodynamics and the well-known gas laws. Since the energy is transported from the star's interior to its surface mostly in the form of radiation, we must also apply the laws of radiation as modified by the quantum theory to take proper account of the interaction of the radiation with the ionized atoms as it flows from the interior to the star's surface.

The basic stellar interior equations (four in number) that incorporate the basic physical laws and describe the changes in the interior stellar conditions as one moves from point to point inwardly or outwardly inside a star were developed by the great British astrophysicist Arthur Eddington, who showed that radiation is the principal mode of energy transport inside stars and wrote the equation that describes this mode. This formula was the final equation that astrophysicists needed to develop theoretical astrophysics into the precise analytical tool that it is today for probing the structure, the birth, and the evolution of stars, but one more important bit of information was needed before detailed theoretical models of stars could be developed: the stellar energy generation mechanism. From the known luminosities of stars, Eddington correctly surmised that stars produced their energy by fusing protons together to form helium nuclei, but not knowing just how the stars do this, since nuclear physics was unknown when Eddington did his work, he assumed very general relationships among the interior parameter, which enabled him to solve the interior equations and obtain models of stars with radii, masses, and luminosities close to those of the sun. These assumptions are not a satisfactory solution to the astrophysical problem, but they were important at the time because they demonstrated that the equations, even without the energy mechanism, were on the right track.

Sir Arthur Stanley Eddington (1882–1944) was born in Kendal, England, the son of a Quaker schoolmaster who passed away when Arthur was two years old. Soon thereafter, the Eddington family moved to Som-

Sir Arthur Stanley Eddington (1882–1944)

erset, where Arthur spent the early years of his life. Although the Eddington family had little money, Arthur's mother managed to send Arthur to Brynmelyn School, where he was fortunate to have several outstanding teachers who instilled in him an appreciation of classical literature and a solid foundation in mathematics. Although shy by nature, Arthur earned high marks in his subjects and won a scholarship to what is now the University of Manchester. While at Manchester, Eddington was especially influenced by the physicist Arthur Schuster, one of his most distinguished instructors, who encouraged young Eddington's interest in the sciences. In 1902, Eddington won a scholarship to study at Trinity College, Cambridge, and he began his studies there that autumn.

Eddington's first two years at Cambridge involved concentrated studies of mathematics; his rigorous preparation enabled him to obtain the prestigious first wrangler position in the tripos examinations in 1904, the first time that a second-year student had ever won such an honor.[1] The following year, he received his degree in mathematics, and he made his living for a short time as a mathematics tutor. In 1906, he accepted an appointment as chief assistant at the Royal Observatory at Greenwich, where he spent the next seven years, learning to become an astronomer. The Royal Observatory required Eddington to perform a variety of duties that enabled him to hone his skills in practical astronomy while helping him to stay abreast of the most recent advances in the subject. Soon after assuming his duties at Greenwich, he was elected a fellow of the Royal Astronomical Society. He began the theoretical astronomical investigations on the internal constitution of stars that established his reputation, and he went on several astronomical expeditions to Malta and Brazil. In 1913, he accepted the Plumian professorship at Cambridge and moved into the Observatory House as director of the observatory, where he lived for the rest of his life.[1]

The training Eddington had received at the Royal Observatory was very helpful in his researches on stellar structures. He was particularly interested in Karl Schwarzschild's theory of the radiative equilibrium of a star's atmosphere with respect to a star's interior.[2] Eddington showed that the outward-flowing stellar radiation produces a gas pressure that just counterbalances the weight of the outer envelopes of the star's mass and that there is a direct relationship between the mass of a star and the rate at which it radiates energy. In other words, a star several times the mass of the sun would have a correspondingly shorter lifetime: "Eddington concluded that relatively few stars would exceed ten times the sun's mass and that a star of fifty times the solar mass would be exceedingly rare."[3] This belief was confirmed by observations that showed that there are relatively few stars of such large masses in the observable universe. Eddington also calculated the diameters of several red giant stars and applied his calculations to the dwarf companion of the star Sirius, "obtaining a diameter so small that the star's density came out to 50,000 gm/cc, a deduction to which he said most people had mentally added 'which is absurd!' "[3] Although Eddington's estimate raised a few eyebrows, his findings were confirmed by astronomers at the Mount Wilson observatory, who found that the red shifts in spectral lines of the star approximated the results required by Albert Einstein's theory of relativity.[3]

No less revolutionary was Eddington's conclusion that the rate of radiation of stars of one solar mass necessitated an evolutionary time scale of several trillion years if the Hertzsprung–Russell sequence of stellar colors and luminosities were to be retained.[4] Since Eddington saw that no known sources of chemical energy could keep the stellar fires burning for such a long time, he suggested in 1917 that the stars are fueled by nuclear processes. Since Eddington's proposal was offered more than 20 years before the discovery of nuclear fission by Otto Hahn and Lise Meitner, it was greeted with widespread skepticism. However, Eddington continued to argue that there were no viable alternative energy sources that could power a star, and he was finally vindicated in 1938 when Hans Bethe published his carbon-cycle theory of stellar energy.

At the same time, Eddington was becoming one of the foremost experts on the theory of relativity. He was the first in Great Britain to receive a copy of Einstein's famous paper on the general theory in 1916, and he soon mastered its mathematical intricacies. Eddington was largely responsible for the acceptance of the general theory by the British scientific community owing to his elegant *Report on the Relativity Theory of Gravitation,* which he wrote for the London Physical Society in 1918. He also wrote a popularized version of the general theory in his *Space, Time and Gravitation,* which increased the public interest in Einstein's work. In 1919, Eddington provided the needed observational confirmation for one of the central predictions of the general theory—that the path of light is bent by gravity—when he led an expedition to the isle of Principe in the Gulf of Guiana to photograph a solar eclipse. Eddington was so busy changing photographic plates during the eclipse that he could not observe it firsthand, but once the plates were developed and the displacement of starlight passing near the rim of the sun was documented, he announced that the general theory had indeed been confirmed. It is reported that after Einstein was notified of the experimental confirmation of the theory by Eddington, he was asked how he would have felt had the predicted bending of the light not been found. Einstein replied, "Then I should have been very sorry for the Lord. The theory is correct." In 1923, Eddington published his *Mathematical Theory of Relativity,* which Einstein himself thought to be the finest presentation of the subject ever written.

In the 1920s, Eddington became obsessed with the idea of formulating a grand theory that would unify the theory of relativity with quantum theory, and he saw the fundamental constants of nature such as the speed of light and Planck's constant as the key to that effort. He derived an estimate for the number of particles in the universe that is not too different from the values commonly accepted today and evaluated over 25 physical constants. Unfortunately, he became bogged down owing both to the difficulties of reconciling the two theories and his reliance on generalizations about the significance of these physical constants that were not supported by empirical data. He also found time to write a succession of books about physics and cosmology that established him as the foremost popularizer of science of his time. Although he developed a theory of matrix mechanics equivalent to that offered by Paul Dirac, he became increasingly convinced that truth could come only through some sort of mystical revelation instead of from scientific theories. Although Eddington's subjectivist philosophy is generally ignored today, it was offered with a conviction that it was as pure as a syllogism and that no exceptions could be tolerated. While such confidence is generally a precedent to a theory being thrown on the scrap heap, the imagination that Eddington brought to his work on stars and relativity also pervades his fundamental theory.

With the rapid growth of nuclear physics after the discovery of the neutron, the theory of the thermonuclear fusion process for the production of energy in stars was developed in 1936 by Hans Bethe (see Chapter 18), who analyzed two different processes for the thermonuclear fusion of hydrogen into helium: The first is the proton–proton chain, in which four protons, in a series of steps, are fused into a helium nucleus. The second is the famous carbon–nitrogen cycle, in which four protons are also fused into a helium nucleus, but not directly; carbon acts as a catalyst. The proton–proton chain goes as the 4th power of the absolute temperature; the carbon cycle goes as the 20th power of the temperature. Owing to these different temperature dependencies, the proton–proton chain operates in solar-mass stars and in stars of lower mass, whereas the carbon cycle dominates in the more massive stars. This phenomenon accounts for the very large luminosities of blue-white stars like Rigel in Orion, which is about 65,000 times as luminous as the sun.

Astrophysicists had just begun to incorporate Bethe's equations about thermonuclear energy generation into Eddington's stellar interior equations when World War II began, so everything in this exciting area of physics was suspended until after the war, when it expanded rapidly owing to the availability of high-speed electronic computers. The stellar models thus obtained were in amazing agreement with the observed properties of stars over a wide range of stellar parameters (masses, radii, surface temperature, and chemical composition). These models clearly indicate the importance of mass and chemical composition in stellar structure and prove that these two parameters uniquely determine the star's structure. This is one of the most beautiful examples in the history of physics of the correlation of a physical system (a star) with a set of equations (physical laws expressed mathematically). The model of the sun deduced from these equations shows that the sun's central temperature is about 15 million degrees Kelvin and its central density is 150 grams per cubic centimeter.

The vast increase in nuclear physics data that followed the war enabled astrophysicists to go beyond models of individual stars and to develop theories of the evolution of groups of stars that are in excellent agreement with observations. The star's mass is the critical parameter for its evolution: The more massive a star is, the more rapidly it "burns" its nuclear fuel and the more rapidly it evolves. The star's mass also determines the star's final state and whether it ends its life as a white dwarf, a neutron star, or a black hole. All stars evolve by burning their hydrogen first, to produce helium, and then, when they reach the red giant stage, their helium, to produce carbon. A star of one solar mass does not evolve beyond the carbon stage because its mass is not large enough to force its central temperature to rise above a few hundred million degrees, which is essential for the nuclear transformation of carbon nuclei into the more massive nuclei. The sun and other such stars thus settle down to the white dwarf stage after their red giant stage. They are prevented from collapsing any further by the pressure of the free electrons, which are in a degenerate state: They move about freely among the relatively fixed, closely packed nuclei. The white dwarf material thus behaves like a metal.

Stars more massive than the sun cannot remain in equilibrium as white dwarfs: They continue collapsing beyond the white dwarf stage until the free electrons are forced into the heavy nuclei and the nuclear protons are changed into neutrons. Since such nuclei are unstable, they emit neutrons until the star consists almost entirely of neutrons; it has become a neutron star, a few miles in diameter with a density of billions of tons per cubic centimeter. Such a star is also spinning very rapidly and is surrounded by a very intense magnetic field. It is supported against further gravitational collapse by the pressure exerted by its degenerate neutrons.

If a star is very massive (10–15 times the solar mass), it continues to collapse beyond the neutron star stage and becomes a "black hole," which is the most remarkable state of bulk matter known. The theory of black holes was developed to its most advanced degree by John Archibald Wheeler, who gave this state of the ultimate gravitational compression of matter its very appropriate name. Because space–time in the neighborhood of a black hole is extremely curved, one must apply the general theory of relativity to the study of black holes as Wheeler did. Without going into the complex details of black-hole theory, one can understand its essential features by applying Newtonian gravitational theory that is appropriately altered to take space–time curvature into account. As a massive sphere such as a massive star collapses, the force of gravity on its surface becomes so great that the speed of escape from the surface equals the speed of light. It thus becomes invisible, since light itself cannot leave the surface. This can be understood in a general way from Newtonian theory, but general relativity is required for a full analysis and complete understanding of black-hole physics. Since a black hole cannot be seen, its existence must be inferred from the behavior of a visible star that is revolving around it. Thus, a very tiny invisible x-ray-emitting region is observed in Cygnus X-1, about which a massive star is revolving every 5.6 days. We conclude that the invisible x-ray source is a black hole into which matter from the visible massive star is falling.

One of the most imaginative and versatile physicists of his generation, John Archibald Wheeler was born in 1911 in Jacksonville, Florida. Both his parents were librarians, and their regard for books and learning influenced John's decision to become a scientist. He showed his promise at an early age and pursued his studies so diligently that he earned his Ph.D. in physics from Johns Hopkins University at the age of 21. After completing his doctoral research, Wheeler spent one year with Gregory Breit at New York University and a second year with Niels Bohr in Copenhagen, concentrating on nuclear research. He joined the faculty at the University of North Carolina, where he taught for three years and "described [in 1937] the concept of resonating group structure in light nuclei and supplied the mathematical formalism to construct nuclear wave functions, taking this structure into account; he showed how this approach supplied among other results a means to evaluate an effective potential—which turned out to be velocity dependent—for the interaction between one alpha-particle and another."[5] That same year, Wheeler formulated the concept of the scattering matrix and outlined its principal characteristics, creating a theoretical tool that contributed greatly to the work in elementary particle physics being done by Werner Heisenberg and others.[5]

In 1938, Wheeler left Chapel Hill for Princeton, where he continued his research on the atomic nucleus, which culminated in 1953 with the introduction of "the collective model of the atomic nucleus," which distinguishes between the states of the individual nucleons and the nucleus as a whole.[5] He also introduced the resonating group structure into atomic physics, but his most newsworthy work may have occurred in the wake of the political crisis triggered by the Soviet Union's explosion of an atomic bomb in 1949 (which ended the American monopoly on nuclear weapons), when he joined Edward Teller at Los Alamos to see whether it was

John Archibald Wheeler (1911–   )

possible to build a hydrogen bomb. Wheeler returned to Princeton to supervise a secret effort to work out the nuclear physics of a variety of hydrogen-fueled triggering devices. Wheeler's work on the so-called Project Matterhorn came to an end with the successful testing of a hydrogen bomb in 1952. Although the design and technological features of nuclear weapons have changed greatly since the time of Project Matterhorn, Wheeler and his co-workers developed many codes, later refined by physicists at the national laboratories, that continue to be a foundation for the design of nuclear weapons.[6]

Soon after completing his work for the government, Wheeler shifted his attention to Einstein's general theory of relativity to understand better the relationships between particles and fields. Working with his graduate student Richard Feynman, Wheeler decided that a geometrical interpretation of nature as envisioned by Einstein's general theory of relativity meant that the particle was an improper basis for considering nature because all matter is simply the by-product of space–time curvature. To try to understand nature in terms of particles would be like trying to understand the ocean by viewing it from above the surface. In any event, Wheeler's geometrical conception of nature led him to try to tie together quantum theory and Einstein's general theory of relativity by developing a probabilistic scheme in which space is not described by any single geometry, but resonates from one geometry to another.[7] He envisioned space as foamlike on the microscopic level, characterized by violent fluctuations on the order of the Planck length (1.6×10^{-33} cm). Wheeler coined the phrase "superspace" to refer to this strange arena in which each point contains an entire three-dimensional geometry. For Wheeler, the fluctuations in the superspace are responsible for the manifestations of the particles; the particles are merely the images created by the undulating foam of superspace.

Wheeler's theoretical investigations of general relativity theory led him to wonder about the dynamics of gravitational collapse. Drawing on an idea introduced by Pierre Laplace and John Mitchell more than a century earlier, Wheeler calculated that once the thermonuclear furnace of a star having at least three solar masses (a mass three times that of the sun) stopped burning, the star would collapse to what Wheeler called a "black-hole state." In short, the star would no longer be able to withstand the imploding force of its atmosphere and would consequently collapse on itself until it was no more than a few miles in diameter. The intensity of the gravitational field on the surface of such an object would be so great (owing to the vastly increased density of the black-hole star) that not even light could escape, making the star invisible. While doing much of the pioneering work in the physics of black holes, Wheeler suggested that the universe itself might lie inside a gigantic black hole because its mass might curve space sufficiently so that a light signal could never escape it. Wheeler also posited that if we live in an expanding universe, then the outward rush of galaxies might someday be reversed, hurtling all matter and energy back to a single nodule of matter and space which would be squeezed out of existence. Although a number of physicists disagree with his geometrical view of nature and believe that his attempt to unite quantum theory and relativity theory is fundamentally flawed, Wheeler has done more than any other person, save Einstein, to cast all phenomena on the microscopic and macroscopic levels of nature in terms of a geometrical framework, although the usefulness of such an approach continues to be hotly debated.

An important consequence of the evolution of massive stars is that they build up all the heavy elements now observed in the universe. The temperatures in the cores of these giant stars reach billions of degrees, so that the light nuclei in their interiors are moving just fast enough to merge and to form heavier nuclei when they collide. Such massive stars, at the end stage of their evolution, when they have built up iron cores, collapse violently and then explode to become supernovae. In this process, the iron core, compressed enormously by the collapse, becomes a rapidly spinning neutron star, and the exploded material, expanding rapidly away from the core, injects heavy nuclei into the interstellar clouds of matter, enriching them as raw material for the formation of second-generation stars.

Two more features of the universe are important for our understanding of its dynamics. First, extremely luminous objects have been discovered at vast distances; indeed, these objects, the famous quasars, at distances of billions of light-years, are the most distant objects ever observed. Since quasars are starlike objects on photographic plates, their vast distances indicate that they are also the most luminous concentrations of matter known. The most distant known quasar is some 10 billion light-years away from us, and we find the number of quasars per given volume of space increases with increasing distance. Quasars therefore tell us something about the early universe (the universe billions of years ago). For a quasar to look like an ordinary star image on a photograph, despite its vast distance, means that it is a superluminous object; from all the observational evidence pertaining to quasars, astronomers estimate that a typical quasar is as luminous as 100 galaxies. A quasar's energy source is still a great mystery; nothing that we know about force fields and particles today can account for such concentrated sources of energy.

The second remarkable feature of the universe that contains important information about cosmological dynamics is the background (cosmic) radiation that permeates all of space. This radiation was detected by chance with a radio telescope in 1965 by Arno Allan Penzias and Robert Woodrow Wilson. This long-wavelength radiation, which comes to us uniformly from all directions in space, is very cold thermal black-body radiation; it has all the characteristics of thermal radiation emitted by a furnace at a temperature of 2.7°K. This background radiation is a very important feature of the present universe, because it confirmed a theoretical deduction about the state of the early universe and enabled cosmologists to choose correctly between two models of the universe concerning which two groups of adherents had been feuding for years. This result leads us to the dynamics of the universe and the question of whether it is static or expanding and, if it is expanding, whether it will expand forever or stop at some distant future time and then begin to contract.

We consider this question from the theoretical and observational points of view simultaneously, since theory and observation went hand in hand in this exciting area of physics, although this congruity was not by design but by happenstance. In 1912, V. M. Slipher had discovered that the galaxies have red or blue Doppler shifts in their spectral lines, which indicate that galaxies beyond ours are either approaching or receding and that beyond a certain distance the galaxies are all receding from us at increasing speeds. Slipher's observational evidence, however, was not definitive enough to lead one to any spectacular conclusion about the behavior of the universe as a whole. Some ten years later, Edwin Hubble began a systematic study of the spectra of the distant galaxies and obtained indisputable observational evidence that the distant galaxies are receding at speeds that increase linearly with the distances of the galaxies. He stated this phenomenon in the form of a law (Hubble's law) that gives the rate of recession of the distant galaxies per unit of distance. This number is called "Hubble's constant"; its magnitude is about 16 kilometers per second per million light years, but some uncertainty is still associated with this value. Hubble's work, which spans a period of about a dozen years, is universally accepted as evidence for the expansion of the universe.

Edwin Powell Hubble (1889–1953) was the greatest astronomer ever produced by the United States. He was largely responsible for the development of the study of extragalactic astronomy and collected the observational evidence that demonstrated that the universe is expanding. Hubble's determination that the galaxies are moving apart from each other made it possible for cosmologists such as Abbé Georges Lemaître and George Gamow to postulate that the universe began as a superdense clump of matter that for unknown reasons exploded, hurtling across space the material that became the stars and galaxies.

Despite the impact Hubble has had on modern astronomy, little in his background indicated that astronomy would be his chosen profession. He was born in Kentucky, the son of an insurance salesman. The family eventually moved to Chicago, where Edwin attended high school. An outstanding scholar and a fine athlete, Edwin won a scholarship to the University of Chicago, where he became acquainted with Robert Millikan and the astronomer George Hale.[8] Hubble was undoubtedly influenced by these two men, because he majored in mathematics and astronomy. He also boxed and was considered so promising a professional prospect that a promoter tried to arrange a fight between Hubble and the then reigning

Edwin Powell Hubble (1889–1953)

heavyweight champion, Jack Johnson.[8] Any plans that Hubble might have had to pursue a boxing career were interrupted by his being awarded a Rhodes Scholarship in 1910 to study at Queen's College, Oxford.

Despite his academic background, Hubble chose to study jurisprudence while at Oxford. His interest in the evolution of the Anglo-American legal system caused him to consider seriously a legal career when he returned to the United States in 1913.[8] After being admitted to the bar and opening an office in Kentucky, Hubble soon found that there was a great deal of difference between the academic study of jurisprudence and the difficulties of running a legal office and dealing with the problems of clients. As Hubble had never been very much interested in the actual practice of the law, he decided that a legal career was not for him and returned to his original field of astronomy.

In 1914, Hubble returned as a graduate student to the Yerkes Observatory at Chicago, where he investigated the classifications of nebulae.[9] His work impressed George Hale, who invited Hubble to join him at the Mount Wilson Observatory, where the 100-inch reflector was under construction. Fully aware of the many valuable discoveries that would come once the new telescope became operational, Hubble accepted Hale's offer. Before he could begin his astronomical career, however, the United States entered World War I and began mobilizing what became a 2-million-man Army for duty in France. Hubble enlisted in the American Expeditionary Force and saw action in France. He remained overseas for two years before returning to the United States in 1919, when he finally joined Hale at the Mount Wilson Observatory.[9]

Hubble's early researches in galactic formation with the 60-inch telescope at Mount Wilson prompted him to offer a classification system for distinguishing between galactic and nongalactic nebulae.[9] "He discovered many new planetary nebulae and variable stars, but the most important result of his early researches concerned the origin of the radiation from diffuse galactic nebulae."[9] Once the 100-inch telescope became operational, however, Hubble began an intensive research effort to determine the nature and constitution of the neubulae outside the Milky Way galaxy. His first discovery of a Cepheid variable star in the Messier 31 nebula was invaluable to his later efforts to map the observable universe, because Cepheid variables can be used to determine the distances to the extragalactic nebulae.[9] By studying the photographic plates of several Cepheid variable stars, Hubble calculated that the Messier 31 nebula lies nearly 1 million light-years outside the Milky Way galaxy and is hence a separate "island universe." The formal announcement of Hubble's discovery in 1924 established the modern view of the universe as a collection of hundreds of billions of such "island universes." Previously, a spirited debate had divided astronomers, with some arguing in favor of a universe populated by many "island universes" and others asserting that the Milky Way is the only significant star system in space. But with the development of such instruments as the 100-inch reflector at Mount Wilson, it became clear that the former group's views were correct.

In 1925, Hubble unveiled his galactic classification scheme, which remains the standard guide to galactic formations. Hubble's investigations had revealed that most galaxies have some degree of rotational symmetry, so he divided all galaxies into regular and irregular classes, the former of which he further divided into spirals and ellipticals.[10] Hubble's classification scheme brought considerable order to the chaos that had frustrated previous efforts to understand the structure of galaxies.

Having shown that the galaxy is the fundamental unit of the observable universe and offered a universally accepted "map" of galactic types, Hubble next sought a reliable method for calculating distances to the edge of the observable universe. Initially, he returned to his reliable

Cepheid variables, which enabled him to extend the extragalactic distance scale out to a distance of only about 6 million light-years.[10] In the next few years, he relied on the brightnesses of galactic clusters to measure distances and ultimately expanded the observable universe to a distance of about a quarter of a billion light-years.[10]

Hubble's investigations of galactic distances made possible his greatest achievement—his law of the proportionality of the distances and radial velocities of galaxies. His previous work had shown that the farther a galaxy is from the Milky Way, the faster it appears to be receding. Hubble estimated that "velocities increased at the rate of roughly 100 miles a second for every million light-years of distance."[11] Further investigations revealed that this relationship appeared to hold out to a distance of over 100 million light-years.[11] This discovery that all the galaxies are moving apart from each other at steadily increasing speeds was the most important astronomical discovery since Copernicus had suggested that the earth revolves around the sun, because Hubble's law shows that the universe is a dynamic entity and not a static star system, as had been imagined by most scientists from Galileo to Einstein. This discovery also gave astronomy Hubble's constant, which is the ratio of velocity of recession of a galaxy to its distance and has the dimensions of an inverse time—its reciprocal is the age of the universe; according to Hubble's original determination, it is approximately 2 (since revised to 15) billion years.[11] Hubble's detailed investigations of the extragalactic realm suggested that the distribution of galaxies throughout space is relatively uniform. But recent, more detailed studies of the distribution of very distant clusters of galaxies show that these are not uniformly distributed in space; rather they form a kind of lacy pattern on the surfaces of huge bubbles of what appear to be empty spaces.

In the 1930s, Hubble devoted himself to mapping the distribution of galaxies in the observable universe and investigating the rotational dynamics of galactic spiral arms. He also engaged in an ongoing debate with a number of theoretical cosmologists regarding the proper interpretation of the red shifts of the light from the receding galaxies. Hubble felt that these red-shift measurements were not trustworthy owing to the adjustments that had to be made to compensate for the diminished energy of the light from observed galaxies, which makes the light appear fainter than it would otherwise be. Because of Hubble's rejection of the red-shift interpretation that the galaxies are moving outward at a significant fraction of the velocity of light, he concluded in 1936 that the galaxies are actually stationary.[12] This conclusion was attacked by many theorists even though Hubble's observational work remained unchallenged.[12] In any case, this assumption represented one of the few times that Hubble disregarded his own observations simply because a theory that gave credence to those observations was not acceptable to him.

The outbreak of World War II took Hubble from astronomy, and he became the chief of ballistics and a director of the Supersonic Wind Tunnel at the Aberdeen Proving Ground, where he remained until 1946.[12] After leaving government service, Hubble returned to California to help supervise the manufacturing of the Hale 200-inch reflector telescope planned for the Palomar observatory. When the new telescope became operational in 1949, Hubble was the first one to use it. The Hale telescope expanded the volume of the observable universe by a thousandfold and reinforced the already impressive evidence favoring Hubble's original vision of an isotropic, expanding universe. Although the cosmic frontiers were further expanded with Walter Baade's determination that all extragalactic distances had been underestimated by a factor of 2 (so that the Cepheid variable Hubble used in the Messier M31 nebula was actually about 2 million light-years away), Hubble's findings remained essentially unchallenged.[12] He spent the last few years of his life continuing to observe galactic formations and enjoying trout fishing as well as the many honors conferred on him, such as his election as an honorary fellow at Queen's College, Oxford, and numerous honorary degrees.

While observational astronomers during the first three decades of this century were beginning to interpret the data presented by the Doppler shifts of the distant galaxies as indicating that the universe is expanding, cosmologists were turning to Einstein's theory of gravity (the general theory of relativity) as the pathway to an understanding of observations and a guide to the development of a correct model of the universe, since Newtonian theory led to a dead end. This work was begun by Einstein himself in a famous paper on cosmology in 1917, in which he applied his gravitational field equations to the universe as a whole. These equations treat gravity as the space–time curvature in the neighborhood of a point produced by the mass and energy in the neighborhood. To apply his equations, Einstein pictured all the individual masses in the universe smeared out into a very rarefied, isotropic, uniform (homogeneous) fog filling all of space. Each point of the universe would then be characterized by the density of matter at that point (the same for all points) and the curvature of space–time (also the same at each point). Einstein was looking for a solution of his field equations of the universe that corresponds to a static, homogeneous, isotropic model of the universe, in accordance with the famous cosmological principle that states that the universe must look the same to all observers no matter where they are in the universe.

When Einstein did his cosmological work, the recession of the distant galaxies was not known, so he had no reason to believe that the universe is not static. But a static model of the universe is untenable because it requires that the galaxies remain separated from each other, suspended in space, despite the force of gravity (curvature of space–time) that would tend to bring them together. Einstein therefore altered his field equations somewhat by adding another term, called the ''cosmological constant,'' to keep the universe from collapsing. With that additional term, he obtained a static solution of his field equations that corresponds to a spherical, closed universe with a finite radius.

Here, we must define the ''radius of the universe'' carefully, because the universe must be pictured as a three-dimensional spatial surface (a hypersurface) embedded in a higher-dimensional manifold, one dimension of which may be pictured as time. The only real dimensions are the three surface dimensions (our three-dimensional space) of the hypersphere, which we may compare to the surface of a balloon. The galaxies and clusters of galaxies are to be pictured as physical objects pasted on this surface, and the real distance between any two of them is measured along the surface that corresponds to depth (radial distance) in our real space. The radius R of such a balloon is not a physically measurable quantity, but merely a scale parameter that determines the distance between points on the balloon's surface; *if* R doubles, all distances double. Einstein's static universe is just such a ''space-balloon,'' and its ''radius'' is a scale parameter that lies along an imaginary direction at ''right angles'' to space itself and therefore not measurable. The expansion of the universe must now be understood in the following sense: The galaxies are not receding from any central point in space; no such point exists. All the galaxies are separating from each other, so that the expansion appears the same from every point in space. This means that the volume of the universe is increasing as the expansion continues, just as the surface area of a balloon increases as it is blown up.

Einstein's static model of the universe is not the only one that can be obtained from Einstein's cosmological field equations, as was demonstrated by a number of different cosmologists, in particular by the Russian mathematician Alexander Friedmann, who recast Einstein's field equations into a form in which the basic parameters of the universe, its radius R and its mean mass density, depend on time. These time-dependent cosmological equations thus require a changing universe (expansion or collapse), rather than a static one. In Einstein's static cosmological model, the radius R does not change, so that distances between galaxies remain constant, but in the Friedmann models, R does change, so that both the distances between galaxies and the mean density of the universe change. This brings us back again to Hubble's discovery of the recession of the distant galaxies, which is thus a direct consequence of Einstein's law of gravity, as incorporated in Friedmann's time-dependent form of the cosmological equations. But these equations do not limit the universe to an expansion; they also permit the universe to collapse. In addition, they tell us how the geometry of the universe—whether it is Euclidean (flat) or non-Euclidean (curved inwardly like the surface of a sphere, and therefore closed, or curved outwardly like a huge dish, and therefore open)—is related to its rate of expansion (Hubble's constant) and the mean density of matter in the universe.

To explain all this, we consider the two time-dependent cosmological equations more carefully. The first of these equations is merely a statement of the conservation of energy for the entire universe—that is, it relates the total kinetic energy of the universe, which is given by the square of the rate of increase of R (how rapidly the universe is expanding), to the mass energy, the gravitational potential energy, and the radiation energy of the universe. The universe is closed, and its geometry is therefore non-Euclidean in a spherical way (elliptical) only if its total energy is negative; its kinetic energy, which is positive, must be small enough so that the speed of recession of the distant galaxies remains below the speed of escape from the universe as the universe expands. This statement means that the mean mass density of the universe must always exceed a certain critical density that is proportional to the square of Hubble's constant (the square of the rate of increase of R) divided by the universal gravitational constant. The numerical value of the critical density thus depends on Hubble's constant—that is, on the present rate of expansion of the universe. Accepting the value of 16 kilometers per second per million light-years for this very important constant, we find that the critical density is 4.5×10^{-30} g/cm^3, or 2.7 millionths of a nucleon/cm^3 of space, or 1 nucleon/400,000 cm^3 of space. If we neglect the hidden mass in the universe (the so-called "missing mass") and use only its observable matter (stars, galaxies, and other matter) to calculate the universe's mean mass density, its value is less than the critical density, which points to an open, hyperbolic universe that will expand forever. But the indirect evidence that most of the mass in the universe is undetectable at present is so persuasive that we must suspend our decision as to the geometry of the universe. If all the estimated hidden mass does indeed exist, the mass density of the universe exceeds the critical density and the universe is a closed, elliptical space–time manifold (Riemannian non-Euclidean geometry), the expansion of which will halt in time; this event will be followed by a collapse. Such a condition points to a pulsating universe with alternate expansions and contractions.

We consider now the second of Einstein's time-dependent cosmological equations, from which we should, in principle, be able to deduce the nature of the universe's geometry without using its mass density. This equation relates the rate at which the expansion is slowing down (the deceleration parameter) to the square of Hubble's constant and the square of the radius R of the universe. This parameter is a pure number (no space–time units). If its value is larger than $\frac{1}{2}$, the universe is elliptical and closed; if it equals $\frac{1}{2}$, the universe is flat and infinite; if it is smaller than $\frac{1}{2}$, the universe is open and hyperbolic. In principle, we should be able to calculate this parameter by comparing the present rate of expansion of the universe (from the red shifts of the nearby galaxies—those within a hundred million light-years) with its rate of expansion billions of years ago (from the red shifts of very distant galaxies—those billions of light-years away). But this calculation is very difficult owing to our great uncertainty about the distances of the remote galaxies. For the present, we must leave open the question of the universe's geometry.

We close this chapter with a discussion of the very early universe and the conditions that existed at, or very nearly at, the moment the universe was born, if we can speak of such a moment at all. Until recently, this area of science was the exclusive domain of a few theoretical cosmologists; most physicists left it strictly alone, for they felt that the observational evidence formed too fragile a base on which to construct an acceptable model of the universe. But with the recent rapid development of high-energy particle physics, physicists see the early universe as a laboratory in which the energies that were present far exceeded (by many orders of magnitude) the energies that can be produced in our earthbound laboratories. Particle physicists have therefore become greatly interested in cosmological models to see whether such models can reveal some important features of such basic building blocks as quarks.

What were the conditions in the universe shortly after its birth, which we generally identify as the "Big Bang" or the initial "fireball?" To discover these initial conditions in our universe, we again go to Einstein's time-dependent cosmological equations, but apply them to the past rather than the future by reversing the direction of the flow of time—that is, by replacing t (time) in the equations by $-t$ (negative time, flowing from the future to the past). The equations then tell us that as we go back in time, the universe gets smaller, the mass and energy density and the temperature increase, and Hubble's constant (the rate of expansion) increases. We can get a good picture of the overall conditions in the early universe just by applying general thermodynamic principles to a contracting universe, keeping in mind that the contraction must be adiabatic, since no energy can enter or leave the universe. Thermodynamics tells us that the temperature, hence the internal energy, of a gas increases if the gas is compressed adiabatically by an external force; the increase in the internal energy of the gas comes from the work done on it by the external force. This phenomenon is also true of the universe, except that the external force is replaced by gravity; as the universe contracts under the pull of gravity, the gravitational potential energy is transformed into internal energy (the kinetic energy of the matter in the universe and the energy of the radiation).

As we go back in time, the radius of the universe decreases, and its temperature increases in the same proportion that the radius decreases; that is, if the radius decreases to 1/10th of its present value, the temperature increases to 10 times its present value, and so on. The densities of the matter and the radiation also increase, but not at the same rate; the matter density increases as the inverse cube of the radius (proportional to $1/R^3$), but the radiation density increases as the inverse 4th power of the radius (as $1/R^4$). The reason for this is that as the universe gets smaller, more and more photons are squeezed into a cubic centimeter (increase in matter density); however, not only are more and more photons squeezed into a cubic centimeter with decreasing universe size, but also each photon's energy is increased (photons become bluer). We see that as we go back in time, radiation becomes ever more important relative to matter in the past history of the universe, not only because each photon becomes more energetic by itself but also because the number of photons exceeds the number of nucleons by a factor of about 10 billion.

Continuing with our backward journey in time, we see the present organization of the universe give way to increasing disorganization as the temperature rises. Stars, planets, and galaxies are torn apart into their constituent atoms by the very hot radiation surrounding them. In time, these atoms are torn apart into electrons and nuclei; finally, the nuclei themselves are dissociated into nucleons. This situation existed when the temperature of the universe was about 1 trillion degrees Kelvin and its radius was about 1 trillionth of its present radius. But particle physicists have speculated about even higher temperatures and earlier times, and so they looked further into the past. As one goes back to these very early epochs, to almost 1 trillionth of a trillionth of a trillionth of a second after the initial moment (after the universe's radius R was 0), the cosmological equations tell us that the temperature and density of the universe will keep on growing without limit as we approach the zero moment, finally becoming infinite. This state is known as the "initial singularity," which has no physical meaning; the equations break down, and so we have no way, with the theory as it is, of understanding the "birth of the universe." Owing to this breakdown, theoreticians begin their cosmological studies when the universe was 10^{-35} second old and its temperature was of the order of 10 thousand trillion trillion degrees Kelvin. We may consider this state as the "Big Bang," because the universe was a very hot "fireball," smaller than its present size by a factor of 10^{28}.

The radiation-dominated universe was then a mixture of very hot radiation of baryons and antibaryons of all kinds, of every variety of meson, and of leptons and antileptons. As the universe expanded and cooled, all the baryons except nucleons and all the leptons except electrons disappeared; the temperature was then of the order of a billion degrees, cool enough for helium nuclei to be formed from the nucleons, and 25% of the protons were indeed fused into helium nuclei. The temperature continued dropping as the universe expanded until it had become cool enough for neutral hydrogen and helium atoms to be formed. As the temperature continued to fall, the radiation became too cool to interact with the neutral helium and hydrogen; radiation and matter thus became decoupled, and matter (gravity) became dominant, with the present state of the universe evolving.

We have described above the standard theory, which is clearly unsatisfactory as it stands because it cannot eliminate the physically impermissible initial singularity. But one of the authors (Motz) has shown a way out of this predicament that involves a drastic change in our picture of the structure of nucleons. He has deduced the existence of very massive, basic particles that he calls "unitons," and has identified these with the Gell-Mann quarks. Owing to the large masses of unitons (of the order of 10^{-5} grams), the binding force of the quarks in nucleons is gravity, and nucleons themselves are pictured as linear rotators with three unitons lying in a line: two at the ends and one in the center. The "Big Bang" is now accounted for by picturing the universe initially as consisting only of unitons, which combined in triplets to form nucleons with a vast release of energy; this was the "Big Bang," the birth of the universe.

This model of nucleons eliminates the initial singularity of the universe by eliminating its initial moment; the uniton model allows for no initial moment of creation. If we go back in time again to when the temperature of the universe was 10^{32} °K, the photons now in the universe were energetic enough to dissociate the nucleons into unitons. Thus, the radiation now in the universe was sopped up and the universe became cold, with all its energy present in the form of the uniton masses and their mutual gravitational energy. The only way this state of the universe could change was by the collapse of the unitons into nucleon triplets. This phenomenon occurred repeatedly in the past and will continue doing so in the future.

Epilogue

We have presented the evolution and growth of physics on two levels—the lives of the scientists who discovered and developed the basic physical principles that constitute physics, and the way these principles were discovered and proposed. In studying the lives of these discoverers, one is impressed by their dedication to an unquenchable inner drive, their need to probe nature and find out what governs it. In the early years of physics, no guiding precepts were available so that the early physicists were like explorers lost in an uncharted world. Not only did they have to discover how nature "works," but they had to discover and develop a new logic, that of their own science whose very nature was only faintly perceived. This task required a certain minimal amount of confidence and arrogance—the confidence that nature (the universe) is rational (not whimsical) and the arrogance to believe that the intellectual abilities of these explorers were equal to the tasks they faced in solving the problems presented by nature.

In accepting the rationality of nature and, therefore, its describability in terms of basic principles, the early physicists began an intellectual revolution that has expanded far beyond anything they could have imagined and has greatly influenced every phase of modern society. Starting along such a revolutionary road required considerable courage because very few observable events in the universe pointed to regularity, as demanded by a well-ordered and rational universe. The great diversity of matter and form indicated to the early Greeks that nature was not governed by universal laws but by an array of gods who directed events according to their own whims. That certain events in the heavens, such as the diurnal rising and setting of the constellations, the apparent motions of the planets, the phases of the moon, and eclipses occurred with great regularity was not taken as evidence of the operation of universal natural laws but rather as evidence of the power of the all-mighty Creator who imposed His will on the heavenly bodies forcing them to move in regular patterns as befitted supernal objects.

To depart from such a simple theistic picture and to suggest that a god or gods have nothing to do with the dynamics of the solar system which behaves according to immutable laws that even the "gods had to obey" was heresy of the worst kind, the promulgation of which required great courage. The burning of Giordano Bruno at the stake in 1600 and the imprisonment of Galileo are evidence of just how dangerous proclaiming scientific principles was at that time. But being a trailblazer in science required courage on a more parochial level—courage to persist in views even if they are not in accord with the prevailing scientific views of the day. To be ridiculed by one's peers for proposing off-beat ideas and concepts can be a powerful deterrent to the presentation of new ideas. Thus Copernicus confessed that he was reluctant to publicize his heliocentric theory of the solar system for "fear that he might be hissed off the stage of history" and Julius Mayer was driven almost to suicide by the contempt heaped on him by his 19th-century contemporaries for proposing that heat is a form of energy and thus discovering the first law of thermodynamics. The need for courage to propose new theories was most eloquently expressed by Albert Einstein on learning of Niels Bohr's quantum model of the atom. Einstein stated that he had the same idea a year earlier but did not have the courage to publish it. Thus running through the story of physics is the story of courage which was an outstanding characteristic of all the persons in this story.

When modern science first began with the works of Johannes Kepler, Galileo Galilei, and Isaac Newton, it developed with no underlying theme; all that was attempted was the discovery of the correct descriptions of specific phenomena without seeking universal principles which can account for many phenomena at the same time. The first step in achieving this goal was taken by Newton in his discovery of the law of gravity and his proposal that the force of gravity is the "prime mover" in the universe, governing not only how "apples fall to the ground" but also how the moon revolves around the earth, how the planets revolve around the sun, how the tides rise and fall, and how the stars themselves are governed in their motions. The importance of this great generalization of Newton's concept cannot be overemphasized for it projected physics onto a new plane—to discover universal principles or laws and not just specific explanations of particular events. This is the thread that ties together physics from the time of Newton to the present and manifests itself in various ways. Most recently, it has become known as "the theory of everything."

The motive to unify physics was generated by the discovery that all the structures in the universe are related to the forces in the universe. This idea may at first have seemed to lead to diversity rather than to unification since the variety of structures appears to be numberless so that, at first sight, an infinitude of different forces seems to be required to account for all the structures. But this difficulty in achieving unification in physics by relating structures to forces is only apparent and not real since a single force can produce many different structures. In principle, a few forces can account for all the structures in the universe.

Just how a force can produce a structure is evident from Newton's laws of motion which relate the way a body moves to the forces acting on it. If forces did not exist in the universe, all the particles in the universe would move in straight lines and they could never combine into groups, as they do now owing to their material attractions, so that structures could never originate. Newton's discovery that a force acting on a body causes the body to alter its state of motion continuously was the key to understanding structure and the beginning of the drive toward unification in physics.

It is worth considering briefly the application of this concept to the solar system as a structure governed by gravity, the first force that was studied in detail, using Newton's laws of motion and his law of gravity. If the force of gravity were suddenly eliminated, the planets would all move away from each other and from the sun in straight lines and the solar system, as a structure, would disappear. Fortunately for us, the sun's gravitational pull on the planets and their natural tendency to maintain their uniform motion along straight lines (their inertia) combine to keep them moving around the sun in nearly circular orbits. The solar system, as a dynamical structure, is thus preserved.

Newton made a remarkable generalization by proposing the concept that gravity is universal and accounts for all the large structures in the universe (planets, stars, and stellar aggregates). Indeed, he went so far as to suggest that the structure of the universe itself is governed by gravity and then attempted to calculate the observed distribution of the stars using his law of gravity; this effort marked the beginning of modern cosmology.

That gravity is the prime cosmological and astronomical force was quickly accepted by physicists and astronomers, and based on this belief, the great mathematicians of the 18th and 19th centuries developed one of the most magnificent edifices of theoretical physics, namely, celestial mechanics, which represented the peak of Newtonian gravitational theory. These mathematicians introduced new concepts which have continued to play important roles in physics and which have guided physicists in their evaluation of new theories. Physicists insist that a theory must be mathematically beautiful to be correct. This means that the mathematics in which the theory is expressed must be elegant and simple and introduce as few basic concepts as possible. This concept of beauty has prevailed in physics to the present.

When expressed in 19th-century mathematical elegance, Newtonian physics produced other concepts which enabled physicists to develop their subject without specific references to forces. Most important among these concepts are the conservation principles which state that in all interactions among particles, regardless of the nature of the interactions, certain physical entities must be conserved. Three such entities—momentum, energy, and angular momentum (rotational motion)—in any isolated system of particles (hence, the entire universe itself) are of particular importance. These conservation principles, which still dominate physics, are universal in the sense that they apply everywhere and at all times.

The conservation principles are related to another concept, that of symmetry, which has played a very important part in guiding physicists in their search for the correct physical laws. Symmetry, as used by physicists, means that the mathematical form of a law must be the same when expressed by different observers; this view encompasses the concept of invariance. The principle of the conservation of momentum thus means that space must be symmetrical in the sense that the laws of nature must not change if we move from one part of space to another and the conservation of energy means that the laws must not change from moment to moment (the temporal symmetry of the universe). Conservation of angular momentum means that the laws must not change if we orient ourselves differently and look at the universe from different directions.

All of these concepts, which represent the continuity of physics from Newton to the present, were derived from Newtonian dynamics when gravity was still the only force that was understood. At that time it was clear that although gravity controls the dynamics of the planets and the stars, it plays no role or a very minor role in the internal dynamics of matter or in the interactions among particles all around us. From a superficial point of view it appears that numerous different forces must be operating to account for the vast variety of materials ranging from the hardest and strongest, such as diamond and steel, to the softest, such as living cells. If this were indeed so, attempting to find all such forces would be fruitless. Happily, this state of affairs does not exist, for the early Greeks, without knowing it, had discovered electricity, the force that accounts for all the ordinary observed properties of matter and for living organisms and their life processes.

Some 2000 years elapsed after the discovery of the electric force before physicists demonstrated that this force accounts for the structure of atoms and molecules and for all of chemistry. All the forces, other than gravity, that we encounter in our daily lives (friction, elastic forces, muscular forces, etc.) are different manifestations of the electric force. But this force has another important feature which the early Greeks did not know and was discovered only accidentally in the 18th century—it is associated with magnetism. The early Greeks had discovered magnetism, but they did not know that the electric and magnetic forces are intimately related to each other and are two different aspects of a single force, called the electromagnetic force.

The discovery of electromagnetism began a new phase of physics which we may call "unification physics," and which during Einstein's lifetime was pursued as the search for a "unified field theory." The aim, as best expressed by Einstein, who led this search, was to show that the electromagnetic field and the gravitational field can be derived from a single, more basic field. Einstein's search, though not successful, was highly inspirational and stimulating because it generated today's attempts to unify the four basic forces now recognized by physicists: gravity, electromagnetism, the nuclear (or strong) force, and the weak interaction.

The unification concept, which now permeates physics, was strongly rooted in the mathematics of Newtonian physics. The same general mathematical formulas that are used in treating Newtonian gravity can be applied with very little change to treating electromagnetic interactions among electrically charged particles. That the mathematical formulations of Newtonian gravitational theory and electromagnetism are very similar, strongly supported the drive toward unification of the two.

Although this unification was not achieved, unification of another sort occurred in Maxwell's remarkable discovery of the electromagnetic nature of light. His famous equations of the electromagnetic field show that this field is propagated through empty space at the speed of light and the great experimentalist Heinrich Hertz, near the end of the 19th century, demonstrated experimentally that the Maxwellian electromagnetic waves have all the properties of light waves, thus completing the unification of electromagnetism and light. This accomplishment in no way altered the basic Newtonian conservation principles which were carried over bodily to electromagnetism.

But unification was not limited to electromagnetism and light; other physicists demonstrated that the laws of thermodynamics and the laws of gases can be deduced from the Newtonian laws of motion; if one accepts the molecular concept. Indeed, Newton's laws of motion reduce the pressure in a gas to molecular collisions and the temperature of the gas to the mean (average) kinetic energy of a molecule of the gas. The most spectacular and dramatic unification of all, however, was achieved by Einstein in 1905 in his special theory of relativity which unifies space and time into a single physical entity called space-time. This concept destroyed the separate Newtonian notions of absolute space and absolute time and replaced them by a single absolute four-dimensional space-time manifold. But Einstein's discovery went beyond the unification of space and time, which stemmed from the observed constancy of the speed of light for all observers moving with constant (but different) velocities through empty space; it unified the energy, momentum, and mass of a particle into a single mass-energy-momentum entity. Thus the separate principles of the conservation of energy, momentum, and mass were replaced by a single conservation principle of energy-mass-momentum. Thus the equivalence of mass and energy, without which contemporary high-energy particle physics could not be understood, was established.

All of these remarkable consequences of Einstein's special theory of relativity can be deduced from the basic principle of invariance which states that a law of nature must have the same mathematical form in all uniformly moving frames of reference. In other words, the mathematical expression of the law, as formulated by any uniformly moving observer must be the same as that formulated by any other such observer. This is not true of any statement about nature that is not a law; the principle of invariance is thus a powerful analytical tool that enables physicists to separate the basic universal laws from all possible statements about nature. The principle of invariance is, and has been, an unfailing beacon in the physicist's search for universal truths; it underlies all of contemporary physics.

Einstein unified physics further in his general theory of relativity by replacing the gravitational force by the curvature of the space-time manifold. According to this theory, a mass curves the four-dimensional space-time manifold near it, thus causing a particle to move along this space-time curvature rather than in a straight line just as though it were being pulled by the first particle. Einstein thus replaced the force of gravity by a curved space-time and, in doing so, made geometry part of physics. Einstein applied this theory of gravity to the entire universe, which was the beginning of modern cosmology; an important deduction from this cosmology is that the universe is expanding and that the expansion began about 15 billion years ago in a vast explosion (the "Big Bang"). The merging of geometry and physics opened up vast new domains to the physicist, which encompass not only the structure of the universe but also the structures of the most elementary particles.

As the 19th century was drawing to a close, the universe appeared to consist of two distinct entities which interact with each other but which are qualitatively quite different: particles and fields. The particles (e.g., electrons and protons) are localized in space and account for all matter in the universe, whereas fields are not localized but spread out as waves and may be pictured as pure energy. But this concept of a dichotomy of the content of the universe into particles and waves was shattered by Max Planck and Albert Einstein who discovered that the electromagnetic field behaves like a stream of particles (photons) as well as a train of waves. This discovery was the beginning of the wave-particle dualism that now permeates all of physics. This wave-particle dualism picture was completed in the 1920s with the discovery that particles (e.g., electrons and protons) have wave-like properties. Thus the circle of ideas was completed and a single mathematical and physical discipline, called quantum mechanics, was developed to treat both particles and fields.

Our story of physics must end at this point, not because physicists have nothing more to do; they have many more exciting discoveries to make, but the two great theories that will guide them in their unending search—the theory of relativity and the quantum theory—are known. These theories are at the very peak of humanity's achievement and creativity and they have united the microscopic world of elementary particles with the macroscopic world of stars and galaxies. When we refer to the unity of the universe, we have these two theories in mind because they teach us how the expansion of the universe itself can be deduced from the properties of the elementary particles that constitute the matter and energy in the universe.

Notes

CHAPTER 1

1. Herbert Westren Turnbull, *The Great Mathematicians* in *The World of Mathematics*. James R. Newman, ed. New York: Simon & Schuster, 1956.
2. Will Durant, *The Story of Philosophy*. New York: Simon & Schuster, 1961, p. 41.
3. *Ibid.*, p. 44.
4. G. E. L. Owen, "Aristotle," *Dictionary of Scientific Biography*. New York: Charles Scribner's Sons, Vol. 1, 1970, p. 250.
5. Durant, *op. cit.*, p. 44.
6. Owen, *op. cit.*, p. 251.
7. Durant, *op. cit.*, p. 53.

CHAPTER 2

1. William H. Stahl, "Aristarchus of Samos," *Dictionary of Scientific Biography*. New York: Charles Scribner's Sons, Vol. 1, 1970, p. 246.
2. *Ibid.*, p. 247.
3. G. J. Toomer, "Ptolemy," *Dictionary of Scientific Biography*. New York: Charles Scribner's Sons, Vol. 11, 1975, p. 187.

CHAPTER 3

1. Stephen F. Mason, *A History of the Sciences*. New York: Abelard-Schuman Ltd., 1962, p. 127.
2. Edward Rosen, "Nicolaus Copernicus," *Dictionary of Scientific Biography*. New York: Charles Scribner's Sons, Vol. 3, 1971, pp. 401–402.
3. Mason, *op. cit.*, p. 128.
4. Rosen, *op. cit.*, p. 403.

5. David Pingree, "Tycho Brahe," *Dictionary of Scientific Biography*. New York: Charles Scribner's Sons, Vol. 2, 1970, p. 401.
6. *Ibid.*, p. 402.
7. *Ibid.*, pp. 402–403.
8. *Ibid.*, p. 413.
9. Owen Gingerich, "Johannes Kepler," *Dictionary of Scientific Biography*. New York: Charles Scribner's Sons, Vol. 7, 1970, p. 289.
10. *Ibid.*, p. 290.
11. Mason, *op. cit.*, p. 135.
12. *Ibid.*, p. 136.
13. Gingerich, *op. cit.*, p. 305.

CHAPTER 4

1. W. L. Reese, "Galileo Galilei," *Dictionary of Philosophy and Religion*. Atlantic Highlands, New Jersey: Humanities Press, Inc., 1980, p. 186.
2. Chet Raymo, *The Soul of the Night*. Englewood Cliffs, New Jersey: Prentice-Hall, Inc., 1985, p. 163.
3. G. Szczesny, *The Case Against Bertold Brecht: With Arguments Drawn from His Life of Galileo*. New York: Frederick Ungar Publishing Company, 1969, p. 68.
4. Dietrich Schroeer, *Physics and Its Fifth Dimension: Society*. Reading, Massachusetts: Addison-Wesley Publishing Company, p. 81.
5. *Ibid.*, p. 84.

CHAPTER 5

1. "Sir Isaac Newton," *Encyclopaedia Britannica*. Chicago: Encyclopaedia Britannica, Inc., Vol. 13, 1974, p. 17.
2. Isaac Asimov, *Asimov's Biographical Encyclopedia of Science and Technology*. Garden City, New York: Doubleday & Company, Inc., 1982, p. 148.
3. "Sir Isaac Newton," *op. cit.*, p. 17.
4. E. N. da Costa Andrade, "Isaac Newton," *The World of Mathematics*. Ed. James R. Newman. New York: Simon & Schuster, 1956, p. 256.
5. "Sir Isaac Newton," *op. cit.*, p. 17.
6. John Maynard Keynes, "Newton, the Man," *The World of Mathematics*. Ed. James R. Newman. New York: Simon & Schuster, 1956, p. 278.
7. Asimov, *op. cit.*, p. 148.
8. *Ibid.*, p. 232.
9. "Sir Isaac Newton," *op. cit.*, p. 18.
10. Asimov, *op. cit.*, p. 152.
11. *Ibid.*, p. 153.
12. Andrade, *op. cit.*, p. 270.
13. Henry A. Boorse and Lloyd Motz, *The World of the Atom*. New York: Basic Books, 1966, p. 89.

CHAPTER 6

1. Henry A. Boorse and Lloyd Motz, *The World of the Atom*. New York: Basic Books, 1966, p. 54.
2. *Ibid.*, p. 55.
3. *Ibid.*, p. 38.
4. *Ibid.*, p. 40.
5. *Ibid.*, p. 65.
6. Isaac Asimov, *Asimov's Biographical Encyclopedia of Science and Technology*. Garden City, New York: Doubleday & Co., Inc., 1982, p. 155.
7. "James Bradley," *Encyclopaedia Britannica*. Chicago: Encyclopaedia Britannica, Inc., Vol. 3, 1974, p. 101.
8. "Edmond Halley," *Encyclopaedia Britannica*. Chicago: Encyclopaedia Britannica, Inc., Vol. 8, 1974, p. 556.
9. *Ibid.*, p. 557.

CHAPTER 8

1. "Sir William Rowan Hamilton," *Encyclopaedia Britannica*. Chicago: Encyclopaedia Britannica, Inc., Vol. 8, 1974, p. 588.
2. Henry A. Boorse and Lloyd Motz, *The World of the Atom*. New York: Basic Books, 1966, p. 1027.
3. "Sir William Rowan Hamilton," *op. cit.*, p. 588.
4. Boorse and Motz, *op. cit.*, p. 1028.
5. "Sir William Rowan Hamilton," *op. cit.*, p. 589.
6. Boorse and Motz, *op. cit.*, p. 1028.
7. "Sir William Rowan Hamilton," *op. cit.*, p. 589.
8. Herbert Westren Turnbull, *The Great Mathematicians* in *The World of Mathematics*. Ed. James R. Newman. New York: Simon & Schuster, 1956, p. 163.
9. Boorse and Motz, *op. cit.*, p. 1029.
10. W. R. Hamilton, *Dublin University Review*, 1833, pp. 795–826.
11. Turnbull, *op. cit.*, p. 153.
12. "Joseph-Louis Comte de Lagrange," *Encyclopaedia Britannica*. Chicago: Encyclopaedia Britannica, Inc., Vol. 10, 1974, p. 598.
13. Turnbull, *op. cit.*, p. 154.
14. "Joseph-Louis Comte de Lagrange," *op. cit.*, p. 598.
15. Turnbull, *op. cit.*, p. 155.

CHAPTER 10

1. Henry A. Boorse and Lloyd Motz, *The World of the Atom*. New York: Basic Books, 1966, p. 319.
2. *Ibid.*, p. 320.
3. *Ibid.*, p. 321.

4. *Ibid.*, p. 263.
5. *Ibid.*, p. 264.
6. *Ibid.*, p. 265.
7. *Ibid.*, p. 266.
8. *Ibid.*, p. 267.

CHAPTER 11

1. R. Steven Turner, ''Julius Robert Mayer,'' *Dictionary of Scientific Biography*. New York: Charles Scribner's Sons, Vol. 9, 1974, p. 237.
2. *Ibid.*, p. 238.
3. James F. Challey, ''Nicolas Léonard Sadi Carnot,'' *Dictionary of Scientific Biography*. New York: Charles Scribner's Sons, Vol. 3, 1971, p. 81.
4. Edward E. Darb, ''Rudolf Clausius,'' *Dictionary of Scientific Biography*. New York: Charles Scribner's Sons, Vol. 3, 1971, p. 303.
5. *Ibid.*, p. 306.
6. *Ibid.*, p. 307.
7. *Ibid.*, p. 309.
8. Sir Isaac Newton, *Optiks*. New York: Dover, 1952, p. 400.
9. F. W. Magie, *The Source Book in Physics*. New York: McGraw-Hill, 1935, p. 247.
10. Henry A. Boorse and Lloyd Motz, *The World of the Atom*. New York: Basic Books, 1966, pp. 213–214.
11. Stephen G. Brush, ''Ludwig Boltzmann,'' *Dictionary of Scientific Biography*. New York: Charles Scribner's Sons, Vol. 2, 1970, p. 266.
12. *Ibid.*, p. 262.
13. *Ibid.*, p. 263.
14. *Ibid.*, p. 264.
15. Martin J. Klein, ''Josiah Willard Gibbs,'' *Dictionary of Scientific Biography*. New York: Charles Scribner's Sons, Vol. 5, 1972, p. 388.

CHAPTER 12

1. James Murphy, ''Introduction,'' in Max Planck, *Where is Science Going?* New York: W. W. Norton, 1932, p. 18.
2. *Ibid.*, pp. 19–20.
3. Albert Einstein, ''Prologue,'' in Planck, *supra* note 1, p. 12.
4. *Ibid.*, p. 20.
5. Max Planck, ''Scientific Autobiography,'' *Scientific Autobiography and Other Papers*. New York: Philosophical Library, 1949, p. 14.
6. *Ibid.*, p. 15.
7. *Ibid.*, p. 16.
8. *Ibid.*, p. 20.
9. *Ibid.*, p. 21.

10. Max Planck, *The Universe in Light of Modern Physics*. New York: W. W. Norton, 1931, p. 82.
11. *Ibid.*, p. 30.
12. *Ibid.*, p. 32.
13. *Ibid.*, pp. 33–34.
14. Max von Laue, "Memorial Address," in Planck, *supra* note 5, p. 8.
15. James Murphy, "Introduction," in Planck, *supra* note 1, p. 37.

CHAPTER 13

1. *Nobel Lectures: Physics 1901–1921*. New York: Elsevier Publishing Co., 1967, p. 97.
2. "John William Strutt, Lord Rayleigh," *Encyclopaedia Britannica*. Chicago: Encyclopaedia Britannica, Inc., Vol. 15, 1978, p. 538.
3. *Nobel Lectures, op. cit.*, p. 97.
4. *Ibid.*, p. 98.

CHAPTER 14

1. *Nobel Lectures: Physics 1901–1921*. New York: Elsevier Publishing Co., 1967, p. 31.
2. *Ibid.*, p. 32.
3. Alfred Romer, "Henri Becquerel," *Dictionary of Scientific Biography*. New York: Charles Scribner's Sons, 1970, Vol. 2, p. 558.
4. *Ibid.*, pp. 558–559.
5. Adrienne R. Weill, "Marie Curie," *Dictionary of Scientific Biography*. New York: Charles Scribner's Sons, Vol. 3, 1971, p. 500.
6. *Ibid.*, p. 501.
7. Henry A. Boorse and Lloyd Motz, *The World of the Atom*. New York: Basic Books, 1966, p. 430.
7. *Ibid.*, p. 502.
8. Albert Einstein, *Out of My Later Years*. New York: Philosophical Library, 1950, pp. 227–228.
9. Boorse and Motz, *op. cit.*, p. 438.
10. *Ibid.*, p. 439.
11. *Ibid.*, p. 449.
12. *Ibid.*, p. 450.
13. *Ibid.*, p. 451.
14. *Ibid.*, p. 641.
15. *Ibid.*, p. 701.
16. *Ibid.*, p. 702.
17. *Ibid.*, p. 703.
18. *Ibid.*, pp. 704–705.
19. *Ibid.*, p. 804.
20. *Ibid.*, p. 805.
21. *Ibid.*, p. 806.

CHAPTER 15

1. Ronald W. Clark, *Einstein: The Life and Times*. New York: Avon Books, 1984, p. 25.
2. *Ibid.*, p. 27.
3. Henry A. Boorse and Lloyd Motz, *The World of the Atom*. New York: Basic Books, 1966, p. 534.
4. Clark, *op. cit.*, p. 66.
5. *Ibid.*, p. 87.
6. Boorse and Motz, *op. cit.*, pp. 535–536.
7. Clark, *op. cit.*, p. 252.
8. *Ibid.*, p. 313.

CHAPTER 16

1. Leon Rosenfeld, "Niels Henrik David Bohr," *Dictionary of Scientific Biography*. New York: Charles Scribner's Sons, Vol. 2, 1970, p. 240.
2. *Ibid.*, p. 241.
3. *Ibid.*, p. 244.
4. *Ibid.*, p. 248.
5. *Ibid.*, p. 250.
6. Henry A. Boorse and Lloyd Motz, *The World of the Atom*. New York: Basic Books, 1966, p. 739.

CHAPTER 17

1. Henry A. Boorse and Lloyd Motz, *The World of the Atom*. New York: Basic Books, 1966, p. 1047.
2. *Ibid.*, p. 1105.
3. *Ibid.*, p. 1106.
4. *Ibid.*, p. 1107.
5. *Ibid.*, p. 1065.
6. *Ibid.*, p. 1066.
7. *Ibid.*, p. 1529.
8. *Ibid.*, p. 1530.
9. "Richard Phillips Feynman," *McGraw-Hill Modern Men of Science*. New York: McGraw-Hill, 1966, p. 170.

CHAPTER 18

1. Henry A. Boorse and and Lloyd Motz, *The World of the Atom*. New York: Basic Books, 1966, p. 1319.
2. *Ibid.*, pp. 1319–1320.

3. Emilio Segrè, "Enrico Fermi," *Dictionary of Scientific Biography*. New York: Charles Scribner's Sons, Vol. 4, 1971, p. 577.
4. Boorse and Motz, *op. cit.*, p. 1320.
5. Segrè, *op. cit.*, p. 577.
6. *Ibid.*, p. 578.
7. *Ibid.*, p. 579.
8. *Ibid.*, p. 580.
9. "Enrico Fermi," *Biographical Encyclopedia of Scientists*. New York: Facts on File, Inc., 1981, p. 258.

CHAPTER 19

1. Henry A. Boorse and Lloyd Motz, *The World of the Atom*. New York: Basic Books, 1966, p. 1760.
2. *Ibid.*, p. 1761.
3. "Robert Hofstadter," *McGraw-Hill Modern Men of Science*. New York: McGraw-Hill, 1966, p. 239.
4. Isaac Asimov, *Asimov's Biographical Encyclopedia of Science and Technology*, 2nd ed. Garden City, New York: Doubleday & Co., Inc., 1982, p. 854.
5. "Robert Hofstadter," *op. cit.*, p. 239.
6. Asimov, *op. cit.*, p. 854.
7. "Robert Hofstadter," *op. cit.*, p. 239.
8. "Murray Gell-Mann," *McGraw-Hill Modern Men of Science*. New York: McGraw-Hill, 1966, p. 188.
9. *Ibid.*, p. 189.
10. *Ibid.*, p. 190.

CHAPTER 20

1. A. Vibert Douglas, "Arthur Stanley Eddington," *Dictionary of Scientific Biography*. New York: Charles Scribner's Sons, Vol. 4, 1971, p. 278.
2. James R. Newman, *The World of Mathematics*. New York: Simon & Schuster, 1956, p. 1069.
3. Douglas, *op. cit.*, p. 279.
4. *Ibid.*, p. 280.
5. "John Archibald Wheeler," *McGraw-Hill Modern Men of Science*. New York: McGraw-Hill, 1968, p. 590.
6. *Ibid.*, p. 591.
7. *Ibid.*, p. 593.
8. G. J. Whitrow, "Edwin Powell Hubble," *Dictionary of Scientific Biography*. New York: Charles Scribner's Sons, Vol. 5, 1972, p. 528.
9. *Ibid.*, p. 529.
10. *Ibid.*, p. 530.
11. *Ibid.*, p. 531.
12. *Ibid.*, p. 532.

Recommended Readings

Walter Baade, *Evolution of Stars and Galaxies*. Cambridge: MIT Press, 1963.

W. W. Rouse Ball, *A Short Account of the History of Mathematics*. New York: Dover, 1960.

E. T. Bell, *Men of Mathematics*. New York: Simon & Schuster, 1937.

J. D. Bernal, *Science in History* (4 vols.). Cambridge: MIT Press, 1971.

Arthur Berry, *A Short History of Astronomy*. New York: Dover, 1961.

Hermann Bondi, *Rival Theories of Cosmology*. London: Oxford University Press, 1960.

Henry A. Boorse and Lloyd Motz, *The World of the Atom* (2 vols.). New York: Basic Books, 1966.

Daniel J. Boorstin, *The Discoverers*. New York: Vintage, 1983.

Max Born, *Physics in My Generation*. New York: Springer-Verlag, 1969.

Max Born, *The Restless Universe*. New York: Dover, 1951.

Percy W. Bridgman, *The Logic of Modern Physics*. New York: Macmillan, 1927.

Norman R. Campbell, *Physics: The Elements*. London: Cambridge University Press, 1920.

Fritjof Capra, *The Tao of Physics*. New York: Bantam, 1977.

Ronald W. Clark, *Einstein: The Life and Times*. New York: Avon, 1984.

I. Bernard Cohen, *Revolution in Science*. Cambridge: Harvard University Press, 1985.

Frederick Copleston, S. J., *A History of Philosophy* (3 vols.). Garden City, New York: Doubleday & Co., Inc., 1985.

Richard Courant and Herbert Robbins, *What is Mathematics?* Oxford: Oxford University Press, 1941.

A. D'Abro, *The Evolution of Scientific Thought*. New York: Dover, 1950.

A. D'Abro, *The Rise of the New Physics* (2 vols.). New York: Dover, 1951.

Paul Davies, *The Runaway Universe*. New York: Harper & Row, 1978.

Dictionary of the History of Ideas (4 vols.). Ed. Philip P. Wiener. New York: Charles Scribner's Sons, 1968.

J. L. E. Dreyer, *A History of Astronomy from Thales to Kepler*. New York: Dover, 1953.

Will Durant, *The Story of Philosophy*. New York: Simon & Schuster, 1961.

Arthur Eddington, *The Expanding Universe*. London: Cambridge University Press, 1933.

Arthur Eddington, *The Nature of the Physical World*. Cambridge: Cambridge University Press, 1928.

Arthur Eddington, *Space, Time and Gravitation*. London: Cambridge University Press, 1920.

Albert Einstein, *Ideas and Opinions*. New York: Crown, 1954.

Albert Einstein, *Relativity: The Special and the General Theory*. Trans. R. W. Lawson. New York: Bonanza, 1961.

Albert Einstein and Leopold Infeld, *The Evolution of Physics*. New York: Simon & Schuster, 1938.

Albert Einstein: Philosopher-Scientist (2 vols.). Ed. P. A. Schilpp. Chicago: Open Court, 1951.

Richard Feynman, *The Feynman Lectures on Physics* (3 vols.). Reading, Massachusetts: Addison-Wesley, 1963.

Kenneth Ford, *The World of Elementary Particles*. Waltham, Massachusetts: Blaisdell, 1963.

Philipp Frank, *Modern Science and Its Philosophy*. Cambridge: Harvard University Press, 1949.

George Gamow, *The Creation of the Universe*. New York: Viking, 1947.

Edward R. Harrison, *Cosmology, The Science of the Universe*. New York: Cambridge University Press, 1981.

Werner Heisenberg, *Physics and Philosophy*. New York: Harper, 1958.

Gerald Holton, *Foundations of Modern Physical Science*. Reading, Massachusetts: Addison-Wesley, 1958.

Fred Hoyle, *Astronomy and Cosmology*. San Francisco: W. H. Freeman, 1975.

Aaron J. Ihde, *The Development of Modern Chemistry*. New York: Dover, 1964.

James Jeans, *The New Background of Science*. New York: Macmillan, 1934.

James Jeans, *Physics and Philosophy*. New York: Dover, 1981.

James Jeans, *The Universe Around Us*. Cambridge: Cambridge University Press, 1930.

Horace Freeland Judson, *The Eighth Day of Creation*. New York: Simon & Schuster, 1979.

Edward Kasner and James R. Newman, *Mathematics and the Imagination*. New York: Simon & Schuster, 1940.

Morris Kline, *Mathematics and the Physical World*. New York: Dover, 1981.

Morris Kline, *Mathematics in Western Culture*. New York: Oxford, 1953.

Thomas Kuhn, *The Structure of Scientific Revolutions*. Chicago: University of Chicago Press, 1962.

T. Z. Lavine, *From Socrates to Sartre: The Philosophic Quest*. New York: Bantam, 1984.

Ernst Mach, *The Science of Mechanics*. Chicago: Open Court, 1907.

Stephen F. Mason, *A History of the Sciences*. New York: Macmillan, 1962.

Ernst Mayr, *The Growth of Biological Thought*. Cambridge: Harvard University Press, 1982.

Lloyd Motz, *The Universe: Its Beginning and End*. New York: Charles Scribner's Sons, 1976.

Lloyd Motz and Jefferson Hane Weaver, *The Concepts of Science: From Newton to Einstein*. New York: Plenum Press, 1988.

Ernest Nagel, *Freedom and Reason*. Glencoe, Illinois: The Free Press, 1951.

Ernest Nagel, *The Structure of Science*. Indianapolis: Hackett, 1979.

James R. Newman, *The World of Mathematics* (4 vols.). New York: Simon & Schuster, 1956.

Isaac Newton, *Mathematical Principles of Natural Philosophy*. Berkeley: University of California Press, 1960.

J. Robert Oppenheimer, *The Open Mind*. New York: Simon & Schuster, 1955.

Heinz R. Pagels, *The Cosmic Code*. New York: Simon & Schuster, 1982.

R. E. Peierls, *The Laws of Nature*. New York: Charles Scribner's Sons, 1956.

Max Planck, *Physics and World Philosophy*. Trans. W. H. Johnston. London: Allen & Unwin, 1936.

Max Planck, *A Survey of Physical Theory*. Trans. R. Jones and D. H. Williams. New York: Dover, 1960.

John Herman Randall, Jr., *The Making of the Modern Mind*. New York: Columbia University Press, 1976.
Bertrand Russell, *A History of Western Philosophy*. New York: Simon & Schuster, 1945.
Bertrand Russell, *Mysticism and Logic*. New York: Norton, 1929.
Bertrand Russell, *Our Knowledge of the External World*. London: Allen & Unwin, 1929.
George Sarton, *A History of Science* (2 vols.). Cambridge: Harvard University Press, 1970.
Erwin Schrödinger, *What is Life?* New York: Macmillan, 1945.
Harlow Shapley, *Beyond the Observatory*. New York: Charles Scribner's Sons, 1967.
D. E. Smith, *History of Mathematics* (2 vols.). New York: Dover, 1958.
Carl Friedrich von Weizsäcker, *The Unity of Nature*. Trans. Francis J. Zucker. New York: Farrar, Straus and Girous, 1980.
Jefferson Hane Weaver, *The World of Physics* (3 vols.). New York: Simon & Schuster, 1987.
Steven Weinberg, *The First Three Minutes*. New York: Basic Books, 1977.
Hermann Weyl, *The Open World*. New Haven, Connecticut: Yale University Press, 1932.
Hermann Weyl, *Space, Time, Matter*. Trans. Henry L. Brose. New York: Dover, 1922.
Hermann Weyl, *Symmetry*. Princeton, New Jersey: Princeton University Press, 1952.
Alfred North Whitehead, *Science and the Modern World*. New York: Macmillan, 1925.
Norbert Wiener, *Cybernetics*. New York: John Wiley & Sons, 1948.
Chen Ning Yang, *Elementary Particles*. Princeton, New Jersey: Princeton University Press, 1962.

Index